EINSTEIN AND CULTURE

EINSTEIN AND CULTURE

Gerhard Sonnert

Humanity
Books

an imprint of Prometheus Books
59 John Glenn Drive, Amherst, New York 14228-2197

Published 2005 by Humanity Books, an imprint of Prometheus Books

Inquiries should be addressed to
Humanity Books
59 John Glenn Drive
Amherst, New York 14228–2197
VOICE: 716–691–0133, ext. 207
FAX: 716–564–2711

09 08 07 06 05   5 4 3 2 1

Library of Congress Cataloging-in-Publication Data

Sonnert, Gerhard, 1957–
    Einstein and culture / Gerhard Sonnert.
        p. cm.
    Includes bibliographical references.
    ISBN 1–59102–316–5 (alk. paper)
    1. Einstein, Albert, 1879–1955—Biography. 2. Scientists—Social conditions. 3. Science—Social aspects. 4. Science—Philosophy. I. Title.
QC16.E5S646 2005
530'.092—dc22
                                                                                2004027131

Printed in the United States of America on acid-free paper

For my family

CONTENTS

INTRODUCTION

Albert Einstein is probably the only scientist with "face recognition" among a sizable fraction of the general public. Dubbed the *Time* magazine Person of the Century in December 1999, on top of countless other accolades, the physicist not only was the most celebrated scientist of the twentieth century but also became the pop icon of the maverick genius. If not for his theory of relativity, people fondly remember him for his unruly hairstyle and his steadfast refusal to wear socks. He has been described as "a profoundly simple man," yet his façade of simplicity merely hid layers of complexity (Hoffmann and Dukas 1972, 3). Despite his celebrity, he has remained a somewhat puzzling figure, and this book aims to help us better understand Albert Einstein, the scientist and the man.

This is not an Einstein biography. There seems to be no urgent need for another one because an ample crop of biographies already exists.[1] The scholarly literature about Einstein has been steadily growing in volume, spurred on to a considerable degree by the laudable efforts of the Einstein Papers Project, which has embarked on publishing most of Einstein's extant documents at Princeton University Press. But some important aspects of Einstein's life story and achievements remain to be explored in depth. What distinguishes this study from the bulk of the Einstein literature is its particular emphasis on culture, or, more precisely, its detailed attention to nineteenth-century and early twentieth–century German culture. We shall investigate how the culture that surrounded Einstein in his formative years and during most of his life shaped his physics and his outlook on life in general.

Examining the cultural determinants of scientific activity has been a well-established line of inquiry in the sociology and history of science. Robert K. Merton's pioneering study of the connection between the religious beliefs of the Puritan period and the science of the day immediately comes to mind. It concluded in the sentence, much-quoted ever since, that "the cultural soil of seventeenth century England was peculiarly fertile for the growth and spread of science" (1970 [1938], 238). The cultural roots of Einstein's science and worldview have not yet been probed in an exhaustive way, although such an endeavor appears particularly worthwhile in his case. First, the staggering magnitude of his scientific achievements by itself would already be enough to justify thorough study from all angles. In addition, the study is intended to clear up persistent misconceptions that cloud this particular area. Some people—including Einstein's own secretary, Helen Dukas—viewed him as a wondrous genius who was not quite of this world and essentially detached from his surroundings, and his widespread rebel image depicts him as a countercultural nonconformist (e.g., Feuer 1974). Yet contrary to Einstein's portrayal as a spectacular metacultural singularity or as a young countercultural revolutionary, this study will demonstrate that Einstein was deeply and decisively influenced by the cultural traditions in which he found himself. Einstein, it will argue, was both: a rebel *and* a cultural traditionalist. We are going to explore this fascinating paradox, to show how the two sides are interrelated, and thus to contribute to a more complex, and accurate, picture of Albert Einstein.

Thus, one of the principal reasons that compel us to delve, in quite some detail, into German culture (*Kultur*), some of its central notions, and the concept of culture itself, is to enable us to understand vital motivations that guided Albert Einstein and his contemporary cohort of scientists. Yet this is not the only reason. By the time Einstein had reached middle age, both he as a person and his physics became targeted as prime foes by a radically mutated strand of German culture, and, in a great cataclysm, he and many like him were ejected from Germany. Our account of German culture must go beyond the point where it provided a lasting orientation to Einstein's life and physics. It must follow the further development of Kultur and its central concepts, tracing how some variants diverged more and more from the contents treasured by Einstein, until they were on a collision course with Einstein's formative Kultur. In this book, therefore, Einstein's life history and the history of major components of Kultur form two equally important strands whose multiple intertwinings we set out to document.

Furthermore, a study of culture also promises insights that are relevant for the present day. Samuel Huntington's (1996) widely discussed book titled *Clash of Civilizations* testifies to the growing importance of the cultural dimension. Culture appears to become an increasingly salient criterion for drawing the battle lines in international conflicts. In this situation, it behooves us to know what *civilization* and *culture* mean and how the concepts developed. Too few people seem to realize the history, rich in twists and turns, behind those notions.

In our survey of several key elements of the conceptual landscape of culture, a central question is how the concept of culture itself has emerged and evolved within shifting cultural contexts. We shall see how the very concept of culture became the explicit embodiment of fundamental aspirations of a culture. In a sense, we will study the self-understanding of cultures by studying their respective concepts of culture. It will become obvious that it makes a big difference if one has allegiance to *Kultur* or to *culture*.[2]

In examining fundamental cultural ideas, I like to speak of *conceptual tectonics* because this (originally geological) term suggests that potent underground forces—such as demographic and socio-economic ones—tend to influence those cultural concepts, but that they do not totally determine everything that happens in the conceptual landscape. A number of additional factors are also at work here. The major cultural concepts, I think, form an interdependent system of meaning that at times might evolve quite independent of those underlying basic forces. Our approach thus stays clear of axiomatic "materialist" or "idealist" positions (assuming, respectively, that culture is entirely determined by socioeconomic factors, or that it is totally autonomous from them). Generalizations are less useful here than detailed historical investigations. In the course of our discussion, we will encounter some instances where the development of cultural concepts appears clearly driven by socioeconomic forces, and others where conceptual developments proceeded largely independent of those forces.

PLAN OF THE BOOK

Chapter 1 sets up two seemingly conflicting images: Einstein the rebel and Einstein the cultural traditionalist. To resolve this puzzle, we take a closer look, in the subsequent chapters, at the culture that surrounded Einstein. Thus chapter 2 traces the development of the key concept of Kultur in the German lands during the nineteenth and early twentieth centuries, tying it to underlying social and national constellations.[3] For to understand German culture, and its impact on Einstein, it is essential to examine how Germans themselves defined culture—in the peculiarly German notion of Kultur, which entered German thought around the beginning of the nineteenth century. Chapter 3 then explores the concepts of Weltanschauung and Weltbild. These were truly dominant notions within Kultur—and also in Einstein's thinking. The chapter documents the at times dramatic evolution of these concepts. Chapter 4 examines the uneasy relationship between Kultur and science, focusing on how German scientists attempted to secure a place for themselves among the bearers of Kultur, as well as on how Kultur shaped German science in the process. Chapter 5 relates the general concepts of Weltanschauung and Weltbild to science, examining how German scientists adapted them and made them central to their understanding of science.

After these general chapters on central cultural concepts, the sixth chapter turns more specifically to Einstein's background. It starts with describing the position of the Jewish population in nineteenth-century Germany socioeconomically and vis-à-vis Kultur and science and then focuses on Einstein's family history as a case in point. Thus the stage is set for discussing, in chapter 7, Einstein's physical Weltbild and resolving our initial puzzle of conflicting Einstein images. The chapter shows how he was shaped by the cultural traditions that surrounded him, how he combined them with his rebellious streak, and how his amazing achievements in physics can be understood as a synthesis of these two elements and as the fulfillment of profound cultural directives. The eighth chapter examines Einstein's Weltanschauung, his philosophy and religion, in detail. The topic here is no longer how his adherence to the surrounding culture shaped and aided his scientific efforts but how it motivated and sustained his decision to lead a life of science in the first place.

The concluding chapter draws comparisons between the nineteenth-century German intellectual situation and the current American one, and it focuses on the blend of German Kultur and American culture within American science that Einstein and his fellow immigrants helped bring about. Thus, I hope that this book not only is of historical interest but also might inform current debates about culture, about science, and about the role science should play in society and culture.

What makes this book distinctive is that it is not only about Einstein but also about a culturally grounded, specifically German style of scholarship—which, it is claimed, Einstein wholeheartedly espoused. The reader may notice that the book's style mimics that approach to scholarship as well as Einstein's own predilections. Using a Mertonian term, one might say that the book methodologically self-exemplifies one of its topics. For the book is atypically wide ranging and thus replicates a key feature of the German approach that rejected narrow specialization in favor of synthesis and comprehensiveness.

ACKNOWLEDGMENTS

In this undertaking, I owe the largest debt of gratitude to Gerald Holton, with whom I have had the privilege of working, over many years, on several research projects. His sage and generous advice was also pivotal throughout the process leading to this book. The study originated in the mid-1990s with research and drafts for Professor Holton's 1997 Rothschild Lecture, on which he and I collaborated. Earlier versions of major parts of sections 1.2, 1.3, and 7.2 of this book, as well as a small part of section 5.2, previously appeared in the published version of the Rothschild Lecture (Holton 1998).

The Einstein Papers Project was the most valuable archival resource for this study. Whenever I visited that project, then housed at Boston University, I received efficient and gracious assistance from Robert Schulmann and his helpful staff. They all have my gratitude. Archival materials are cited by their call numbers. The shorthand *CP* in a citation refers to *The Collected Papers of Albert Einstein*, published by the Einstein Papers Project at Princeton University Press.

I thank Robert Spaethling for reading, and commenting on, the chapter about Weltanschauung and Weltbild and Sabine Schenk for helping me access some rare material from the Universitäts- und Landesbibliothek Sachsen-Anhalt in Halle, Germany.

Numerous German language sources have been used in this inquiry. When an originally German text is quoted in English, the translation is usually mine, but, in a few cases, the passage is taken from a published English translation (recognizable as such in the reference list).

NOTES

1. E.g., Bernstein 1973; Brian 1996; Clark 1971; Fölsing 1993; Frank 1947, 1949; Hermann 1994; Hoffmann and Dukas 1972; Kondo 1969; Pais 1982; Seelig 1954; Wickert 1972. A bibliography of books on Einstein, compiled by David Cassidy, is available on the Web site of the American Institute of Physics at http://www.aip.org/history/einstein/bibliog.htm.

2. Reasoning about culture often poignantly illustrates the preferred mode of how a culture reasons. A representative of Kultur might describe what happened in the development of the term *Kultur* around 1800 as Kultur becoming "aware of itself" and thus speak of the emergence of "Kultur in and for itself." Such highly abstract reasoning at the metalevel was treasured within the mind frame of Kultur, but within our present-day culture may seem quaint at best, bizarre at worst.

3. When we talk about the culture, or Kultur, that constituted Einstein's formative environment, we have a period in mind that began roughly around 1800 and continued to about the middle of the twentieth century, and an area that included the German-speaking lands of Central Europe—i.e., not only the area of post-1871 Germany but also Austria and most of Switzerland. (As long as the Hapsburg monarchy existed, Kultur was of normative significance also for many non-German-speaking ethnicities in that empire.) In the following, *Kultur* or *nineteenth-century German culture* should be read as shorthand for the cumbersome *nineteenth-century and early twentieth-century culture of the German-speaking areas*.

I.
REBEL OR TRADITIONALIST?

Even if Albert Einstein had lived "at the North Pole," Helen Dukas, Einstein's trusted assistant and secretary, once told Lewis Pyenson, he would have made his great scientific discoveries (1985, 60). She apparently believed that Einstein's genius was entirely independent of any social and cultural determinants. This study must take issue with Helen Dukas on that account; its major goal is to demonstrate how the Kultur in which Einstein lived helped shape his thinking in general and his physics—or, to use two key notions of Kultur itself, how it nurtured Einstein's general *Weltanschauung* as well as his physical *Weltbild*. Because we contend that these concepts—Kultur, Weltanschauung, Weltbild, and several others of equal centrality in the German intellectual landscape—played a crucial role for German science and especially for Einstein, we shall examine them in subsequent chapters with a degree of detail not routinely applied in the sociology of science. When we return from this foray into the conceptual tectonics of nineteenth-century German culture, not the least of our rewards will be a deeper understanding of Albert Einstein's dazzling accomplishments in physics and of his wider outlook on life.

1.1. DETERMINANTS OF SCIENTIFIC GENIUS

Helen Dukas has not been alone in her view, of course. Genius is often seen as something supracultural and perhaps even supra-human, as something that cannot be understood through the causes that drive and explain the actions of lesser mortals. Yet various disciplines have, from different angles, tried to get a scientific grip on the elusive subject of genius or—more specifically to our topic—scientific genius and to ferret out some of its determinants. To start with, ever since Sir Francis Galton's (1869, 1875) classic studies, people have been fascinated with the genetic component of genius. The genetic hypothesis has been vigorously challenged by views emphasizing environmental factors, and "nature-nurture" controversies have periodically flared up around this issue (as in other fields). The present study can certainly not contribute to, let alone adjudicate, these controversies. Suffice it to note that, in Albert Einstein's case, believers in hereditary genius might point to certain mathematical and technical inclinations that can be found on both the paternal and maternal family branches. Furthermore, Albert Einstein's oldest son, Hans Albert, became a distinguished professor of engineering. His second son, Eduard, was highly intelligent but mentally unstable and a long-term patient in a psychiatric institution. The latter's affliction perhaps interests those who, with Cesare Lombroso (1891), believe the much-disputed idea that but a fine line separates genius from insanity. Nonetheless, evidence for a hereditary component of scientific genius must be considered rather limited in Einstein's case. In general, it seems safe to say that heredity is not a sufficient explanation of scientific excellence, at least not in any straightforward way. Various additional factors, notably environmental ones, are needed to bring about exceptional scientific achievement.

Within psychology, there is now an extensive research literature about explanations of scientific excellence that focus on the cognitive, motivational, and personality traits, as well as on other characteristics of superior scientists. One of the many interesting findings in Anne Roe's (1952) landmark study *The Making of a Scientist*, for instance, was that eminent scientists tended to be somewhat introverted, socially awkward, and aloof—"nerds," as we might call them today—during their childhood. More recently, symptoms of nerdiness and social ineptitude were classified as Asperger's syndrome, a mild form of autism. Autism expert Simon Baron-Cohen even claimed that Einstein, as well as Newton, exhibited signs of Asperger's syndrome (Muir 2003).

Roe also found, as did other observers (e.g., Clark and Rice 1982; Galton 1875; Sonnert and Holton 1995; Visher 1947), that firstborns were substantially overrepresented among scientists. Frank Sulloway (1996) gave the birth-order hypothesis a fresh twist by showing that the protagonists of novel scientific theories tended to be later-born, whereas the defenders of scientific orthodoxy tended to be firstborn—the firstborn Einstein being one of the inevitable exceptions. There are, of course, numerous other psychological explanations of scientific excellence, which we cannot survey here.[1] Most of the psychological studies tend to neglect the social and cultural environments of excellence. It has long been noticed, however, that geniuses are not spread evenly across time and space, but that curious clusters of geniuses emerge at the same time in the same locale (e.g., Kroeber 1944). This is a strong indication that social and cultural milieux contribute to producing those outstanding individuals. Howard Gardner's (1993) psychological theory of creativity is exceptional in its strong attention to the role of the environment. It identifies some kind of a mismatch between an individual and his and her environment as the major stimulus of creativity. Geniuses, according to Gardner (1993), tend to be marginal persons within their discipline and larger environment. This idea will prove very helpful when we explore the roots of Einstein's scientific achievements.

Furthermore, it seems obvious that any successful theory of genius will have to incorporate an element of chance (e.g., Austin 1978; Simonton 1988), which by itself will limit the explanatory power of the theory. This is probably true not only for the psychological theories of scientific excellence but also for the sociological theories to which we turn now.

Sociological approaches to science have focused less on the individual genius and more on the characteristics, determinants, and consequences of science as a social activity. Sociologists of science have been studying the workings of what could be called the social system of science, or the institutional structure of science (e.g., Ben-David 1984, 1991; Cahan 1985; Merton 1973; Turner 1971; Zloczower 1981). This kind of research has investigated the organizational features of science departments and research institutes, the social background and typical career paths of scientists, the funding mechanisms for research, the reward structures and internal stratification of the scientific community, and similar topics. Here, scientific excellence appears to result, at least in part, from conditions within the social system of science, for instance, from the availability and utilization of resources and networks, or, to use a key Mertonian term, from the "accumulation of advantage and disadvantage" over the course of a career. (See Zuckerman 1989.) Nonetheless, many sociologists working in this tradition have assumed that the validity of scientific knowledge itself remains uncompromised by the social circumstances of its production, and they have thus retained an "objective" dimension of scientific excellence in their theoretical frameworks.[2]

In addition to what one might consider an "intermediate level" sociological analysis, which focuses on the social system of science and its structures, there has also been interest in the microlevel and the macrolevel. At the former level, researchers have examined the social construction of scientific knowledge by highly detailed investigations. They have, for instance, examined the discussions and negotiations that go on among scientists while they come to grips with new data, as well as other similar aspects of scientific work (e.g., Latour and Woolgar 1979; Traweek 1988).

At the latter level, sociologists and historians of science have examined the impact on science of broader social and political forces (e.g., Borscheid 1976), and of fundamental cultural values. Erwin Schrödinger (1994 [1932]), the eminent physicist, aptly phrased the central question for that kind of investigations in 1932: To what extent is the pursuit of science *milieubedingt?*—where *bedingt* can have the strict causal sense of "dependent on" or the more gentle and useful meaning of "to be in *resonance* with."[3]

We do have major studies of such milieu-resonance for earlier scientists. First of all, one thinks of Robert K. Merton's (1970 [1938]) classic study of the connection between science and the religious beliefs of the Puritan period, which was already mentioned. Work has also been done, for example, on the effect of the neo-Platonic philosophy on the imagination of seventeenth-century figures, such as Johannes Kepler and Galileo, on the theological interests of Isaac Newton that affected his work, and on the philosophical base that supported the discoveries of Hans Christian Oersted, J. R. Mayer, and André Marie Ampère. Similar studies of more recent scientists have been sparser, although there have been notable exceptions. Paul Forman (1971) made the—somewhat controversial—attempt to interpret aspects of some scientists' presentations of quantum mechanics chiefly as their response to the sociopolitical malaise in the Weimar Republic.[4] Max Jammer (1966) and Gerald Holton (1973) investigated the extent to which Niels Bohr's wider philosophical interests, especially his delight in Søren Kierkegaard's writings, prompted him to introduce the complementarity principle into physics. More directly on our subject, Lewis Feuer (1974) argued that the physics of Einstein and his cohort was influenced by the spirit of countercultural rebellion that pervaded the young generation in Einstein's student days. Lewis Pyenson (1985) thoroughly examined the social and cultural environment of the young Albert Einstein.

This study will primarily focus on the interaction of culture with a scientist's research program (a "macro-micro" interaction, one might say), but we must emphasize that the broad forces of culture, the more specific elements of the institutional structure of science, and the even more individualized modes of knowledge-production interact in multiple ways. We should particularly note that the "intermediate level" institutional structure of the social system of science provides powerful opportunities and constraints for a scientist's research program (Merton 1970 [1938]), and that it is clearly advantageous for individual scientists if the cultural values and motivations to which they themselves subscribe are also articulated in the structure of the science system.[5] The Prussian educational reforms of the early nineteenth century, for instance, are a relevant example of how cultural values can become expressed and "institutionalized" in the social system of science. The model of the German university, created by Wilhelm von Humboldt, clearly institutionalized certain core ideals of Kultur and *Bildung*, although, over time, the state bureaucracy that was in charge of the universities appeared to have subverted those ideals to some extent (Sorkin 1983). In the secondary schools (*Gymnasien*), the implementation of neohumanism in the curriculum was the vehicle by which those central cultural values were institutionalized at that level of the educational system (see, e.g., Loewe 1917).

Studies that focus on larger cultural conditions cannot claim to explain scientific excellence fully. Although Einstein's cultural roots are, as we intend to show, a *necessary* component for understanding his scientific achievements, it is clear that they can never amount to a *sufficient* explanation. Einstein's discoveries were unique; the surrounding culture, pervasive. As will become obvious, Einstein shared his allegiance to central elements of Kultur with numerous other scientists, and his physical research program—the quest for a unified Weltbild—embodied a chief imperative of Kultur, but only he came up with his particular trailblazing discoveries. To be sure, he was more deeply committed to certain cultural ideals than was the average scientist of his cohort, yet, doubtlessly, many additional factors had to contribute to Einstein's genius and success, be they science-internal, biological, psychological, sociological, or random.

Nevertheless, cultural effects had a crucial part in Einstein's achievements. He did not live at the North Pole; he lived from earliest childhood in, and soon identified with, a cultural milieu that both shaped and nourished his science. We will see that Einstein was thoroughly acquainted with Kultur; he had a conscious and out-spoken allegiance to it; and his physics research resonated with it. His achievements thus were the fulfillment of a major cultural directive to which both Einstein and large parts of the science community were committed. Hence, Einstein's cultural milieu legitimized, validated, and supported his quest.

We are going to carry out a thematic analysis of Einstein's research program (Holton 1973), which took as its starting point the precise formulations of major physical laws that had been known for a long time (e.g., Newtonian mechanics) or had more recently become general knowledge in the physics community (e.g., Maxwell's electrodynamics). This analysis will reveal that Einstein's physics was not driven, to any overriding extent, by the latest experimental results becoming available at the time. Rather, he was motivated by a few basic ideas, which he called categories and Holton called themata, and these themata strongly resonated with the culture in which Einstein was embedded.

Before embarking on our study of the impact of culture on a scientist's work, we should note that the reverse relationship—science, and particularly technology, influencing culture—is also a well-established phenomenon. For instance, people of very diverse convictions, from Karl Marx to Ludwig Büchner to the Darwinists, have claimed that their societal projects and cultural schemes would follow logically from scientific findings.[6] We shall encounter various attempts at constructing a comprehensive worldview based on scientific results. It is beyond the scope of this study to investigate the influences that emanated from Albert Einstein's science into the cultural realm, but we should point out, at least in passing, that his physics is a fascinating example of how a scientific theory (or, in this case, the popular distortion of a scientific theory) can have a deep cultural impact.[7] Einstein's theory of relativity was widely noticed in a variety of fields (including ethics, anthropology, religion, and literature) where it commonly served as a prop for some brand of relativism because it had proved, supposedly, that "everything is relative." Einstein was perturbed by the popular misunderstanding of his theory. He would have preferred his theory—which Max Planck and Max Abraham, not Einstein himself, had named the "theory of relativity" in 1906—to have become known as the "theory of invariance" instead.[8]

Although Einstein himself was fond of proclaiming simplicity as one of his major goals in life, C. P. Snow fittingly noted, in concert with many of Einstein's biographers, "there were some singular paradoxes in Einstein's career" (1980, 4; see 1979). These might include, for instance, internationalist and Zionist, pacifist and "father of the atomic bomb," humanist and loner, famous and indifferent toward fame, rationalist and mystic, believed in both individual justice and collective guilt, wise and with childlike naiveté, and, of course, complexity and simplicity.

In the following, we outline two sides of Albert Einstein that, at first view, seem starkly juxtaposed and even contradictory: his popular rebel persona and his less conspicuous character of a cultural traditionalist. Our analysis of nineteenth-century German culture will then enable us to explain this seeming paradox and to understand how the disparate elements fit together. We will see that Einstein's type of rebellion was far from the modern image of the twentieth-century countercultural rebels in art, poetry, politics, parts of academe, or music—rebels who typically reject the cultural canon of the bourgeoisie along with its social-political conventions.

1.2. THE REBEL

The public image of Albert Einstein has been that of a rebel. This perception, rudimentary as it often is, is also backed up by weightier testimonials. An Einstein biography by the mathematician Banesh Hoffmann (who once worked with Einstein) and none other than Helen Dukas is titled *Albert Einstein—Creator and Rebel* (1972). As mentioned, Lewis Feuer (1974) placed Einstein in the countercultural milieu of young revolutionaries who lived in Zürich and Bern around the turn of the century. And to the *New York Times*, the observational confirmation of Einstein's general relativity theory appeared to be gravely unsettling. On November 16, 1919, under the title "Jazz in Scientific World," the newspaper reported at length that Charles Poor, a professor of celestial mechanics at Columbia University, thought Einstein's success demonstrated that the spirit of unrest of the period had "invaded science," and the *Times* added its own warning, "When is space curved? When do parallel lines meet? When is a circle not a circle? When are the three angles of a triangle not equal to two right angles? Why, when Bolshevism enters the world of science, of course" (p. 8).

We will now document traits, incidents, and events that support this image of Einstein the rebel. Our account can be brief because most of the facts are well known. When Albert was a small child, he was notorious for his bad temper. His sister Maja remembered that she was often at the receiving end, for instance, when Albert hit her on the head with a hoe. These outbursts subsided when Albert reached school age, but then he launched his first attempt at a more sustained rebellion. In opposition to his thoroughly secular home environment, Albert decided to become a religious Jew (Winteler-Einstein 1987, lix).[9] Whereas he attended classes in Catholic religion at a public elementary school, he received private instruction in Jewish religion from a distant relative, and this, his sister thought, sparked Albert's devotion to traditional Judaism (ibid.).[10] At home, Albert made an effort to obey a few of the kosher dietary laws by refusing to eat when pork was served at the family table.

After elementary school, Albert was promoted to the Munich Luitpold Gymnasium, where he encountered the state-prescribed instruction in Jewish religion. Under the guidance of Oberlehrer Heinrich Friedmann and a rabbi, he also prepared for the Bar-Mizwa. By the time he turned thirteen (the traditional age for Bar-Mizwa), however, Albert was no longer interested in Judaism. At the age of twelve, Einstein still remembered vividly in his later years, he rebelled again, this time against the Jewish faith (1979a [1949], 1). He became convinced that organized religion was a fraud in which "the young people are deliberately lied to by the government" (ibid.). Note how Einstein identified religion with the government, presumably because religious instruction was a part of the public school curriculum. He now turned to "fanatical free-thinking" and formed a deep "distrust of any kind of authority" (ibid.).

Thus was lost the "religious paradise of youth" (ibid., 2). What prompted that transformation? The twelve-year-old Albert found himself enchanted with a "holy little book of geometry" that inspired a sense of awe at the certainty and clarity of geometric conclusions (ibid., 4). This enthusiasm for mathematics, which, as the adjective *holy* indicates, apparently absorbed some of Albert's religious impulses, went hand in hand with an interest in the scientific Weltbild. Albert encountered it, under Max Talmey's guidance, in Aaron Bernstein's *Naturwissenschaftliche Volksbücher*—"a work I read with breathless suspense" (ibid., 5)—and in Ludwig Büchner's *Kraft und Stoff*.[11] Einstein reminisced that he, "at about thirteen years of age, read Büchner's 'Kraft und Stoff' with enthusiasm, but later that book seemed to me rather child-like in its naïve realism" (Seelig 1954, 14). In addition, Maja remembered, Talmey introduced Albert to Alexander von Humboldt's *Kosmos* at that time (Winteler-Einstein 1987, lxii). When the boy's mathematical skills rapidly surpassed those of his mentor, Talmey recalled that he shifted the subject of their conversations to Kant's *Critique of Pure Reason* (Talmey 1932, 164).

Whereas Einstein's home thus nourished his precociously burgeoning intellect, his memories of the Gymnasium were not happy; he hated the drill and the authoritarian discipline.[12] Dr. Joseph Degenhart, Einstein's *Klassenlehrer* (homeroom teacher) in the 7. *Klasse*, scolded him, "your mere presence undermines the classroom's respect for me" and suggested that he leave the school (Winteler-Einstein 1987, lxiii n. 58). In December 1894, the fifteen-and-a-half-year-old Albert did Dr. Degenhart the favor; he decided to leave the Gymnasium without *Abitur* and to join his parents who had moved to Italy.[13] Some sources report that, around this time, Albert also formally left the Jewish community (Frank 1949a, 34–35; Jammer 1995, 26; Snow 1980, 6; also see Fölsing 1993, 56). When his father applied for Albert's release from Württemberg citizenship, he did describe Albert as "konfessionslos" (without denomination) (*CP*, 1: 20, n. 1).[14] The release was granted shortly before Albert's seventeenth birthday, just in time to avoid the military draft (ibid.).

At the Kantonsschule in Aarau, Switzerland, we find Albert as what one might consider a thoroughly alienated youth, having left his school, his country, his religion, and his family and even having failed at his first attempt to enter the Swiss Polytechnic Institute at Zürich, known as Eidgenössische Technische Hochschule (ETH) after 1911. Once Albert obtained his Matura at Aarau—a school he really liked—and thus gained entry into the Polytechnic Institute, he continued in many ways his in-your-face rebelliousness. He did not get along well with physics professor Heinrich Friedrich Weber, on whom his career might well depend. Einstein obstinately insisted on calling him "Herr Weber" instead of the more respectful "Herr Professor" (Seelig 1954, 35). In turn, Weber once told Einstein, "You are a smart fellow, Einstein, a very smart fellow. But you have one big flaw: You let nobody tell you anything!" (ibid.). And Weber did nothing to help Einstein in his job search later.

At the institute, Einstein was less than diligent; in hindsight, he described himself as a "mediocre student" (Einstein 1956, 10). Because he preferred studying the classics of physics on his own and neglected the prescribed courses, he was in trouble when the final examination approached. He was saved by the "rescue anchor" (ibid., 11) of his friend Marcel Grossmann's careful notes and course outlines. Hermann Minkowski, who taught mathematics at the institute, remembered Einstein as a lazy student: "[The special theory of relativity] came as a tremendous surprise to me. For, in those earlier days, Einstein was a real lazy bone. He paid no attention at all to mathematics" (Seelig 1954, 33).[15]

His father apparently had hoped that Albert would become an electrical engineer and join the family business. According to Einstein's son Hans Albert, "At the age of sixteen, his father [Hermann] urged him [Albert] to forget his 'philosophical nonsense,' and apply himself to the 'sensible trade' of electrical engineering" (quoted in Clark 1971, 21). To Albert, however, choosing a practical profession "was simply unbearable" (ibid., 24). Here we find the rebellion of the "pure" scientist against the more technological and applied pursuits of his family. In any case, the nature of the Polytechnic Institute allowed both father and son to find something positive in an enrollment. Because it was a "technical" school, Hermann could still hope that Albert would eventually become an engineer, while Albert, of course, availed himself of the opportunity to concentrate on basic science during his studies.

Between Einstein's graduation from the Polytechnic Institute (1900) and his entry into the Bern patent office (1902) lay the most bohemian phase of his life.[16] He lived on the margins of bourgeois society—economically, socially, and, by the standards of the place and time, also morally. Einstein's applications for a number of academic positions were rejected. He had to take various temporary teaching jobs, and those tended to end abruptly and noisily.[17] He lived in impoverished circumstances, as his old mentor Talmey observed on a visit (1932, 167). His dissertation project went under.[18] Against the vigorous opposition of his parents, he conducted an affair with Mileva Marić, whom he had met when they both were students at the institute.[19] She gave birth to a daughter, Lieserl, out of wedlock; the child was apparently given up for adoption. Amid these vicissitudes, Einstein's general attitude was one of defiance of the philistine world.

Albert found several prime specimens of the despised philistines in his own family, first of all, his aunt Julie Koch. To his girlfriend Mileva, Albert described Aunt Julie—who nonetheless financed his studies through a monthly allowance—as "a veritable monster of arrogance and insensitive formality" (Renn and Schulmann 1992a, 11). In a letter he sent to Mileva from his 1900 summer vacation with family and friends, he complained, "The people here and their way of life are hopelessly empty" (ibid., 20). In the same letter, Einstein proudly told Marić his war story about the family crisis that ensued after he announced his plans to marry her. "[Mama] asks me quite innocently: 'So, what will become of your Dollie now?' 'My wife,' I said just as innocently, prepared for the proper 'scene' that immediately followed. Mama threw herself onto the bed, buried her head in the pillow, and wept like a child" (ibid., 19). A few days later, Albert reported back to Mileva, "[Mother] has given up on open warfare and will probably wait to let loose the big philistine guns when she is joined by Papa" (ibid., 24).

A letter from Milan in the fall of 1900 (dated October 3) reiterated Albert's favorite theme of two against the (philistine) world: "You don't like the philistine life anymore either, do you? He who has tasted freedom can no longer wear chains. I'm so lucky to have found you, a creature who is my equal, and who is as strong and independent as I am! I feel alone with everyone except you" (ibid., 36).

Even in Einstein's great paper on relativity in 1905, one can find many touches of that seeming arrogance and self-confident defiance— not only of accepted ideas in science but also of the accepted style and practice. The paper contained none of the expected footnote references or credits, except for a mention of a close friend, Michele Besso, a person who would be unknown among research physicists.

Throughout his life, Einstein considered himself different from the regular, solid *Bildungsbürger* (educated bourgeois—a term whose significance will be explained later) and characterized himself as a gypsy (e.g., in Seelig 1954, 46).[20] When the Aarau class celebrated their twenty-fifth Matura reunion in 1922, the former students sent a letter to their most famous classmate and received a friendly reply, in which Einstein called them "happy people, grounded in the soil, who remember the gypsy" (ibid., 16).

Early in 1908, Einstein succeeded on his second try to gain his *Habilitation* at Bern University and was thus able to teach at that university as a *Privatdozent* (the lowest faculty rank).[21] During the winter semester of 1908–1909, Maja Einstein, who studied Romance languages at Bern, once wanted to sit in on her brother's lecture. Although Einstein had managed to step onto the first rung of the academic career ladder, Maja quickly discovered that his appearance still was deemed less than professorial and dignified. When she asked the custodian for directions to the lecture hall, he exclaimed, "That slob is your brother? I would have never thought that" (ibid., 105).

Yet Einstein's quickly growing renown in the physics community soon made him sought after by a number of universities.[22] In quick succession, he held professorships at Zürich University, the German University of Prague, and at his alma mater, which by that time had changed its name to Eidgenössische Technische Hochschule (ETH). In 1913, he entered the inner sanctum of physical research when he accepted a call to Berlin—to the Prussian Academy of Sciences and Berlin University.[23] In a letter to his colleague and friend Paul Ehrenfest, he said he took this position because it allowed him to focus more on research and afforded independence from teaching duties: "I accepted this curious sinecure, because lecturing gets so strangely on my nerves, and I don't have to lecture there at all" (quoted in Fölsing 1993, 376).[24] Although Einstein gained freedom from teaching, other obligations loomed in Berlin, for he was appointed director of the Kaiser Wilhelm Institute for Physics in 1917. As one might have expected, he did not relish administrative duties (Castagnetti and Goenner 1997). A discussion Einstein had with John D. Rockefeller Jr. serves to illustrate his generally laissez-faire style of organizational leadership. Einstein argued that the strict regulations Rockefeller had laid down for his educational foundations stifled genius. "'Red tape,' the Professor exclaimed, 'encases the spirit like the bands of a mummy!' Rockefeller, on the other hand, pointed out the necessity for carefully guarding the funds of the foundations. . . . 'I,' Einstein said, 'put my faith in intuition.' 'I,' Rockefeller replied, 'put my faith in organization'" (quoted in Nathan and Norden 1960, 157).

When war broke out in August 1914, ninety-three of the chief cultural luminaries published a manifesto backing the German war effort, with the significant title "Appeal to the World of Kultur." Einstein supported a pacifist counterdeclaration ("Appeal to the Europeans"), which was never published because it attracted a grand total of only four signatures. During the war, Einstein never made a secret of his pacifist and cosmopolitan attitude, and in an increasingly hostile Germany he took care to express publicly his support for the founding of a Jewish state in Palestine.[25]

After the war, Einstein undertook trips abroad to help end the isolation of German science. He did this probably out of an overriding feeling of fairness and out of his customary solidarity with the underdog. (See Weizsäcker 1979, 168.) During the war, Einstein had unequivocally condemned German militarism, but he also considered it unfair that postwar Germany and German science were treated as pariahs in the international community. Characteristically, Einstein felt closest to Germany in the immediate postwar years marked by internal unrest, economic hardship, and international isolation.

In the winter of 1930–31, Einstein visited the United States' West Coast. There he met Upton Sinclair, about whom he wrote, "He is in disrepute here, because he relentlessly highlights the dark sides of American life" (quoted in Fölsing 1993, 719). To his host Robert Millikan's dismay, Einstein gave Sinclair an interview for a socialist weekly and delivered a speech to Caltech students in which he was rather ambivalent about technological progress. One of the reasons for Einstein's settling at the Princeton Institute for Advanced Study was that he realized that he would be constantly surrounded by social events, all arranged by Millikan, if he went to California. Yet Einstein's stay at the Institute for Advanced Study also turned out to be less than ideal; it was marred by conflicts with the director of the institute.

Einstein had signed the famous letter to President Roosevelt that gave rise to the atom bomb project, but because the FBI and the military secret service considered him a security risk, he was not allowed to participate in the nuclear research, although he would have liked to (ibid., 802–803). After the war, Einstein used his celebrity status to support the burgeoning anti–nuclear arms movement. Throughout his stay in the United States, he was carefully monitored, and the FBI files on Einstein are voluminous (Jerome 2002).[26]

A central element of the popular Einstein myth is his unconventional dress code. And it is true: Albert Einstein was a casual dresser and preferred old clothes. When his second wife, Elsa, felt he needed a new suit, she had to take one of his old jackets to the store for size because Einstein would protest that the old suit was perfectly fine and would refuse to shop for a new one (Herneck 1978, 40–41). Einstein's housekeeper in Berlin also remembered that Einstein did not use a professional barber; instead, his wife did his hair. The extremely short-sighted Elsa had to lay aside her lorgnette and was thus visually impaired while cutting hair (ibid., 23–24). This practice resulted in Einstein's distinct hair style. Antonina Vallentin reported that Einstein told her why he did not wear socks—socks only develop holes (1954, 28).[27] In general, Einstein explained, "My dress and the way I wear my hair stem solely from my desire for simplicity. It is my feeling that the less I can get along with in daily life, such as automobiles and socks, the freer I am from these drudgeries" (Bucky 1992, 111).

Einstein paid relatively little attention to conventional manners. Both Thomas L. Bucky and Peter A. Bucky, sons of Einstein's friend and family physician Gustav Bucky, inform us that Einstein found the rules of etiquette inordinately comical (Thomas Bucky 1990, 447; Peter Bucky 1992, 7). "Invariably, when we heard loud laughter resounding through the house and ran upstairs to check, we would find him with this Emily Post book [on etiquette] out, thoroughly amused. Often he would share with us whatever section of the book he considered outlandishly funny" (Bucky 1992, 7).

In sum, Einstein's personality contained a vigorous and persistent strand of independence as well as a large amount of mistrust of, sometimes even resistance against, authorities—best symbolized perhaps by his radical turn against organized religion at the age of twelve. This basic tendency manifested itself in several domains of Einstein's life. In the social realm, it appeared as a rebellion against the social conventions of "philistine" life. Einstein typically preferred outspokenness to tact. He also was the very opposite of the "organization man." At the political level, his leanings toward pacifism, socialism, and cosmopolitanism several times brought him into conflict with dominant doctrines. The described deep-seated trait in Einstein's personality had two sides: withdrawal and rebellion.

Despite his activism for humanity at large, Einstein had little interest in establishing relationships with actual human beings, as he observed himself. "My passionate sense for social justice and social responsibility has always been in a peculiar opposition to my distinct lack of the need to connect with humans or human communities" (Einstein 1955a, 8). He always felt like a stranger among humans and had a yearning for solitude. In this tendency toward social withdrawal, he certainly fit the psychological profile of distinguished scientists that Anne Roe's (1952) study discovered. Not without reason, thus, L. Pyenson viewed Einstein as an "outlier" or "stranger" who was marginal and wanted to be left alone to follow his own goals (1985, 60–61).

Yet, in addition to those reclusive "outlier" tendencies, Einstein was, of course, also a rebel—not only in his high-visibility activism for unpopular social and political causes but also because his flaunting of social conventions contained certain elements of provocation and in-your-face rebellion vis-à-vis the despised philistines. Although Einstein emphasized, "One should draw his attention from others not by his outer appearance but rather by his inner qualities," drawing attention was precisely the effect of some of Einstein's nonconformist behaviors (Bucky 1992, 111). Following conventions in minor points would probably simplify life more than would breaking them—and having to deal with the ensuing fallout. In his old age, Einstein still remembered that "when I was a little boy I was being spanked for not having on my Sunday clothes at the proper time and also because I did not conform to the usual way of saying, 'How do you do?' to our guests" (ibid.). This suggests that Einstein's quest for simplicity, also in daily life, was not pragmatic, but principled. He was not primarily after simplicity in the sense of convenience or freedom from hassles; he pursued the ideal of simplicity, even at the cost of creating real-life complications. Einstein thus was clearly both: an outlier and a rebel, or, one might say, a lonely rebel. This curious brand of rebellion is but a first level of complexity in the persona of our paragon of simplicity.

1.3. THE TRADITIONALIST

In contradistinction to the image of Einstein the rebel, we shall now assemble a catalog (equally one-sided) of incidents, traits, and characteristics that portray Einstein as a traditionalist. We start by drawing up a list of examples indicating that Einstein's rebelliousness had limits. For instance, it does not fit in the picture of the unruly child that Albert's nanny called him "Pater Langweil," or "Father Boring" (Reiser 1930, 27). Furthermore, Einstein was called *Biedermann* as a young boy.[28] This is particularly ironic because the word *Biedermeier* (or *Biedermann*) is almost synonymous with philistine—the very lifestyle Albert loved to denigrate as a young adult.

Moreover, Elsa furnished their Berlin home in *Biedermeier* fashion; the salon was called the "Biedermeierzimmer" and, for a while, sculptured busts of both Johann Wolfgang von Goethe and Friedrich von Schiller adorned the home (Herneck 1978, 14, 47–48).[29]

Thus Einstein's rebelliousness against the philistine lifestyle clearly was not extreme. He gladly took advantage of the material support and comfort his family provided and preferred grumbling about a philistine environment to living outside of it. His second marriage to Elsa—who had grown up in his extended family—provided a continuation of his familiar surroundings. Yet Philipp Frank wrote about Einstein in his Berlin home "that Einstein always remained a stranger in such a 'bourgeois' household: a wanderer through the world who is resting for a moment, a bohemian as guest in a bourgeois home" (Frank 1949a, 219).

While he was rebelling against his family in the "Dollie affair" with Mileva, he stopped short of forcing the issue by marrying her against his parents' wishes. Only after father Hermann, on his deathbed, gave his blessing—and thus, in fact, preempted the more fervent opposition of mother Pauline—did Albert proceed.

Although Albert rebelled against joining his family's engineering business and preferred "pure science," he nevertheless maintained a lifelong interest in experimentation, technology, and invention—even after he had left the patent office (Galison 1987, 2003).

Whereas Emily Post's etiquette greatly amused Einstein, he did not entirely reject the *bürgerliche* formality of social interaction. Einstein and Gustav Bucky, for instance, although close friends of long standing, never abandoned addressing each other formally as "Professor Einstein" and "Doktor Bucky." "The formality appeared to be agreeable to both," commented Bucky's son Thomas (Bucky 1990, 446). Einstein also was cognizant of the nuances of social address. At first, he called his Berlin housekeeper "Herta." After she located a book that nobody else could find in his study, he started calling her "Fräulein Herta"—as a "sign of respect," he said (Herneck 1978, 32). She knew where the books were because she dusted them weekly.

Einstein indeed belonged to the tiny group of German pacifists and democrats during World War I who stood up against overwhelming popular opinion, but he was far from being a political activist, understandably so because he always regarded himself as a physicist first. He also had no problems accepting rank, honors, and money from a regime he detested. Moreover, his political opinions were characterized as unsophisticated by some commentators. In their study of Einstein's politics during World War I, Hubert Goenner and Guiseppe Castagnetti concluded: "Nevertheless, in contrast to his work in science, his political thinking, as far as he expressed it during the period studied, lacked a rational discussion of the points of view of the warring parties and of their possible aims. Einstein never considered social and economic forces at work" (1996, 51–52). The fact that Einstein did not typically view political and social issues in accordance with any of the common ideological frameworks available at the time might be yet another sign of his independence—and of the dominance of an ethical (as opposed to political) point of view.[30]

More significantly, a deep reverence for tradition characterized Einstein's assessment of his own place in physics. Whereas large segments of the public were quick in hailing him as the great scientific revolutionary, he took great pains to deflect this label.[31] Instead, he emphasized that his work was firmly embedded in the tradition of physics and that it had to be considered an evolution, not a revolution. Together with Leopold Infeld, Einstein published a book with the telling title *The Evolution of Physics* (1938), which stressed the slow and cumulative progress of physics.[32] In 1927, on occasion of the bicentenary of Newton's death, Einstein unequivocally stated, "What has happened since Newton in theoretical physics is the organic development of his ideas" (quoted in Jammer 1995, 75). Einstein never wavered in his allegiance to the grand project of modern physics, and he saw himself as the heir of the great physicists who went before him, rather than as their dethroner. After all, Einstein himself characterized his work as the "Maxwellian Program," and it is no accident that among the framed portraits he kept in his Berlin study there were three scientists, each of whom pursued a great synthesis in physics—Newton, Michael Faraday, and James Clerk Maxwell.[33]

For our depiction of Einstein's traditionalist side, an additional aspect is much more important than those specific limitations to his rebelliousness we listed above: his lifelong allegiance to German Kultur. At an early age, he made the acquaintance of the classic poets and philosophers that formed the core of *Bildung* (education—another central concept to be explained), and they became his constant companions.

There was, of course, an extensive compulsory exposure to German Kultur that nobody could avoid who passed through the Gymnasium, the training ground of future *Kulturträger* (carriers of Kultur—see below). The students in this neohumanistic secondary school for ages roughly from ten to eighteen or nineteen were expected to be quite thoroughly acquainted with the great German poets and other thinkers, the *Dichter und Denker*, and classics from other cultures, especially from antiquity. The team preparing the publication of Einstein's *Collected Papers* found the curricula of Einstein's Munich schools and the Aarau school. A quick scan of a few of the mandatory parts of the canon gives us a good impression of how the young minds of Einstein and his cohorts were meant to be shaped. We find, at the start, readings from the Bible; Latin enters at age ten; Greek at age thirteen. Early on, the students were exposed to Latin literature in the shape of Caesar's *Gallic Wars* and Ovid's *Metamorphoses*. Then, under the supervision of Einstein's only beloved teacher, Ferdinand Ruess, there were poems by Ludwig Uhland, Schiller, Goethe, and others; then Goethe's prose poem "Hermann and Dorothea," Xenophon's "Anabasis"; next year, more Schiller, Johann Gottfried von Herder, Cicero, and Virgil. The Aarau Kantonsschule proffered a similar fare: more of the classics in German, French, and Italian. The 1896–97 German curriculum for the top grade, for example, included the "History of literature from Lessing to the death of Goethe," the reading of Goethe's *Götz von Berlichingen*, as well as his *Iphigenia* and *Torquato Tasso*, which students were supposed to read on their own (*CP*, 1: 360).

To some students, it is true, these classics of Kultur were merely force fed. (There was no choice in selecting courses.) They gladly turned their back on Kultur as soon as they received their official certificates or diplomas. Others went on to engage in the superficial practice of incessantly citing the classics to bolster their social status and demonstrate their membership in the educated elite (Frühwald 1990). They were greatly helped by Georg Büchmann's *Geflügelte Worte* (1926), a best-selling compilation of classic quotations and lengthier excerpts, which was first published in 1864 and went through twenty-seven editions by 1926. These were the philistines who attracted the scorn of Einstein and his like-minded cohorts.

For Einstein, however, the Bildung that was offered him in secondary school led to a sustained self-refinement through knowledge of the best works, rather analogous to Matthew Arnold's (1994 [1869]) concept of culture. Einstein's outreach in the traditional cultural environment was enormous. He had a great love for classical music—an interest stimulated and supported by his musically talented mother. He played the violin from an early age, and quite well. Mozart was his favorite composer (Hoffmann and Dukas 1972, 250; Saz 1990, 504).[34] Like Einstein, many Kulturträger considered classical music an essential part of Bildung, cherishing the aesthetic autonomy of music and disdaining its utilitarian functions (Dahlhaus 1990).

But above all, Einstein loved books (Holton 1995). Starting as a child with his reading of popular-level science and Euclid's "holy" geometry, his books remained companions of enormous importance, and their variety was astounding. Einstein appreciated the core group of the classic canon, but he also had a special admiration for some more marginal figures of Kultur, such as Baruch Spinoza and Arthur Schopenhauer.[35] There exists a list of books in the Einstein household.[36] If we regard only those published up to 1910, we find, among the wealth of volumes, the works of Aristophanes, Ludwig Boltzmann, Ludwig Büchner, Cervantes, William Clifford, Dante, Richard Dedekind, Charles Dickens, Fyodor Dostoyevsky,[37] Frederick Hebbel, the *Collected Works of Heine* (two editions), Hermann Helmholtz, Homer, Alexander von Humboldt (both his collected works and an English translation of his *Kosmos*); many books by Kant, Gotthold Ephraim Lessing, Ernst Mach, Nietzsche, Schopenhauer, Sophocles, Spinoza, and for good measure, Mark

Twain.[38] But above all, there is lots of Goethe, a thirty-six-volume edition and one of twelve volumes, plus two volumes on his optics, the exchange of letters between Goethe and Schiller, and a separate volume of *Faust*.

At the Polytechnic, where, after all, he was being prepared to be a high school physics teacher, he took all the obvious courses— differential equations, analytical geometry, mechanics, and so forth, although not what he most wanted to learn about, Maxwell's elec- tromagnetism. That was not yet taught there, and he read it on his own. But even the training of physics teachers resonated with a wider conception of Bildung. In his first year, Einstein enrolled also in two optional courses, one on the philosophy of Kant and one titled *Goethe, Werke und Weltanschauung* (*CP*, 1: 364).

During his scientifically most creative and intense period in Bern, Einstein and two friends founded the *Olympia-Akademie* for the self-study of scientific, philosophical, and literary classics. We have a list of the books they read from Maurice Solovine, one of the members of the Akademie: Spinoza, Hume, Mach, Richard Ave- narius, Karl Pearson, Ampère, Helmholtz, Bernhard Riemann, Richard Dedekind, William Clifford, Henri Poincaré, J. S. Mill, Gustav Kirchhoff, as well as Sophocles and Jean Racine, Cervantes, and Dickens (in Einstein 1987, 8–9).

During Maja's later years, the siblings did a lot of reading together. Einstein reported in a letter to his friend Besso (April 21, 1946) that he read "Herodotus, Aristotle, Russell's *History of Phi- losophy* and many other interesting books" together with Maja, who lived with him in Princeton (Einstein and Besso 1972, 377). When Maja's health deteriorated to the point that she was bedridden, Albert sat by her side: "In her last years, I read to her every evening from the finest books of old and new literature" (quoted in Fölsing 1993, 819). From the wide range of the litera- ture Einstein appreciated, it already becomes obvious that Ein- stein's intellectual scope was comprehensive and cosmopolitan— not only physics but also philosophy and works of literature, not only German-language writers but also authors of other cultures, and not only the most acclaimed poets and thinkers but also those who were less well known.

When we pick up our discussion of Einstein's outlook again, we will be able to glimpse a synthesis of what at this point may seem disparate elements of rebellion and traditionalism. But to do so, we first have to survey some major elements of the conceptual landscape and tectonics of nineteenth-century German culture.

NOTES

1. For an overview, see Simonton 1988. On creativity in general, see Glover, Ronning, and Reynolds 1989; Rothenberg and Hausman 1976; Runco 1997; and Weisberg 1993. On psychological aspects of Einstein's life, see Erikson 1982 and Wertheimer 1945, 213–33.

2. From a wider perspective, one might say that, of all forms of knowledge, scientific knowledge was the last one to be suspected of social and cultural biases. In the early twentieth century, Karl Mannheim, one of the founders of *Wissenssoziologie* (sociology of knowledge), still exempted scientific knowledge from the social biases that, in his view, permeated other kinds of knowledge. He appeared to adhere to one of the typical notions of nineteenth-century German philosophy—setting up the humanities (*Geisteswissenschaften* or *Kulturwissenschaften*) as a separate branch of *Wissenschaft*, next to the sciences (*Naturwissenschaften*)—when he argued, "It is only because natural science, especially in its quantifiable phases, is largely detachable from the historical-social perspective of the investigator that the ideal of true knowledge was so construed that all attempts to attain a type of knowledge aiming at the comprehension of quality are considered as methods of inferior value. For quality contains elements more or less intertwined with the Weltanschauung of the knowing subject" (Mannheim 1936, 290–91). Ludvig Fleck (1980 [1935]), however, who in 1935 published his long-forgotten *Entstehung und Entwicklung einer wissenschaftlichen Tatsache*, already investigated the social determinants of scientific facts themselves. More recently, various schools of constructivist social science have ventured into the realm of epistemology and challenged the epistemological privilege of scientific knowledge, emphasizing that it, too, is socially constructed and hence not universally valid (e.g., Collins 1981; Collins and Pinch 1993; Keller 1985). In this mode of thinking—which has been extensively challenged—scientific excellence becomes an entirely social phenomenon: No longer predicated on objective advances in scientific knowledge, it

results exclusively from processes of definition and labeling, in which the scientific community and other powerful agents decide which theories and which scientists are to be considered excellent.

3. Also see Meyenn 1994.

4. For critical views, see Brush 1980; Hendry 1980; Kraft and Kroes 1984.

5. Also see Merton's general sociological concept of the "opportunity structure" (1996, 153–61).

6. For a Marxist interpretation of the influence of Einstein's theories on culture, see Kuznetsov 1979b.

7. Szende (1921), for instance, argued that the theory of relativity had social relevance because it represented and supported the wider societal and cultural principle of relativism.

8. Einstein's letter to E. Zschimmer, September 30, 1921; see Holton 1996, 131–32, and Berlin 1980, 10. On the impact of Einstein's work on culture, see Cassidy 1995.

9. Reiser thought the religious phase started even before entry into elementary school, but this seems improbable (1930, 28–29).

10. In addition, Albert, the only Jewish pupil, experienced a heavy dose of anti-Semitism from his schoolmates (Winteler-Einstein 1987, lx n. 44). In this context, too, Einstein's turn to religious Judaism may have been a gesture of defiance and an assertion of his independence. Although Albert's teachers usually did not exhibit any anti-Semitism, Marianoff reported one rather blatant anti-Semitic episode. A teacher one day brought a large nail into class and explained to the students that this was one of the nails the Jews used to crucify Jesus (1944, 30–31). In Marianoff's account, it remains unclear whether this happened in primary school or in the Gymnasium.

11. Max Talmud (later Talmey), a Jewish medical student from Poland, was a weekly dinner guest in the Einstein home and became an informal tutor to young Albert.

12. Einstein's view of the Munich Luitpold Gymnasium may have darkened somewhat in hindsight (e.g., Einstein 1955b, 146; 1956, 9–10; Marianoff 1944, 30–31). Despite his negative memories, this Gymnasium was actually among the more liberal and progressive ones in Germany (Pyenson 1985, 3–6). Through the indictment of his own school, Einstein perhaps expressed his more general disgust at the distortion of *Bildung* that, in his view, the German school systems perpetrated. As Pyenson commented, "We can only speculate on what irreparable damage might have been inflicted on the young genius had Einstein attended a more traditionally inclined secondary school" (1985, 6). Later in life, Einstein campaigned for the abolition of the dreaded final exam. Under the title

"The Nightmare," he published a short essay in the *Berliner Tageblatt* (December 25, 1917; in *CP*, 6: 581) that attacked the *Reifeprüfung* (also called *Abitur* or *Matura*) for traumatizing many students and for lowering the standards of education through placing a premium on useless rote-learning and drill. Looking back on his educational experience in Munich, Einstein often likened the schools to the military, calling his primary school teachers sergeants and his secondary-school teachers lieutenants (Frank 1949a, 24–25).

13. When their Munich business failed in June 1894, the Einstein brothers made a fresh start in northern Italy (which also turned out to be unsuccessful). After Albert's parents had moved to Milan, he stayed in Munich with distant relatives.

14. As a condition of his appointment as a professor at the German University at Prague in 1911, Einstein formally had to declare a religion, which he did, reluctantly, as "mosaisch" (Frank 1949a, 137).

15. Also see Lanczos 1974, 4.

16. Helen Dukas insisted that Einstein's lifestyle in Zürich and Bern was "anything but 'bohemian'" (Pyenson 1985, 77 n. 9). Yet she had of course no firsthand knowledge of that period in Einstein's life. It is also likely that she felt protective of Einstein's "respectability," as she may have understood it.

17. For instance, in 1901, Einstein was hired as a *Privatlehrer* (private teacher) in Dr. Jakob Nüesch's private *Lehr-und Erziehungsanstalt* (Educational Institute) in Schaffhausen, Switzerland. He soon fell out with his employer and left his position early. In a letter to his friend Habicht (February 4, 1902), Einstein commented, "I shipped out from N[üesch] with a big bang" (*CP*, 1: 331).

18. Einstein withdrew the dissertation he had submitted to Zürich University early in 1902. His second attempt, again at Zürich University, succeeded in 1905.

19. On Mileva Marić, see Gabor 1995 and Trbuhović-Gjurić 1983.

20. Einstein expressed his disdain for the philistines fervently in a letter to his friend Besso from Prague (May 13, 1911): "My position and my institute here give me a lot of pleasure. Only the people I find so strange. These are people with no natural feeling at all; without heart, and with a peculiar mixture of snobbery and servility, without any good will toward their fellow human beings. Ostentatious luxury and, right next to it, creeping misery in the streets. Barren of thought, without faith" (Einstein and Besso 1972, 19). In the same letter, he addressed Besso as a fellow gypsy: "You, too, are a bit of a gypsy of this kind" (ibid., 20).

21. *Habilitation*, a special feature of Central European academia, was an academic degree above the doctorate that certified its holder's ability to teach at the university level. It still exists today.

22. It is remarkable how swiftly Einstein's academic career took off after a rather rocky and inauspicious start. Einstein had certainly not accumulated the resources, network connections, early recognition, and other advantages that usually portend successful academic careers. As Merton put it, the social system of science has an "institutionalized bias in favor of precocity"—those who, early on, are considered promising young scientists reap many advantages that are denied those who are judged less brilliant, which sets in motion the process of a self-fulfilling prophecy (1996, 323). Einstein beat this bias and became a "late bloomer," solely on the strength of his dazzling publications, for which he needed little more than his brilliant mind. Had Einstein not been interested in theoretical physics, but in an experimental or observational branch of the sciences, it would have been much more difficult, if not impossible, to do the kind of breakthrough work on his own that would catapult him from the patent office into the top echelons of academe. Nonetheless, the theoretical physics community still had to recognize the caliber of Einstein's work. Here Max Planck's influence was crucial. As editor of the *Annalen der Physik*, he immediately understood the significance of the unknown young physicist's work and supported it.

23. In addition, Einstein became the founding director of the Kaiser Wilhelm Institute for Physics, which opened in 1917.

24. In fact, all his life Einstein managed to avoid turning out more than a single PhD of his own (Seelig 1954, 125). The graduate student's name was Hans Tanner, and Einstein supervised his dissertation when he was a professor at Zürich University.

25. Einstein's interest in the Zionist cause may go back as far as to his stay in Prague in 1911–12. Prague was among the European cities in which national and ethnic tensions were the most virulent. A sharp antagonism existed between the German minority and the Czech majority, and a sizeable Jewish population was caught somewhere in the middle. This atmosphere might have sensitized Einstein to Jewish nationalism. In Berta Fanta's salon, he met the Zionist Hugo Bergmann, Fanta's son-in-law (Fölsing 1993, 321). However, Bergmann could not remember ever to have discussed Jewish problems with Einstein (ibid., 322), and Einstein apparently kept his distance from the Prague Jewish community. According to Gilbert, Einstein formed a greater affinity to Zionism in 1920, under the influence of the prominent German Zionist Kurt Blumenfeld (1980, 22).

26. A collection of FBI files on Einstein is available on the Internet at http://foia.fbi.gov/einstein.htm.

27. The Swiss military medical exam in 1901 found excessive foot perspiration (*CP*, 1: 278). Pyenson surmised that the real reason for Einstein's disdain of socks was that he liked the feel of leather (1985, 71).

28. Fölsing 1993, 29. Or *Biedermaier* (Moszkowski 1922, 220), *Biedermeier* (Hermann 1994, 78; Herneck 1963, 26).

29. The Einsteins gave them to the younger brother of Herta, their domestic servant, as a present.

30. Einstein's ethical point of view was not without what one might consider blind spots. In the 1930s, Einstein supported Stalin's purges as necessary under the special circumstances of the Soviet Union (Fölsing 1993, 727). In a 1937 letter to Born, he dismissed the reports of Stalin's show trials as propaganda (Einstein, Born, and Born 1969, 179). After World War II, he condemned the German people collectively for the Holocaust.

31. See also the characteristic book title *Einstein's Revolution: A Study in Heuristic* (Zahar 1989).

32. The point that Einstein was *no rebel* in physics was also made by Forman (1981) and Pyenson (1985, 60). Infeld (1950) himself, however, later wrote a book that prominently featured Einstein's scientific "revolution."

33. Reiser (1930, 194), Bucky (1992, 52), and Konrad Wachsmann (1990, 144) reported pictures of Faraday, Maxwell, and Schopenhauer. Hoffmann and Dukas, however, mentioned pictures of Faraday, Maxwell, and Newton, noting that the portrait of Newton had been lost (1972, 46). Wachsmann (1990, 144) explicitly denied seeing a Newton portrait in the study and added that he saw one in the library. But a photograph of Einstein's study, reprinted in Hermann, clearly shows a picture of Newton (1994, 355).

34. In addition to Mozart, Kuznetsov mentioned Bach, Haydn, and Schubert as Einstein's favorite composers (1965, 89). Also see Kuznetsov (1979a, 437–58).

35. Most of the "classic" German literature was written in the few decades before and after 1800. Then came a time of canonizing this extraordinary literary wealth. In the latter part of the nineteenth and the early part of the twentieth centuries, it was quite common to assemble "best book" lists of the outstanding works of literature. In 1911, Heinrich Falkenberg compiled a bibliography of such "Listen der besten Bücher" in the *Zeitschrift für Bücherfreunde*. The earliest of its forty-six entries was Johann Neukirch's (1853) *Dichterkanon*. In Neukirch's compilation, as well as in the subsequent ones, Goethe played a dominant role.

Around 1906, the Viennese bookseller Hugo Heller polled a number of intellectuals about their choice of "the ten best books." A selection of the responses was printed in the *Jahrbuch deutscher Bibliophilen und Literaturfreunde* (Feigl 1931, 108–27). As one might expect, Goethe figured prominently in these replies, both explicitly and implicitly. Marie von Ebner-Eschenbach, for instance, who did not list Goethe, noted, "I believe I am in accord with the basic intention of your enterprise when I do not mention that which goes without saying" (ibid., 116). Similarly, Emil Marriot wrote, "I deliberately avoided our classics. I read them time and again; but to recommend these works would be carrying owls to Athens [i.e., totally redundant]" (ibid., 122). These intellectuals considered the canon of classic works so self-evident that they judged it superfluous to endorse these works on a ten-best-books list. At the time, the consensus about the classic literary canon, with Goethe at its head, was so strong that it did not have to be made explicit.

Much has changed since then. At the close of the twentieth century, the German weekly *Die Zeit* again asked a group of intellectuals about the literary canon (reported in *Die Zeit*, no. 21 [May 16, 1997]). This time they were to nominate only three to five works that they thought should be compulsory reading for German *Gymnasiasten*. Goethe and his *Faust* still received numerous nominations, but now many respondents lamented the almost complete erosion of the classic literary canon since the 1970s. Indeed, the *Zeit* project was intended to help resurrect a canon that had clearly faded. The report started with the statement that nowadays "up to 90 percent of those who begin to study German at a university, do not know 'Faust'"—which to the earlier generations of *Bildungsbürger* would have sounded utterly unbelievable.

36. This database for all books remaining after his death was compiled by NHK (Japan Broadcasting Corporation).

37. On Einstein and Dostoyevsky, see Kuznetsov 1972. When C. P. Snow (1979) visited Einstein in the summer of 1937, he found that Einstein's favorite novel was *The Brothers Karamazov* and his second-favorite novel, *Don Quixote*. In addition, Snow mentioned *King Lear* as one of the physicist's favorites. Another reading list can be found in Pais (1982, 16). Auguste Comte, the great French sociologist, whose program of applying the scientific method to the study of society might have interested Einstein, was notably absent from his reading. Comte remained relatively unknown in the German-speaking parts of Europe at the turn of the century. German translations of his works were slow to appear (see chronology in Comte 1994 [1844], xliii–xliv). In 1914, none other than Wilhelm Ostwald edited

Comte's *Prospectus des travaux scientifiques nécessaires pour réorganiser la société,* almost a century after it was first published in 1822. (See Comte 1914 [1822].)

2.
KULTUR

People nowadays like to talk about culture. A recent search for the word *culture* on the Internet revealed 28.7 million hits. Within the narrower purview of the scholarly publications referenced in the Harvard University electronic library catalog, a keyword search for *culture* resulted in the much smaller but still substantial number of 61,385 items. Because of its ubiquitous use, the meaning of *culture* is often taken for granted. Many people, I suspect, may not even realize that today's predominant usage of the concept is relatively new, and they may not know how seminal intellectual developments in Central Europe during the late eighteenth and early nineteenth centuries propelled the notion of *Kultur* to enormous prominence and in the process also changed the meaning of *culture*.

Kultur and *culture* share a common origin in the Latin word *cultura*.[1] At first denoting cultivation in the concrete sense of agriculture, *cultura* later acquired more general and abstract meanings. A milestone in this development was Cicero's famous statement, "Philosophia cultura animi est [Philosophy is the culture of the mind]." The dominant meanings of both *Kultur* and *culture* can be traced back to this notion of *cultura animi*.

When Samuel Baron von Pufendorf, a seventeenth-century legal scholar, introduced the notion of Kultur to the German public, he still used its Latin form *cultura*.[2] It is sometimes said that Johann Gottfried Herder's *Ideen zur Philosophie der Geschichte der Menschheit* (1784–91) first recorded the Germanized form *Cultur* (e.g., Rauhut 1953, 85). Yet Herder's usage can be traced back even further to his *Versuch einer Geschichte der lyrischen Dichtkunst*, written in 1764 (Bayer 1975, 334; Taylor 1938, 16). "And what more can we ask of hymns than that they sing the virtue of their age, the virtue that can be sung most forcefully! But of course: Because it [the virtue] was also related to strong passions, hatred, bias, ambition, and cruelty, we must weigh them on no other scales than those of the stage of Kultur that was reached at that time, and with no other weights than those of political and poetic virtue" (Herder 1968 [1764], 126). Furthermore, "Because of the Great Flood, the human spirit descended considerably from the achieved stage of its Kultur: reduced to the bare necessities of life and depressed by extreme deprivation, it was impossible for the mind to lift itself up and expand" (ibid., 138).

Whereas Pufendorf's understanding of the concept remained close to *cultura animi*, that is, the process of an individual's education and refinement, Herder's *Kultur* was a complex mixture of individual and collective elements and thus foreshadowed a specifically German development.[3] In English, *culture* originally was a synonym for tillage or husbandry, but it early on acquired more metaphorical meanings along the lines of *cultura animi*. In 1510, Sir Thomas More spoke of "the culture and profit of their minds," and Thomas Hobbes, in his *Leviathan*, equated the education of children with "the culture of their minds" (Levin 1965, 1).

2.1. KULTUR AND BILDUNG

Around the turn of the nineteenth century, *Kultur* and *culture* began to diverge, as the meaning of Kultur evolved into a distinctly supra-individual, collective concept, whereas the meaning of culture remained unchanged. Why did this happen? We find a first clue when we consider Kultur in connection with its companion concept *Bildung*. Bildung is not a neologism but stems from a long religious tradition going back to the Christian idea of humans being formed in the image (*Bild*) of God. To German mysticists, notably Master Eckhart (1260–1328), the concept was crucially important. In the eighteenth century, the poet Christoph Martin Wieland (1733–1813) secularized the concept of *Bildung* under the influence of Lord Shaftesbury. The latter's "formation of a genteel character" read *Bildung* in the German translation.[4]

Bildung thus emerged as a synonym for the original meaning of Kultur as individual cultivation. Consequently, it became more easily possible for the term *Kultur* to focus on collective structures, primarily at the level of nations. This then also made the dichotomy, at the collective level, between Kultur and *Zivilisation* more feasible. In the evolving division of labor between Kultur and Bildung, the two terms remained closely connected and were even defined in reference to each other: Bildung was the subjective-individual, Kultur the objective-collective side of the same coin. Bildung was not individual development in a vacuum; it was the process through which individuals acquired the structures and products of Kultur. Kultur, in turn, was sustained and advanced by *gebildete* individuals, who were often called the *Kulturträger*—a German term that has the double meaning of *carriers* and *pillars* of Kultur. (On the one hand, *gebildete* individuals were seen as personally carrying Kultur. On the other hand, they also functioned as the chief supporters—"pillars"—of the nation's *collective* project of Kultur.) Although the term *Kulturträger* itself became most popular only after World War I, it was already a key concept before the war, as the following episode illustrates. In Prussia in 1910, a bill—which eventually failed—proposed a change in the three-tiered electoral law that expressly favored Kulturträger, by which they would be put in a smaller pool of voters above the class for which their wealth would qualify them so that their votes would count more (Gebhard 1962, 3: 305).

Whereas the availability of Bildung as a synonym for the individual-level contents of Kultur certainly facilitated the shift of Kultur to the collective level, it is hardly a sufficient explanation. To shed more light on this development, we must ask what features of late eighteenth– and early nineteenth–century Germany contributed to the emergence of a collective notion of Kultur and how they shaped its content. Following Norbert Elias's classic discussion of this question, we distinguish *national* and *social* roots, which, in their combination, created a uniquely German definition of Kultur (1939, 1: 1-42).[5] We first turn to the social root.

In terms of social stratification, most of the Kulturträger could be identified as belonging to what has been called the *Bildungsbürgertum* (the educated members of the bourgeoisie)—more about an important exception later.[6] Karl Mannheim usefully distinguished two components in the modern bourgeoisie (1936, 156). It had, he wrote, from the beginning two kinds of social roots—"on the one hand the owners of capital, on the other those whose only capital consisted in their education."[7] In nineteenth-century Germany, the latter formed the mass of the Bildungsbürgertum; their social ranks were symbolized by the certificates they had attained during the process of Bildung and often also by being a member within the hierarchies of the civil service. *Bildungsbürger* were people who had received a Gymnasium and a university education and who worked predominantly in professions that required such training: physicians, lawyers, clergy, as well as teachers, professors, and other higher officials in government service (*Beamte*).

It was the Bildungsbürgertum who tended to use the concept of Kultur to legitimize its claims for social position. The *social* root of Kultur lay in its use as a fighting term against the elevated social status of the aristocrats, although it increasingly served other purposes as the nineteenth century went on. It signifies the political and economic weakness of the German Bürgertum that the main attack on the aristocracy was carried out in the domain of Kultur. (In

France, of course, the bourgeoisie took political control, and its successful revolution terminated the aristocrats' privileges. In Britain, the middle classes channeled their energies perhaps more into the pursuit of economic predominance.)[8] As Elias put it, "Here is a stratum [the Bildungsbürgertum] that is being kept away from political activity, that hardly thinks in political categories and only tentatively in national categories, and whose whole legitimation depends on its intellectual, scholarly or artistic accomplishments; there, opposed to it, is an upper class that, in the other's perspective, does not 'accomplish' anything, but whose self-perception and self-legitimation centers on the formation of distinguished and distinguishing manners of behavior" (1939, 1: 8–9). The champions of Kultur and Bildung riled against what they considered the aristocrats' unearned and undeserved status, against their depravity and overall bad behavior. Most aristocrats had, in fact, quite sophisticated manners that sprang from court ceremonials, but the proponents of Kultur despised, and fought against, this supposed apex of accomplishment. To them, it was nothing but a polished surface concealing a rotten core. Here lies one of the seeds of a peculiar Kultur-Zivilisation dichotomy, which we will explore.

As the social configurations evolved in the process of modernization, so did the concepts of Kultur and Bildung. Middle classes often engage in a sort of two-front social conflict: While striving to tear down the barriers that keep them from the classes above, they also try to reinforce the barriers that separate them from the classes below. In our concrete case, Kultur and Bildung initially served as the Bildungsbürgertum's rallying cries in its attack on the aristocracy. That original target of the Bildungsbürgertum's rhetoric indeed began losing ground, but new and equally disliked foes materialized. The emergence of mechanization, large-scale industry, unruly proletarian masses, the social ascendancy of capitalists (*Besitzbürgertum*)—all these developments threatened to nullify the vision of the Bildungsbürgertum. The Bildungsbürger loathed a society they saw transforming into one big machine (Harrington 1996, 20). Some of them succumbed to what Fritz Stern (1974) aptly called "cultural despair," which drove them toward a pathetic irrationalism and mystical nationalism that contributed to the meltdown of Kultur in the twentieth century.

Increasingly, the concepts of Kultur and Bildung were put to defensive use, protecting the status of the Bildungsbürgertum against the newly emerging social strata, proletariat and economic bourgeoisie. Early on (in the late eighteenth and early nineteenth centuries), Bildung, for instance, contained a strong universalist and egalitarian element of general "Menschenbildung" (Vierhaus 1972, 528). It was also strictly nonutilitarian in its focus on the development of an individual's character and well-rounded personality.[9] It was, according to Reinhart Koselleck, "a concept that—like religion—balks at its restriction to social boundaries and always calls for transcending them" (1990, 29). Since the second half of the nineteenth century, however, Bildung was increasingly regarded as a property that bestowed social entitlements as a precondition for entering the privileged class of the Bildungsbürgertum (Vierhaus 1972, 543).

At that point, the very rhetoric of equality and meritocracy that the middle classes employed against the upper classes was hurled back at them by their own critics. Emphasizing the earlier universalistic element of Bildung, the working-class movement condemned what it considered a degeneration of Bildung into a source of social privilege and mounted large-scale efforts to appropriate Bildung and Kultur for the workers. And in the *Gymnasien*, young Albert Einstein and many others rebelled not against Kultur and Bildung as such, but, *in the name of Kultur and Bildung*, against what they considered their corruption. Ulrich Linse (1976) astutely characterized the *Jugendkulturbewegung* (Youth Culture Movement) into which the youthful discontent grew around the turn of the twentieth century (e.g., the "Wandervogel" movement, originating in 1901) as a protest movement *within* the Bildungsbürgertum, as a "Gebildeten-Revolte."[10] Here we find a potential for rebellion woven into the very fabric of Kultur itself—universalistic ideals of Bildung and Kultur that can serve as the basis for an *immanent* critique of societal reality (by which the existing society is criticized for falling short of *its own* ideals). We find Einstein among this group. On the other hand, there was, of course, also the potential for a much more radical and total rebellion that not only condemned certain shortcomings of societal reality but also cast off the ideals themselves and espoused various brands of irrationalism and nihilism.

Fritz Stern emphasized that dislike of the German Gymnasium was widespread and intense: "[W]e have still to understand why so many young Germans felt such revulsion for it. The Youth Movement was only the most dramatic rebellion against what its members called the artificiality and pedantry of these schools. Much of the irrationalism and hatred of 'system' which characterized German youth sprang up in opposition to these schools" (1974, 271). More and more, Bildung was filled with "patriotism, duty, and discipline" (Mosse 1985a, 13). The following are some particularly telling examples from a selection of themes of German essays in the final exam and in the upper grades of the Prussian *Gymnasien* at the turn of the century (Herrmann 1990, 355–56):

> Humans are not born to be free.
> The benefits of the military draft.
> Why does Germany need colonies?
> What thoughts does the recently unveiled monument of Emperor
>   William inspire in us?
> Even war has its positive side.
> Death has a cleansing force.[11]

Here the reversal of the earlier meaning of Bildung is almost complete. Whereas at the beginning of the century, Bildung exhorted individuals to unfold and develop their personality on the path toward personal autonomy, it now prepares them for personal annihilation in the service of a military state. Like many others, young Einstein rebelled against the latter variant of Bildung by embracing the former.

Whereas many of the Kulturträger entered the solidly middle-class formation of the Bildungsbürgertum, some were forced, or chose, to remain at the margins of society. Those were the *free-floating intellectuals*. Karl Mannheim identified them as the carriers of unbiased knowledge. In his view, they were free of the usual ideological distortions a social position imparts because they were "free-floating"—marginal existences that lacked a well-defined place in society. "Such an experimental outlook, unceasingly sensitive to the dynamic nature of society and to its wholeness, is not likely to be developed by a class occupying a middle position but only by a relatively classless stratum which is not too firmly situated in the social order. . . . This unachored, relatively classless stratum is, to use Alfred Weber's terminology, the 'socially unattached intelligentsia' (*freischwebende Intelligenz*)" (Mannheim 1936, 154–55). Not surprisingly, the antipathies between the two segments of the Kulturträger often expressed themselves in arguments over Kultur itself, each group considering the other a cultural threat. For the Bildungsbürger, the intellectuals were shiftless negativists who were chipping away at the very foundations of Kultur. For the intellectuals, the Bildungsbürger were merely pretentious and otherwise clueless "philistines." Young Albert Einstein's struggles with his family gives us a taste of this conflict.

Although the Bildungsbürgertum's initial attack on the aristocracy could have conceivably employed the original meaning of Kultur as individual cultivation, the term took on a collective meaning that attached to the Bildungsbürgertum as a stratum and fostered group cohesion and identity. This was greatly helped by the aristocrats' immediately discernable foreign predilections, which made them stand out clearly from the rest of the population. As late as 1780, Frederick the Great of Prussia would lament, in French of course, that German was merely "une langue à demi-barbare" (quoted in Elias 1939, 1: 13). By that time, Elias remarked, writers such as Goethe, Lessing, Klopstock, and Herder— giants among the Kulturträger—had already published many of the greatest works of Kultur, but the Prussian king apparently had not yet taken note (ibid., 1: 14). In this attitude Frederick was not exceptional. Large parts of the aristocracy were thoroughly Francophile and tried hard, within their often modest means, to emulate the "civilization" emanating from the French court. Those German aristocrats preferred to speak French and were embarrassed, even horrified, by what they saw as barbarism surrounding them in their native lands. Thus, because the original social target of the Bildungsbürgertum's Kultur rhetoric, the aristocracy, was also the class in which foreign (French) influences prevailed most strongly, the social and national roots of the term *Kultur* were closely intertwined. We now turn to the latter.

2.2. KULTUR AND ZIVILISATION

The peace treaty of Luneville in 1801 concluded one of several wars that brought military humiliation to the German states at the hands of the seemingly unstoppable Napoleon. This event inspired Friedrich Schiller, next to Goethe the most revered of the German wordsmiths, to write the outline of a poem in which he would address the question, "At this moment, when the German emerges from his tearful war without glory, may he hold his head high and carry himself with self-confidence in the community of nations?" (1968, 447). Schiller never completed the poem, but the outline already contained his poignant answer, "Yes, he may! He was unfortunate in the fight, but he has not lost that which amounts to his worth. The German Empire and the German nation are two different things. . . . Separate from the political sphere, the German has laid the foundation of his own worth, and even if the Empire went under, the German dignity would remain untarnished. It is a moral entity; it dwells in the Kultur and in the character of the nation" (ibid.).[12]

Schiller's argument illustrates a more general point about the national importance of Kultur. We realize that, for the lack of political unity, military strength, and economic development, Kultur—in the sense of the collective characteristics of the nation—became essential for the nascent German national identity and German nationalism. In his chapter in Kroeber and Kluckhohn's *Culture*, Alfred Meyer put it succinctly: "Kultur theories can be explained to a considerable extent as an ideological expression of, or reaction to, Germany's political, social and economic backwardness in comparison with France and England" (1963, 404). Not only did this backwardness make the concept of Kultur important, it also shaped its contents. As Kroeber and Kluckhohn wrote in the same volume, "Being politically in arrears, their [the Germans'] nationalism not only took solace in German cultural achievement, but was led to appraise culture as a whole above politics as a portion thereof; whence there would derive an interest in what constituted culture" (ibid., 52). This is the national root of the concept of Kultur. We should emphasize again that it was entwined with the social root.

For although nationalists everywhere tend to argue that nationalism is a somehow "natural" aspiration of a "people," German nationalism itself rose under specific historical and social conditions and reflected the interests and values of certain social classes, most especially the Bildungsbürgertum under the leadership of the professorate, whom Fritz Ringer called the "mandarins." In his words, "the nation and, through it, the state were defined as creatures and as agents of the mandarins' cultural ideals" (1969, 117). The crucial role that Kultur played in founding the German nation contributed to the social prominence of the Bildungsbürgertum—as pillars of Kultur—in Germany.

When Kultur first acquired its collective meaning, it retained the notions of development and improvement that had been central to the earlier meaning and applied them to the group level. Kultur now tended to be seen from the perspective of an evolutionary "Kulturgeschichte," in which the idea of stages of Kultur became paramount. As we have seen, such a concept of Kultur goes back to Herder who already in 1764 spoke of a "stage of Kultur" in a collective sense (1968, 138). A major exponent of this evolutionary view of Kultur in the nineteenth century was Gustav Klemm with his *Allgemeine Cultur-Geschichte der Menschheit* (1843–52) and *Allgemeine Culturwissenschaft* (1854–55).

The 1885 edition of the *Brockhaus' Conversations-Lexikon* reflected the shift toward the collective meaning of Kultur, but in a way that testified to the inherent conservatism of enterprises of this sort. While the definition of the entry *Kultur* lagged behind, the new collective meaning had already appeared, somewhat surreptitiously, through the back door of the related entry *Kulturgeschichte*. According to this edition of the *Conversations-Lexikon*, "Kultur denotes the activity that is applied to an object to improve it or make it suitable for certain purposes, or the success of this activity. One therefore speaks of both the Kultur of a field, understood as the clearing and planting of it, and the Kultur (Ausbildung) of the mind, the Kultur (nurturing) of the arts and sciences, etc." But this quite traditional definition was immediately followed by a much longer entry on *Kulturgeschichte*, which presents the newer meaning of

Kultur: "Kulturgeschichte is the historical account of the whole formation process [*Bildungsprozesses*] of humankind, from the beginning of human reasoning to its accomplishments in the present. Kulturgeschichte rests on the fundamental hypothesis that humanity is destined to Kultur, and that Kultur itself is capable of continuous progress, that its end point, especially in the area of conquests of the external nature, cannot be determined, and that the progress of Kultur is tied to the perfection of humanity."

Clearly, the subject of cultural improvement was no longer the individual but the whole of humankind in its constituent groups. Kultur as "Bildungsprozess of humankind" had an immediate evolutionary aspect; one could now order the cultures of particular groups and nations along the path of humanity's cultural progress. This evolutionary sense of Kultur retained a strong flavor of Enlightenment thought. For the champions of the German nation, however, such a concept of Kultur was fraught with problems, which the following discussion of the Kultur-Zivilisation distinction will elucidate. In French, the word *civilisation* was by and large a synonym of *culture*; it was primarily contrasted with *barbarie* (Kroeber and Kluckhohn 1963, 16–17). Although many Germans also used *Zivilisation* and *Kultur* interchangeably, others began to dichotomize these two concepts around the turn of the nineteenth century.[13]

The eminent philosopher Immanuel Kant had already differentiated, in his *Idee zu einer allgemeinen Geschichte in weltbürgerlicher Absicht* (Idea for a Universal History from a Cosmopolitan Point of View), between the states of being *cultiviert*, which was achieved through the arts and sciences, and *civilisiert*, which was achieved through social sophistication and decency (1975b [1784], 44). Wilhelm von Humboldt expanded on this dichotomy (1836, 1: xxxvii). He defined *Zivilisation* as the qualitative improvement (*Veredlung*) of the social condition, the increased control and refinement of elementary human impulses. Kultur was assigned to the domain of the arts and sciences. For the scholars following Humboldt—among them, for instance, Albert Schäffle (1896 [1875–78]) and Julius Lippert (1886–87)—*Kultur* denoted the control over nature by means of science and technology, whereas *Zivilisation* described features of social relationships. Ultimately, however, the Humboldtian distinction of Kultur and Zivilisation went under. Kroeber and Kluckhohn found its last German usage in Paul Barth's *Die Philosophie der Geschichte als Sociologie* (Kroeber and Kluckhohn 1963, 25; Barth 1897). For Barth, *Kultur* referred to the control humans exercise over nature, *Zivilisation*, to their control over themselves.

Humboldt's brand of Kultur-Zivilisation distinctions did not gain predominance in Germany, in part, because Kultur had become the rallying cry of the German nation. The definitions by Humboldt and others led to acceptable consequences as long as one compared "savages" with the European nations, which all could more or less be grouped together as "advanced" in this broad perspective. But once the Germans of the nineteenth century compared themselves to England and, especially, France, difficulties arose. These countries were undeniably more modern and advanced than Germany in many ways—in their technological, economic, and military accomplishments, as well as in their sociopolitical structures and their sophistication of manners. In short, it was difficult, impossible almost, to dispute the French (and English) advantage in both Kultur and Zivilisation, as defined in the Humboldtian tradition, and such a conceptual framework could quickly lead to some bothersome conclusions for German patriots. Here may lie one of the potential causes for the popularity of a peculiarly German dichotomy between Kultur and Zivilisation.[14] In a marked departure from the Humboldtian definition, the new conceptual distinction tied Zivilisation mainly to "external," material and technological aspects of the structures and products of a human group, whereas Kultur referred to the "internal," spiritual ones. This distinction became dominant in the late nineteenth and early twentieth centuries.

In his sociological classic *Gemeinschaft und Gesellschaft*, Ferdinand Toennies (1970 [1887]) associated *Gemeinschaft* (community) with Kultur (traditional customs, religion, arts), and *Gesellschaft* (society) with Zivilisation (law and science). As Gesellschaft is superseding Gemeinschaft in Toennies's scenario, so Zivilisation is superseding Kultur. Later on, Oswald Spengler's (1922–23 [1918–22]) doomsday book *Untergang des Abendlandes*, published in the wake of World War

I, became highly popular among the wider German audience. In it, Zivilisation was the terminal, sterile stage of a Kultur. In a similar vein as Toennies, Alfred Weber (one of whose students was Norbert Elias) defined Zivilisation as intellectual and rational, serving the utilitarian purpose of controlling nature, in his 1912 talk "Der soziologische Kulturbegriff" (1927, 31–47). Kultur, by contrast, was nonrational, based on emotion and values, served no external purpose, and was noncumulative. These scholars and writers availed themselves of the existing terminology for their theoretical frameworks, not necessarily with a nationalist purpose in mind. However, their usage rendered the concepts, especially the dichotomy of Kultur and Zivilisation, even more popular in the wider public, who found them conducive for legitimizing German national aspirations.

For a broad characterization of this pervasive way of thinking, one might point to three major shifts in meaning that created the newer and dominant Kultur-Zivilisation dichotomy. First, the previous (Humboldtian) notions of Kultur and Zivilisation just about traded places, with Zivilisation now referring mainly to the material and external aspects of controlling nature and Kultur referring to the nonmaterial and internal aspects of society—particularly the most central ones. While Kultur was extracting the core fibers that constituted the social domain, it left to Zivilisation merely the empty husk of formal and superficial behaviors. Second, the term *Kultur* took on additional meanings; it turned into something no longer completely within the boundaries of rationality and Enlightenment-style universal progress. A Kultur also became the unique embodiment of the collective soul or spirit of a particular nation, and, in this sense, took a relativistic turn. From that time on, the notion of Kultur acquired a Janus face—looking forward to the qualitative uplifting of human conduct and to progress in intellectual and artistic achievements, and looking backward to ancient manifestations of the national spirit in fairy tales, myths, and rural customs. Third, Zivilisation was typically relegated to secondary importance, while Kultur was what really counted and what became regarded as the cornerstone of the German nation. The Janus-faced ambiguity surrounding the concept of Kultur, in concert with its overriding importance in the conceptual landscape, made Kultur a prime focus of ideological contention. With enormous intensity, German intellectuals, and Einstein among them, as we shall see, argued over which of the varied elements in Kultur should prevail.

From the vantage point of the Kultur-Zivilisation dichotomy, the French superiority in Zivilisation revealed itself as superficial and irrelevant. A focus on Zivilisation was considered a misguided priority prone to sterility and decadence; true greatness could only be found in Kultur. The dichotomy of Zivilisation and Kultur now echoed the one between Western Enlightenment and German Romanticism, or the one between reason and soul. In extreme cases, Zivilisation could even be entirely reduced to something English or French, Kultur to something German. In the later part of the nineteenth century, many Germans embraced some distinction of Kultur and Zivilisation along these lines.

The German national resentment against Zivilisation reached a height during and after World War I (Elias 1939, 1: 7)—"the war of 'Zivilisation' against Germany," as Thomas Mann called it, and reversely, the resumption of the "age-old German struggle against the spirit of the West" (Mann 1920, xxxv, 7). During the war, Mann condemned Zivilisation and its companions, such as democracy, in stark words as essentially un-German: "The difference between spirit and politics includes that between Kultur and Zivilisation, between soul and society, between freedom and enfranchisement, between art and literature; and German-ness [*Deutschtum*]: that is Kultur, soul, freedom, art, and not Zivilisation, society, enfranchisement, literature" (ibid., xxxiii). And, "The spirit of politics, in the guises of democratic Enlightenment and 'human Zivilisation,' is not only psychologically anti-German; it necessarily is also politically hostile to Germans, wherever it shows itself" (ibid., xxxiv). The preceding quotes should not be considered representative of the author's large body of work; they immediately flowed from the deep emotions stirred up by World War I. However, the fact that even a person like Thomas Mann would write such things illustrates powerfully the pervasiveness of these ideas in the German educated public.

When German-American Friedrich Wilhelm von Frantzius (1916) tried to influence the predominantly anti-German public opinion in the United States after the outbreak of World War I, he wrote a book whose German edition was titled *Deutschland, der Träger der Welt-Kultur: Wissenswerte Tatsachen für jeden Deutschen und Deutsch-Amerikaner* (Germany, the Carrier of World Kultur: Facts that Every German and German-American Should Know). In addition to celebrating the German contributions in the realm of Kultur, Frantzius's book also embraced the Kultur-Zivilisation dichotomy: "There is a big difference between 'Zivilisation' and 'Kultur.' Many nations possess a high Zivilisation; but to be at a high stage of Kultur, one needs assiduousness, high notions of morality, and a philosophical outlook on life" (1916, 9).

The dichotomy of Kultur and Zivilisation was also acknowledged in the 1922 *Brockhaus Handbuch des Wissens in vier Bänden*, which incorporated the more recent meaning of Kultur into its main entry and defined Kultur as "clearing, planting, and nurturing (of the soil, food plants, forests, also of arts and sciences); then improvement of humans through formation [*Ausbildung*] of the mind and development of all talents, in the philosophical and historical sense, according to newer usage, the unified, configured style in all life expressions of a human or of a people, in contrast to Zivilisation."

Finally, German race theorists also used, and modified, the notions of Kultur and Zivilisation—giving Kultur a biological underpinning. The anthropologist Walter Scheidt (1930, 1934), for instance, advocated a "biologische[r] Kulturbegriff," in which "Kultur consists of . . . Volkstum [folk characteristics] and Zivilisation. The volkstümliche elements of Kultur are 'own-property,' the zivilisatorische elements of Kultur are 'other-property'" (Scheidt 1930, 2: 32, 35). Kultur was seen as a composite of the hereditary, genetic endowment of a people (Volkstum) and of environmental forces (Zivilisation). In this view, Zivilisation thus was universal and could be transmitted to various races; Volkstum, by contrast, remained linked to a particular race. The race-based concept of culture—making German Kultur the property of the Germanic or Aryan race—became dominant under the National Socialist regime.

As the 1990 *Brockhaus Enzyklopädie* summarized, "the differentiation [of Kultur and Zivilisation] in the nineteenth century and the first half of the twentieth century has been used ideologically (O. Spenger) and has become a fighting term in the 'defense' of antimodern ('German') depth and spirituality (Kultur) against modern, Western ('French') 'superficiality' (Zivilisation)."

2.3. CULTURE

In 1871, Edward B. Tylor's *Primitive Culture* established the collectivist meaning of the word *culture* in the English language: "Culture, or civilization, . . . is that complex whole which includes knowledge, belief, art, morals, law, custom, and any other capabilities and habits acquired by man as a member of society" (1873 [1871] 1: 1). Note the synonymous use of *culture* and *civilization*. Tylor borrowed his concept of culture from Germany, where, by that time, it was already established. Tylor's specific sources were Klemm's *Allgemeine Cultur-Geschichte der Menschheit* (1843–52) and *Allgemeine Culturwissenschaft* (1854–55).

The early, Humboldtian, juxtaposition between Kultur and Zivilisation did not die out without leaving an ephemeral trace in America—in Lester Ward's *Pure Sociology* (1903) and Albion Small's *General Sociology* (1905).[15] Later, however, American scholars who employed this terminology did so in concert with the newer German usage.[16] Robert M. MacIver, for instance, equated civilization with means and culture with ends (1931, 226). In a paper titled "Civilization and Culture" (1936), Robert K. Merton criticized MacIver's distinction for being based on differences in motivation; nonetheless, Merton's own distinction of civilization as impersonal and objective, and culture as personal and subjective, was aligned with the newer definition of the Kultur-Zivilisation dichotomy that had risen to dominance in Germany.

> This civilization is 'impersonal' and 'objective.' A scientific law can
> be verified by determining whether the specified relations uni-

formly exist. The same operations will occasion the same results, no matter who performs them. . . . Culture, on the other hand, is thoroughly personal and subjective, simply because no fixed and clearly defined set of operations is available for determining the desired result. . . . It is this basic difference between the two fields which accounts for the cumulative nature of civilization and the unique (noncumulative) character of culture. (Merton 1936, 109–12)

Talcott Parsons was another sociologist who in the early twentieth century popularized the concept of culture among American social scientists (Kuper 1999). For the academic discipline of anthropology, in particular, *culture* (in the collective sense) has become the most central idea. "Culture was the most important concept that held American anthropology together," as Aram Yengoyan put it (1986, 368). Yet one immediately notices that there is no consensus among anthropologists about this concept, despite—or, perhaps, even because of—its centrality to the discipline. For example, some anthropologists, in the mainstream tradition of *Kultur*, emphasize the nonmaterial parts of culture (e.g., values and systems of meaning). Others focus on the material parts (e.g., artifacts, technologies, economic systems), that is, on *Zivilisation* in the preceding German usage.[17]

Moreover, the once-dominant functionalist tradition that has viewed cultures as comprehensive, integrated systems in which all individual elements are interconnected, has come under increasing criticism.[18] As anthropologists have been reaching toward more multifaceted and multileveled accounts of the concept, they have charted internal divisions and fault lines within a larger culture (for instance, between "high" culture and "low" culture[19]). They have also emphasized that the various cultural elements are not static but constantly changing, in concert as well as in conflict with each other. According to distinguished anthropologist Sally Falk Moore, "The functional, totalizing cultural coherence that Malinowski took for granted theoretically no longer has much plausibility. If there is no totally coherent 'whole,' no grand concept of 'the culture' and 'the society' as an integrated consistent unity, there is also no explanatory logical totality into which one can reliably fit any particular cultural item that comes along."[20]

Yet, in a way, the very intensity with which contemporary anthropologists argue about the concept of culture demonstrates how deeply entrenched and central that concept has become. Moore concluded, "Obviously, even if one wanted to, it would be impossible to trash the culture concept because it is so deeply rooted in the history of ideas and in the discipline of anthropology" (1994b, 373). The amazing popularity of the term *culture* in anthropology, as well as in the social sciences and even in everyday language, might abet the false intuition that it is a concept of relatively long standing. It therefore appears worth stressing that, for a long time after its introduction by Tylor in 1871, only small, specialized circles used *culture* in the collective sense (along the lines of Kultur), and it took a "half-century of lag" before the term *culture*, in its new meaning, became fashionable (Kroeber and Kluckhohn 1963, 65). In 1905, when Albion Small defined the word *culture* in his *General Sociology*, he displayed an acute awareness of imposing a rather exotic concept on his readers:

> The very idea of 'culture,' as the term is used among German scholars, has hardly entered distinctly into American calculations. . . . [I]t is necessary to define words in a way not yet adopted as a rule in English usage. What, then, is 'culture' (*Kultur*) in the German sense? To be sure, the Germans themselves are not wholly consistent in their use of the term, but it has a technical sense which it is necessary to define. In the first place, 'culture' is a condition or achievement possessed by *society*. It is not individual. Our phrase 'a cultured person' does not employ the term in the German sense. For that, German usage has another word, *gebildet*, and the peculiar possession of the *gebildeter Mann* is not 'culture,' but *Bildung*. (1905, 59)

In contemporary English, however, the collective meaning of the term *culture*, as introduced by Tylor, Small, and other scholars, has become predominant.

At least two factors appear to have contributed to this lag in the adoption of the collective meaning of *culture* in the English language. First, English lacked a synonym for the old use of *culture* (i.e., the equivalent of Bildung), and, second, it possessed a synonym for the new use (i.e., civilization). In the absence of the Bildung-Kultur differentiation, *culture* thus retained the old meaning of individual improvement for a long time. A strong humanistic tradition, going back to Matthew Arnold's *Culture and Anarchy* (1994 [1869]), insisted on viewing culture as the quest for individual perfection. For Arnold, culture was "a pursuit of our total perfection by means of getting to know, on all the matters which most concern us, the best which has been thought and said in the world" (ibid., p. 5). Culture thus is understood as a process that improves an individual's mind and spirit—in juxtaposition to outward civilization. Adopting the Kultur-Zivilisation dichotomy to his individualistic framework, Arnold noted, "The idea of perfection as an inward condition of the mind and spirit is at variance with the mechanical and material civilization" (ibid., 34).

Reaching at least to the midpoint of the twentieth century, traces of this tradition were still noticeable in T. S. Eliot's *Notes towards the Definition of Culture* (1949), written after World War II. Eliot differentiated three aspects of *culture*—that of the individual, that of a social group or class (especially the elite), and that of society as a whole. Whereas the first of these definitions derived from Arnold's concept of individual self-cultivation, the third derived from Tylor's anthropological usage of the term. The second, intermediate, definition largely rose from a discussion of Karl Mannheim's sociological theory of elites and underlay Eliot's argument for a nonegalitarian society with different levels of culture.

Having completed our review of the conceptual tectonics and landscape of Kultur and its allied terms—concepts that formed the central landmarks in the intellectual life of Einstein and his cohorts—we now turn more specifically to two notions that lay at the very core of Kultur: Weltanschauung and Weltbild.

NOTES

1. The root *cult* of the word *cult-ura* is that of the perfect participle *cult-us/cult-a/cult-um* of the verb *colere*—to cultivate. (From the same root, the future participle of *colere* is also formed, i.e., *cult-urus/cult-ura/cult-urum*.) For the history of the terms *Kultur* and *Bildung*, see Rauhut 1953. For *Kultur*, see Fisch 1992; Hartman 1997, 205–24; Kroeber and Kluckhohn 1963. For Bildung, see Vierhaus 1972; Vogel 1915; Weil 1930.

2. Note, however, that, according to Levin, the German term *Kultur* was borrowed from the French word *culture*, rather than directly from the Latin (1965, 2).

3. For a detailed discussion of *Kultur*, *Bildung*, and allied concepts in Herder's work, see Taylor 1938.

4. See Cocalis 1978; Nordenbo 2002; Rauhut 1953; and Vierhaus 1972.

5. The following sketches what we consider the major lines of conceptual development. Those interested in a more detailed history of *Kultur* can turn to Kroeber and Kluckhohn (1963) and Fisch (1992).

6. *Bildungsbürgertum* is a retrospective term originating only in the 1920s (Engelhardt 1986, 189). In the late eighteenth century, the term *gebildete Stände* came into usage, replacing the earlier term *gelehrte Stände*. For a detailed history of the concept, see Engelhardt 1986.

7. Already Karl Marx noted in his *Kritik des Hegelschen Staatsrechts* that "money and Bildung" were the main criteria for social differentiation in the bürgerliche society (1957 [1843], 284).

8. Equivalents of the social stratum of the *Bildungsbürgertum* existed in many countries, of course, but its social clout vis-à-vis other segments of the middle classes was particularly strong in nineteenth-century Germany. Among the causes, in addition to the political and economic backwardness, one should mention the relative importance of service for the government (*Staat*) in the multitude of German territories large or small, as well as the crucial role of *Kultur* in providing the basis of the German nation (see below).

9. *Bildung* strongly contrasted with *Ausbildung*—the imparting of knowledge and skills necessary for practicing a profession. Bildung in this respect came close to what we consider a "liberal arts education." In this vein, Friedrich Paulsen stressed that Bildung is the opposite of inculcating useful skills and traits by rote-learning: "This is the new ideal of Bildung [based on Rousseau]: not training to become an obedient subject of the

government, a properly acting member of society, a blind follower of a religious system, but Bildung toward becoming a human being, Bildung toward forming a complete, autonomous personality through the development of all powers with which Nature has endowed this being, Bildung toward humanitarianism" (1906, 98–99).

10. Also see Gay 1968, 77–78. Erik Erikson noted that the anti-Establishment phase of *Wanderschaft* became a "well-institutionalized social niche" within Kultur for adolescents who were trying to find themselves (1970, 741).

11. An official Prussian publication of 1892, which sets forth the plan and aims for the *Gymnasien*, illustrates how Bildung had been instrumentalized in the service of the nation. It announces, "Instruction in German is, next to that in religion and history, ethically the most significant in the organism of our higher schools. The task to be accomplished here is extraordinarily difficult and can be properly met only by those teachers who warm up the impressionable hearts of our youths for the German language, for the destiny of the German people, and the German spiritual greatness. And such teachers must be able to rely on their deeper understanding of our language and its history, while also being borne up by enthusiasm for the treasures of our literature, and being filled with patriotic spirit" (*Lehrpläne und Lehraufgaben für die höheren Schulen* 1892, 20).

12. Similarly, Novalis had written in 1799, "While these [other European countries] are occupied with war, economic speculation, and partisanship, the German diligently forms himself into an associate of a higher epoch of Cultur [*sic*], and this progress is going to give him a great advantage over the others in the course of time" (1960e [1799], 519).

13. Jörg Fisch argued that the German language did not know a significant differentiation between *Kultur* and *Zivilisation* until the turn of the twentieth century (1992, 681, 751). Whereas the differentiation indeed reached the climax of its popularity only in the decades before and during World War I, the roots of the differentiation reach much further back in time, as we show in the text. It is also important to note that, at its inception, the meaning of the differentiation was not fixed, but various authors defined *Kultur* and *Zivilisation* in different ways, until the dichotomy settled into what has become the standard meaning.

14. In English and French, *civilization/civilisation* has been the rough equivalent of *culture*—used in the German collective sense of Kultur (Huntington 1996, 41; Braudel 1980, 205; Eliot 1949, 11; Kroeber and Kluckhohn 1963, 52).

15. E.g., "'Civilization' is the positive outcome, up to a given time, of men's working together. It is the sum and the system of men's attainments and accomplishments, measured by human units rather than physical units. 'Culture,' as we use the term at this point, is the total equipment of technique, mechanical, mental, and moral, by which the people of a given period try to attain their ends" (Small 1905, 344). Or, in a direct reference to the Humboldtian dichotomy, "Again, the Germans distinguish between 'culture' and 'civilization.' Thus 'civilization is the ennobling, the increased control of the elementary *human* impulses by society. Culture, on the other hand, is the control of *nature* by science and art'" (ibid., 59). "To translate the German *Kultur* we are obliged to say material civilization. Culture in English has come to mean something entirely different, corresponding to the humanities" (Ward 1903, 18).

16. In the volume *German Culture*, published in 1915, editor W. P. Paterson explicitly introduced the dominant German distinction—defining *Zivilisation* as material and technological and *Kultur* as spiritual (1915, vii).

17. This division in the concept of culture is often accompanied by methodological differences—the former group preferring "qualitative" and "subjective" methods, the latter, "quantitative" and "objective" ones.

18. Among the major exponents of this tradition were Bronislaw Malinowski and A. R. Radcliffe-Brown. Ralph Linton succinctly stated a central premise of this view: "The function of any culture as a whole is to assure the survival and well-being of the society with which it is associated" (1955, 34).

19. The social differentiations and stratifications within culture have also been a major topic of sociological study (e.g., Bourdieu 1984; Gans 1974).

20. Moore 1994b, 364; also see 1994a, 128–29.

3.

WELTANSCHAUUNG AND WELTBILD

Albert Einstein rarely wrote nasty letters. But he did write an uncharacteristically irate letter of protest in 1934 when he discovered that the publisher of a selection of his occasional papers had given the volume the title *Mein Weltbild*.[1] Einstein blasted that title choice as "tasteless and misleading" (quoted in Holton 1986, 315). His reaction suggests that, to him, the concept of Weltbild was too serious to be debased by this casual use. In his reverence for Weltbild and its companion concept Weltanschauung, Einstein was not alone. Weltbild and Weltanschauung were, without doubt, central elements, if not the pinnacle, of Kultur and Bildung. This section presents a case study of how these concepts of Weltanschauung and Weltbild—central to Einstein's as to German thinking—have developed. We shall see that, because these concepts play a central role in Kultur, their evolution, in its major outlines, reflects that of the concept of Kultur, which we addressed earlier.

In the time of the Weimar Republic, the monumental three-volume compilation titled *Das Akademische Deutschland* included the chapter, written by Protestant theologian Reinhold Seeberg, "Hochschule und Weltanschauung." In it, Seeberg expressed a widespread and long-standing consensus when he called Weltanschauung a primary determinant of Bildung: "Bildung thus is the personally acquired possession of a Weltanschauung, and its routine exercise in one's thought, desire, and emotional life" (1930, 166). And furthermore, "the Weltanschauung is the right of citizenship of the intellectual person in the intellectual world" (ibid., 165). One can easily find similar statements in publications, such as Gottlob Friedrich Lipps's *Weltanschauung und Bildungsideal* (1911) and Theodor Litt's *Wissenschaft, Bildung, Weltanschauung* (1928; see also Litt 1930).[2] Social historian Reinhart Koselleck pointed out that Weltanschauung served to stabilize the religious function of a Bildung that had substituted a "secular religiosity" for traditional religion (1990, 27). By extension, as Kultur formed the collective aspect of Bildung, Weltanschauung was also a central aspect of Kultur.

To strive for complete view of the world, in a general sense, is an old and recurring dream of humankind, a dream that the late Isaiah Berlin (1979) somewhat skeptically called the Ionian Fallacy. In the sixth century BCE, Ionian philosophers first inquired into the ultimate nature of the universe. Thales, the earliest of them, concluded that everything came from water; Anaximander identified the unexperienceable (*apeiron*) as the ultimate source; Anaximenes argued for air, and Heraclitus favored fire. Ever since, the quest for a unified world picture has inspired scholars in many places and in many eras. It would thus be wrong to consider this quest unique to nineteenth-century German culture, but it certainly flowered with extraordinary vigor in that milieu. But why did Kultur focus on Weltanschauung and Weltbild so enthusiastically? One might say that the overriding tendency of Bildung and Kultur toward a holistic synthesis—what Ernst Cassirer called the "'metaphysical' trait of the German intellect"—made the quest for a Weltanschauung or Weltbild central to Kultur itself (1961 [1916], xi).[3] This answer is not entirely satisfying, however, for one immediately begins to wonder whence that holistic tendency came. It is well beyond the scope of this book to delve into that question more deeply, and a cursory speculation must suffice here. It seems to me that the outcome of the Reformation in Germany may have been one of the salient long-range contributing factors. In contrast with other major European nations where the Reformation either triumphed completely or was defeated with equal decisiveness, the German lands ended up split almost evenly into a Catholic part and a Protestant part. Hence there could not exist any religion-based worldview that would have enjoyed near-unanimous institutional support and popular allegiance and that would thus have been almost universally accepted as quasi-natural. This religious schism in Germany may have conditioned German thinkers to focus on the subject of intellectual synthesis and to hone their intellectual acumen in this area: As a reaction against the loss of the unity of a religious Weltanschauung, an extraordinary amount of intellectual energy appears to have been channeled into creating a substitute unity at the metaphysical and philosophical levels.

In its extreme expressions, this tendency was not shared by the thinkers of other countries. Because German intellectuals felt it essential to have a consistent and comprehensive Weltanschauung, they threw themselves zealously into debates for which more pragmatically oriented intellectuals would not care much. This might be illustrated by contrasting S. C. Pepper's "world hypotheses" with the German Weltanschauung or Weltbild. Pepper's work can be viewed as an attempt to construct something close to a Weltbild, but Pepper used a very different approach (1942; 1945; 1967).[4]

The critical issue here is the status of "common sense." Agreeing with common sense was not a main requirement for German philosophical thought—on the contrary, transcending it was the highest goal. German philosophers typically saw it as their calling to strip common sense of its contents so as finally to discover underlying "pure" principles, principles of pure reason. (This can be seen most clearly in Kant's and Fichte's transcendental approaches.) Hegel sternly denounced common sense: "Flowing in the more stagnant riverbed of common sense, natural philosophizing pronounces a rhetoric of trivial truths. . . . One might as well save the effort of coming up with final truths of this kind; for they can already by found in the catechism and in popular adages, etc." (1952 [1807], 55–56). In many German intellectual circles, it seems, calling someone's work commonsensical was the ultimate condemnation.

By contrast, Pepper made common sense the foundation and starting point for all more refined knowledge: "Common sense . . . we thus discovered to be the very secure base of all knowledge. There is no evidence to indicate that common sense will ever fail mankind except as more refined knowledge supplements it" (1942, 320). To refine knowledge, Pepper relied on the scientific method. "By now my old drive for the truth was directed toward the study of evidence and hypothesis—toward a reliable method rather than a reliable creed" (ibid., viii). In a process of expanding the scope of one's hypotheses, one finally reaches the level of world hypotheses of unrestricted scope. "Now, the cognitive strength of a structural hypothesis is in proportion to the scope and the precision of the corroborative material. For maximum cognitive strength, we thus reach the conception of hypotheses of unrestricted scope and maximum precision. Only in such unrestricted hypotheses have we the security that there are no outlying facts which will fail to support the hypothesis and its system of structural corroboration. Such unrestricted systems of structural corroboration I call world hypotheses" (Pepper 1945, 8). Pepper explicitly distanced his empirical method from the more speculative approach, which was favored by many German philosophers: "It is surprising how persistent is the idea that there is some short cut to this thoroughgoing empiricism. Nearly everybody wants to think that he has insight into some ultimate truth or fact which is in no need of improvement, and that the progress of knowledge builds up from this favored incorrigible foundation" (ibid., 11).

This characteristic distinction of Anglo-Saxon and German approaches can also be traced in the attitudes toward government and the state. Whereas the Anglo-Saxon tradition typically trusts a generalized version of Adam Smith's "invisible hand" to achieve, through controlled competition, a functioning balance of rival economic and political interests, the German tradition has turned to the state as the guarantor of the synthesis of the general will. Again, such a simple dichotomy between national traditions has to be taken with a grain of salt. For instance, in Hobbes's *Leviathan*, English thought raised the specter of absolute government in the service of the common good—a concept that would strike few adherents of the German notion of *Staat* as alien. Yet the *Leviathan* did not become a lasting mainstay of the Anglo-Saxon way of thinking about government. German thought was not entirely monolithical, either. In the second of his "Gespräche für Freimäurer," G. E. Lessing, for instance, let one of the discussants, Falk, express a view of the state that was still very close to Western mainstream positions: "States unite humans so that each individual may better and more safely enjoy their part of happiness. The sum total of the individual happiness of all its members is the happiness of the state. Other than that, there is none. Any other happiness of the state, in which some individual members (and be it only few) suffer and *must* suffer is a cover-up for tyranny. Nothing else!" (1981 [1776–78], 50). The proximity to the Benthamite tradition is conspicuous.

Whereas many German thinkers after Lessing focused strongly on the nation or the people (*Volk*), it was Hegel in whom the glorification of the *state* reached new heights. Not merely a tool for individuals to get along with each other, the state itself was considered the realization of reason in the world: "The state is the reality of the idea of morality. The state is the reality of the substantive will . . . [it is] reason in and for itself. . . . The state is the spirit that exists in the world and realizes itself in it with consciousness. It is God's impact on the world that the state exists: its foundation is the force of Reason realizing itself as will" (1928 [1821], 328–36).

This worshipful view of the state was grasped upon as a key legitimation by German governments, especially the Prussian one, and later by the unified German state. It became customary to distinguish between the partisan and antagonistic self-interests of individuals and social groups, which threatened to throw society into chaos, and the common good, which was championed and administered by the state—concretely, by a group of impartial civil servants (Beamte), who felt entitled to, and typically received, the highest esteem. In the political and social realm, thus, the state played the role of the supreme synthesis that Weltanschauung played in the intellectual realm. In neither realm, the Anglo-Saxon tradition shared the German predilection for such syntheses and preferred a more pragmatic, from-the-ground-up approach.

At the outset of our exploration of the terms *Weltanschauung* and *Weltbild*, with their often-neglected conceptual and intellectual twists and turns, we should note that the substantive concept of Weltanschauung may have predated the actual word. For instance, Ronald Calinger's (1972) essay "The German Classical *Weltanschauung* in the Physical Sciences," which covered the period from 1760 to 1790, used the term *Weltanschauung* retrospectively—it did not yet exist at that time. Nonetheless, the classical German philosophers of the period were concerned, according to Calinger, with forming what could be considered a Weltanschauung, on the basis of Leibniz's and Newton's philosophies. Although Leibniz did not use the actual term *Weltanschauung* (or *Weltbild*), he has been regarded as the first to use something like the concept of Weltanschauung in his theory of monads—which, for him, were the basic indivisible substances (Grimm and Grimm 1955). In Leibniz's philosophy, all monads have the same "perceptions" of the world—in fact, each monad includes the whole universe—but they differ in their "apperceptions," that is, in their conscious understanding of these perceptions. A perfect monad would consciously and fully understand its unity with the universe. Less developed monads have a conscious apperception of only part (if any) of their perceptions. Monads understand those things more clearly to which they are more closely related. Hence each monad can be said to reflect the universe from its own specific point of view. Leibniz's theory thus also provided an element of perspectivity in Weltanschauung, which is often considered essential for the concept. Furthermore, the concept of discovering or making conscious the whole universe within oneself could be considered a precursor to Fichte's idea of the I (*Ich*).

3.1. DEFINITIONS

Before surveying the evolution of the concepts of Weltanschauung and Weltbild in detail, we present what we consider current consensus definitions of these concepts. They will provide a summary orientation for those who are less interested in the historical intricacies of conceptual development. The definitions, distilled from numerous sources, focus on three main elements of each of the concepts.

Weltanschauung

The first key element of contemporary definitions is the description of *Weltanschauung* as an intellectual *synthesis of the world as a whole*, a "synthesis of the human knowledge or conjectures about the world and reality" (Eisler 1930a, 506). Karl Jaspers wrote in *Psychologie der Weltanschauungen*, "What is Weltanschauung? In any case something whole and something universal. For instance, when we speak of knowledge: not individual pieces of expert knowledge, but knowledge as a whole, as cosmos" (1919, 1). This element of totality is also expressed in Jonas Cohn's popular definition, which was adopted by A. Götze (1924) and F. Kainz (1943, 237): Weltanschauung is "a kind of view that intuitively comprehends individual realities as parts of a unified cosmos" (Cohn 1908, 504). Albert Gombert defined *Weltanschauung* similarly as a "view (*Anschauung*) . . . that is grand and unfettered because it transcends narrow conceptions and dismal details, and evaluates and seeks to understand all things within the totality of the world" (1901, 259). Thus, Weltanschauung is the very opposite of a collection of unconnected facts about the world. We should add here that the intellectual synthesis of Weltanschauung may occur in very different shapes and forms. Ernst Mach, for instance, was very skeptical about grand theories of the world and focused instead on the careful study of sensory data. For him, the synthesis of Weltanschauung was rooted in the unity of methodology, not in the unity of theory. We will discuss Mach's approach in greater detail below.

Second, Weltanschauung involves the synthesizing action of the *human consciousness, mind, intellect, or spirit (Geist)*. "In every person's Weltanschauung, the totality of all perceptions is conceived as having been raised to the climax of a complete totality of consciousness, including the totality of consciousness of the human condition, without which the Weltanschauung would be nothing. . . . The Weltanschauung . . . presupposes the highest self-reflexive action of the human mind" (Schleiermacher 1911, 3: 456). Similarly, Friedrich Kainz spoke of the "comprehensive intellectual outlook and the conscious way of viewing and evaluating the world, which derives from it" (1943, 237). In this sense, the concept of Weltanschauung also characterizes the individual who holds it because the individual's way of looking at the world expresses the basic essence of the individual. In Franz Austeda's words, "A person's way of making value judgments expresses his/her spiritual individuality" (1979, 331).

The preceding quote by Kainz also introduces the third element: Weltanschauung contains a *value judgment*—"evaluating the world" (Kainz 1943, 237)—and, hence, a *practical guideline for living*. It is more than a purely factual statement of what the world is like; it has emotional aspects, gives a normative meaning to the world as well as to the individual's life, and it can influence an individual's actions through moral imperatives. "At the same time, I recognize Weltanschauung only in the person who not only looks at the world, but knows how to live, act, and make judgments based on this Anschauung" (Hoffmeister 1955, 662). According to Jaspers, "Weltanschauung is not only knowledge, but it manifests itself in value judgments, in the hierarchy of values" (1919, 1). In this sense, the synthesis of Weltanschauung goes beyond what science can state about the world. Jonas Cohn wrote,

> Weltanschauung includes those tasks that the assertion that philosophy is not, or not entirely, an academic discipline (*Wissenschaft*) brings to mind. . . . Within the concerns of essential *Weltanschauung*, there is a limit beyond which the purely personal begins. There, the total personality leads others through its vigor and consistency, through abundance of genius or through ethical strength, aesthetic perfection, or religious devotion. . . . The results of aca-

demic philosophy are certainly also principles of Weltanschauung, but, by themselves, they let co-exist a plurality of types of Weltanschauung and an infinite profusion of variations.[5]

A practical short definition of *Weltanschauung* would be the one given in *Der Große Brockhaus* (1977–82): "[T]he view of the world in its diversity as a meaningful whole, and the application of that view to one's attitude about life."

Weltbild

The previously mentioned three main characteristics of Weltanschauung can also guide our definition of Weltbild, which is again based on a broad contemporary consensus. Our definition distinguishes Weltbild from Weltanschauung: While Weltbild shares the first characteristic with Weltanschauung, it differs on the other two.

First, *Der Große Brockhaus* (1977–82), for instance, defines the term *Weltbild* as "the concept of the whole of experienced reality, which comprises more than just the sum of individual experiences." Just like Weltanschauung, Weltbild provides an intellectual *synthesis of the world as a whole*, not just an agglomeration of unrelated facts—a point also made by Karl Jaspers: "On the one side is the development of a Weltbild with direction and order, on the other side is the welling up of a chaotic mass of contents that merely multiplies without becoming a totality, that merely exists as an assemblage without being capable of a refining process, or of a vigorous force" (1919, 128).

Second, in a marked contrast with Weltanschauung, Weltbild focuses less on the activity of the human mind's looking at the world and more on the outcome or products of this activity (theories and beliefs about the world). It is thus more objective than the more subjective Weltanschauung. This difference was expressed by Jaspers: "As far as the soul exists in the dichotomy of subject and object, psychological investigation sees attitudes in terms of the subject, and Weltbilder in terms of the object. . . . We thus define Weltbild as the totality of the objective contents that a human being possesses" (ibid., 122).

Third, Weltbild is more restricted in scope than Weltan-schauung. It is more narrowly cognitive and excludes the emotional, evaluative, and practical elements contained in a Weltanschauung. Often Weltbild is limited to the physical world and defined as "the total concept of the physical world, its objects, forces, events, and their interactions governed by natural laws. This concept is acquired through research and learning."[6] Sometimes, Weltbild is used in a more comprehensive fashion that includes philosophical and meta-physical aspects. Georg Simmel, for instance, considered the quest for a complete Weltbild an important feature of metaphysics: "Meta-physics has the formal merit of at least striving for a completed Welt-bild according to universal principles" (1905, 82). In such cases, Weltbild draws closer to Weltanschauung. In our definition, we follow the narrower understanding of Weltbild.

A practical short definition of Weltbild is that of *Meyers Enzyk-lopädisches Lexikon* (1979)—"synthesis of the findings of objecti-fi-able knowledge into a comprehensive view of the world." Or, as Gerald Holton put it, "a generally robust, map-like constellation of the individual's underlying beliefs of how the world as a whole operates" (1993a, 157).

3.2. CONCEPTUAL EVOLUTION

Both Weltanschauung and Weltbild have been widely used in German philosophy as well as in the German everyday language, but *Weltanschauung* is clearly intellectually more central and more pop-ular than its companion term. In dictionaries that mention both concepts, the entry for Weltanschauung is typically longer.[7] Other dictionaries mention only Weltanschauung.[8] Weltanschauung has also drawn a number of philological examinations of its origins, whereas no such investigation was found for Weltbild.[9]

Although the word *Weltanschauung* was born as a philosophical term only in 1790, it was, already by the middle of the nineteenth century, so entrenched in everyday language that some writers ridiculed it as a popular fad (Götze 1924, 48, 51). Jakob Burckhardt noted in 1843, "now, however, people consider themselves gebildet, patch together a 'Weltanschauung' and start preaching to their fellows" (Burckhardt 1984). In Hermann Kurz's *Die beiden Tubus*, the spirit of Weltanschauung was that of alcohol. "[Eduard had] once overheard a trace of intelligence in [the ants], which moved the listener to tears, with the help of the Weltanschauung from the bottle" (1873 [1859], 242). The great poet Goethe became important in an indirect way. Much more influential than his own casual and varying use of the word *Weltanschauung* was the subject matter of his magnum opus *Faust*, which became synonymous for an individual's quest for Weltanschauung.[10] Faust ardently desires to know "was die Welt im Innersten zusammenhält" (what holds the world together in its innermost—lines 382–83).[11] After *Faust*, some literary critics appeared to have made it a required standard for a play to contain some element of Weltanschauung (Götze 1924, 49). Reacting against this, Friedrich Hebbel at mid-century decried the "silly hunt for a Welt-Anschauung" (1904 [1851], 405).

It signifies the strong link between the words *Weltanschauung* and *Weltbild* and a specifically German brand of philosophy that they entered the English language untranslated as technical terms—Weltanschauung in particular. Three contemporary English-language dictionaries of philosophy, for instance, carry the entry *Weltanschauung* with short explanations (Blackburn 1994; Honderich 1995; Runes 1983). A fourth (Audi 1995) mentions Weltanschauung only to refer the reader to Wilhelm Dilthey, who, in his landmark *Weltanschauungslehre* (1960), distinguished several major types of Weltanschauung (see below). Weltanschauung also appeared in French dictionaries of philosophy and the Italian *Enciclopedia Filosofica* (Centro di Studi Filosofici di Gallarate 1957).[12] The latter called the term "difficilamente traducibile in italiano," thus concurring with Leo Weisgerber, who, in his *Weltbild der deutschen Sprache*, counted Weltanschauung among the German words that are hard to translate into other languages (1953–54, 2: 209; similarly Merz 1965, 3: 445).

Nonetheless, there are, of course, nearly equivalent English words. Attempts at translation typically yield "worldview." We find such a translation in the *Oxford English Dictionary* (1989): "worldview (G[erman] weltanschauung), contemplation of the world, view of life." Similarly, *Webster's New International Dictionary* (1942) describes its entry *world view* as "Translation of Weltanschauung." *Weltanschauung*, conversely, is "Literally, world view; a conception of the course of events in, and of the purpose of, the world as a whole, forming a philosophical view or apprehension of the universe; the idea embodied in a cosmology." This dictionary also contains another dyad—*Weltansicht* and *world concept.* The latter term is defined as "A philosophical conception of the world as a whole, with reference to its fundamental principle and organization. Cf. Weltansicht." *Weltansicht*, however, is "[a] world view; an aspect in which the universe is regarded; a special view or apprehension of reality as a whole." Thereby, *Weltansicht* ties itself, and consequently *world concept*, to the worldview/Weltanschauung nexus. Hence, though it might have seemed tempting to conclude that the dictionary mirrors the distinction between *Weltanschauung* and *Weltbild* (here in the guise of *Weltansicht*) in the distinction between *world view* and *world concept*, we realize that the two concepts have not been neatly separated.

Although, in the German language, Weltanschauung and Weltbild have sometimes been used as synonyms, they are more commonly considered related, yet not identical, terms. We will first sketch the conceptual development of Weltanschauung, the more prominent term, and then move on to Weltbild. Finally, we will examine how the distinction between Weltanschauung and Weltbild has been shifting over time.

The following survey of how the conceptual landscape evolved will show that our concepts underwent a stage progression not unlike the one typically found. At first, when the concepts were created and shortly thereafter, their meaning vacillated in swift and large swings. A distinct group of intellectuals sustained the discourse that nurtured the development of the concepts and provided a nucleus for their stabilization and popularization. In the end, the terms arrived, their meanings relatively stable, commonly understood, and—some even complained—overused. Nonetheless, even at that last stage, decisive changes in meaning occurred. We shall watch the unfolding drama of how Weltanschauung's rational hubris collapsed. Reason, in the guise of *Wissenschaft*, or, more specifically, philosophy, had given birth to the concept, but it overburdened it so much that it could not sustain its claims, as the concept became mired in conceptual quicksand and slipped away—and even, to some extent, pulled Weltbild with it.[13]

Weltanschauung

Physics and Metaphysics. To a speaker of German, the literal meaning of the word *Weltanschauung* poses no puzzle: "Weltanschauung" refers to "looking at the world." Like many other "-ung" words, Weltanschauung may ambivalently describe both the process and the result of an action; it can thus mean both the process of looking at the world and the resulting perception or image (Betz 1981; see Dornseiff 1945–46). Within this ambivalence, the latter meaning grew stronger without completely eclipsing the former meaning. Alexander von Humboldt's usage succinctly exemplifies the literal understanding of Weltanschauung as visual perception, with an emphasis on the process of perceiving: "Humans perceive the external world through organs. . . . The eye is the organ of Weltanschauung" (1845, 85–86).[14] The literal use of Weltanschauung (visual world perception) has not quite died out yet—at least, that is one possible interpretation of a remarkable sentence in an announcement of USAir's *Viva Europe!* auction for frequent flyers: "You and your companion will find a week in Germany can re-energize your zest for the *Zeitgeist* and widen your *Weltanschauung*" (*USAir News*, March/April 1996). Nonetheless, the literal understanding of Weltanschauung has been minor in comparison with meanings far beyond optics. First, the term occurred as part of philosophical attempts to understand the basic characteristics of human perception. Then the notions of an intellectual synthesis became prevalent, as well as an emphasis on the "ultimate questions."

Immanuel Kant is commonly credited for being the first among scholars and writers to use the word *Weltanschauung* (e.g., Grimm and Grimm 1955, 1530). The term appeared in his *Kritik der Urteilskraft*

(1790): "A suprasensual faculty in the human mind is necessary even to think of the given infinite without contradiction. For only through this faculty and its idea of a thing-in-itself (noumenon), which itself allows no Anschauung, but forms the substrate for the Weltanschauung, as a mere phenomenon, is the infinitude of the world of the senses entirely subsumed under one concept in the purely intellectual estimation, although it can never be entirely conceived in the mathematical estimation through numerical concepts" (Kant 1922 [1790], 254–55).

The word *Weltanschauung* appeared only once in Kant's work, and it definitely paled in comparison with its much more important sibling *Anschauung*. The latter was a cornerstone in Kant's theory of the *transcendental (necessary) properties and preconditions of human perception*.[15] "All things that present themselves as objects to our senses are phenomena. That which contains only the special form of sensibility, without touching the senses, belongs to the pure Anschauung (which is void of sense-intuitions, and hence not in the domain of the intellect)" (Kant 1975a [1770], 43).[16] According to Kant, pure Anschauung a priori consists of some fundamental categories and forms (such as space and time) that necessarily shape all human perceptions of phenomena—which he called *Weltanschauung*.

Although Kant's invention of the word *Weltanschauung* was rather offhand, his philosophy had a seminal impact on the ensuing development of the concept, owing to his towering status in the history of nineteenth-century German philosophy. Kant had, of course, a huge number of followers who called themselves Kantians or Neo-Kantians, but even philosophers who fundamentally disagreed with him typically developed their own ideas through a critique of his writings. In any case, Kant was not to be ignored. As for our topic, Kant's critique of Enlightenment thought, and especially his concept of Anschauung, formed the basis upon which the term *Weltanschauung* would evolve.

Most philosophers of the French Enlightenment had also subscribed to the search for a unified world picture. In his *Philosophie der Aufklärung* (Philosophy of Enlightenment), Neo-Kantian philosopher Ernst Cassirer succinctly summarized, "The rationalistic postulate of unity maintained a total domination over the minds of this age" (1932, 29). The Enlightenment philosophers tended to be rationalists who shared the optimistic belief that human reason was able to explain the whole world without limitation in principle. The Enlightenment's search for a unified world picture took the characteristic form of gathering together all the knowledge of the world. One thinks of the grand *Encyclopédie*, edited by d'Alembert and Diderot, in whose service the most illustrious minds of the epoch were enlisted. Reason ruled, but it had clay feet—a weakness that attacks by empiricists, such as David Hume, exposed. German philosophy's drive toward the Weltanschauung can be viewed as an attempt to reconstitute the Enlightenment program after the loss of its epistemological innocence.

Kant explored the necessary epistemological determinants of human thought. Through the transcendental categories and forms thus revealed, Anschauung bore the indelible imprint of the human mind. After Kant, any quest for cognitive unity had to account for the knower's construction of knowledge. Therein lay a seed of relativism. It did not yet germinate in Kant's transcendental program itself, which aimed at identifying *inalterable* categories in human thinking, but it sprouted further down the line, as the concept of Weltanschauung took on additional burdens, when a synthetic purpose superseded the transcendental one.

In the move toward synthesis, the concept of Weltanschauung both became more central in the intellectual discourse and underwent a crucial change in meaning. It shifted from the "input" to the "output" side of the cognitive process. Whereas for Kant, Weltanschauung had been located in the transcendental epistemology of human perception of the world (phenomena were processed and organized in the human mind by Anschauung), now Weltanschauung itself emerged as the ultimate product of human cognition, the intellectual synthesis that was the most revered goal of reasoning. Kant had been primarily interested in the necessary determinants of all human perception. Many German philosophers after Kant focused on how a comprehensive, unified understanding of the world—Weltanschauung, as they understood it—could be established.

Friedrich Schleiermacher's work on education exemplifies the efforts of German thinkers to extend the Kantian agenda, as well as the concomitant change in the concept of Weltanschauung. Schleiermacher, whose pivotal role in the evolution of the term *Weltanschauung* will receive further attention shortly, asked whether there were a general developmental principle that underlay all stages and branches of education. His answer was: "[T]he general task is the development of the chaos of sense perceptions into a Weltanschauung, and of the chaos of spontaneous action into a world-constructing self-presentation" (1911 [1813–14], 3: 455). According to Schleiermacher, a newborn's sense perceptions of the world are utterly chaotic. As the mind develops, it brings order into these perceptions and finally creates a Weltanschauung as the mental synthesis of the totality of all perceptions. Schleiermacher paralleled the area of perception with the area of action where he saw a similar development. A newborn's actions are chaotic, to a large degree determined by physiological impulses. As the human being matures, the actions become more constructive and "partake in the human spirit's evolving project to form the world" (ibid., 3: 456). Still rooted in the Kantian understanding of Weltanschauung, Schleiermacher's deliberation set up Weltanschauung, now understood as a cognitive synthesis, as an educational goal and thus signaled the departure of the term from its origin.

Following Kultur's fundamental imperative of unity and synthesis, many post-Kantian philosophers and thinkers went beyond Kant's austere transcendentalism and attempted to integrate additional elements to achieve an ever more complete and substantive view of the world. Under the influence of Romantic thought, philosophers identified the individual person and various intermediate groups, such as nations, as the subjects to bring forth that grand synthesis of Weltanschauung. The quest for a Weltanschauung that is comprehensive and includes both cognitive and evaluative (normative) aspects will prove to be an overambitious project that was doomed to founder. The following will retell the struggle of Wissenschaft, in the shape of philosophical reason, trying to hold onto the notion of Weltanschauung it had given birth to, but finally failing and surrendering it to nonrational forces, because the concept had become too grandiose and overreaching to withstand the relativist challenge.

This ultimately futile project of a rational Weltanschauung in the post-Kantian intellectual environment echoes a similarly doomed earlier synthesis—the convergence of (Aristotelian) philosophy and (Augustinian) theology, reason and faith, accomplished by Thomas of Aquinas in the thirteenth century. In Etienne Gilson's (1938, 1955) imposing—but perhaps somewhat one-sided (see Ozment 1980)—view, this was the pinnacle of medieval intellectual history. From this height, Gilson argued, the fourteenth and fifteenth centuries then witnessed a steep intellectual decline, in which Thomas's synthesis fell apart and a widening gap opened between an increasingly skeptical philosophy and a theology increasingly based on nonrational belief, revelation, and mystery. The tragic turning point for Gilson were the condemnations in 1270 and 1277 of Aristotelianism, pronounced by the bishop of Paris, that eroded the common ground and caused philosophy and theology to take off in different directions.

One of the critics of Gilson's interpretation, Damasus Trapp (1956), vigorously defended the fourteenth-century thinkers against the charges of intellectual decline. He pointed out that the great intellectual systems built by Thomas of Aquinas and his thirteenth-century cohorts were actually contradictory within as well as between themselves. Thus the obvious challenge for their successors became to sort out these problems in a critical spirit rather than to attempt yet another synthesis.

Something similar to Trapp's scenario appears to have happened in the nineteenth and early twentieth centuries to the comprehensive concept of Weltanschauung. This synthesis, too, collapsed under its own weight because it contained claims that were simply inconsistent and went too far. A grand synthesis of everything— including both facts and values—was a promise reason could not fulfill. The evolution of the notion of Weltanschauung, to which we will now return, should therefore be read not as a chronicle of intellectual decline but as the gradual exposure of fault lines and cracks that, though hidden, had existed from early on in the conceptual structure. It became increasingly apparent that the synthesis did not hold, and facts and values inexorably separated, just as reason and faith did after the 1270s.

There have been extensive, but nonetheless rather sterile, philosophical debates about whether being determines consciousness or the other way around. I rather doubt that it is useful to deal with this question at the level of general principle, and I am more interested in how the influence of the respective "materialist" and "idealist" factors varies from case to case. As we shall now see, the *collapse* of the project of a rational Weltanschauung serves as an example for a development primarily driven by intellectual forces.[17] On the other hand, we already described how strongly German Kultur and its major concepts (and Weltanschauung among them) depended on a particular socioeconomic and political environment. The tectonic forces of basic socioeconomic conditions helped elevate Weltanschauung to a central goal, but the overdrawn ambition of a rational Weltanschauung was bound to implode by the mere force of reasoning. However, the particular Weltanschauungen that took its place, as well as the extent to which relativism came to dominate, cannot be sufficiently explained by factors internal to the intellectual discourse.

The Romantic turn. Friedrich Wilhelm Joseph Schelling has been considered the first to use the word *Weltanschauung* in its more modern meaning (in 1799) (Götze 1924, 42–43; Kainz 1943, 237).[18] This standard view needs three major modifications. First, Schelling's role in bringing about the modern usage of Weltanschauung was rather indirect. (See Meier 1967.) He should be more accurately described as the inspirer, rather than the creator, of the modern understanding. A closer look at the passages in which Schelling used the term will reveal a strong continuity of Kant's epistemological project, albeit with particular Fichtean overtones. Second, Fichte himself influenced the early evolution of Weltanschauung more strongly than is usually acknowledged. Third, I suggest, following H. J. Sandkühler, that Schleiermacher's speech "Über das Wesen der Religion" (On the Nature of Religion, 1799) deserves recognition as a veritable milestone on the way toward the modern meaning of Weltanschauung (Sandkühler 1990, 784).

Around 1800, the small provincial town of Jena was arguably the intellectual hub of all German lands. The eminent philosophers Fichte, Schelling, and Hegel were professors at Jena University. Friedrich von Schiller, who taught history at the university until 1793, lived in Jena until 1799, and Goethe frequently visited the town from nearby Weimar. These were only the most famous of an array of brilliant scholars, intellectuals, and poets who all inhabited this small town together. (See Strack 1994.) Jena at that time was also the hotbed of early Romanticism, which formed in a circle of friends around Ludwig Tieck, Novalis (Friedrich von Hardenberg), and the brothers August and Wilhelm Schlegel. These people were young, and they were rebels (rebelling, first and foremost, against Enlightenment-style rationalism but also against the stifling conventions of "philistine" respectability). To some degree, thus, Jena in 1800 resembled Einstein's Zürich in 1900. (See Feuer 1974.) The intense social and intellectual interactions of this group of early Romantics were the crucible in which the concept of Weltanschauung was transformed.

Before turning to Schelling's contribution, it is both appropriate and important here to acknowledge the role Johann Gottlieb Fichte played in the evolution of the term *Weltanschauung*. His use of the term has been generally neglected in the accounts of the conceptual history—with the notable exception of Helmut Meier's (1967) dissertation—although, in his work, one finds the second documented occurrence of the word. Two years after being introduced by Kant, it appeared in Fichte's discussion of natural and moral laws: From the human point of view, Fichte argued, these two laws are very different; they cannot be reconciled and be made the basis of a unified Weltanschauung. Yet there is a superior law that coordinates natural and moral laws. God alone knows this superior law; it is the basis of his Weltanschauung. "According to the postulates of reason, God must be considered the being that determines nature in keeping with moral law. In him, thus, is the union of both laws; and the principle, on which both of them together depend, is the foundation of his Weltanschauung" (Fichte 1845 [1792], 108). Compared to Kant's transcendental approach, Fichte's conceptualization of Weltanschauung contained at least two new aspects: The element of a supreme synthesis first appears, and Weltanschauung is being located in the context of religion. As we shall see below, Schleiermacher went on to elaborate the religious dimension of Weltanschauung (although in a different way).

A few years after Fichte's treatise of 1792, the concept of Weltanschauung reappeared in the context of his philosophy of the I (*Ich*). In the *Grundlage des Naturrechts nach Prinzipien der Wissenschaftslehre* (The Foundation of Natural Law according to the Principles of the Study of Wissenschaft), first published in 1796, Weltanschauung was contrasted with the direct reflection of the I on itself. Weltanschauung, according to Fichte, looks at the objectified I—which is the world, in Fichte's understanding—and therefore is not free. "[T]he activity in Weltanschauung is that free activity in the state of constraint; and reversely, free activity is the activity engaged in Weltanschauung, if the constraint is absent" (Fichte 1979, 19). In Fichte's connection of Weltanschauung to restraint, we encounter the concept of limitation so prominent in Fichte's (1970) philosophy of the I. And it is this idea of limitation that appeared to be taken up in Schelling's work.

In his *Erster Entwurf eines Systems der Naturphilosophie* (First Draft of a System of Naturphilosophie), written in 1799, Schelling pointed out, when discussing the artistic drive in animals,

> Just as human reason conceptualizes the world only according to a certain type, whose visible expression is human organization, each organization is an expression of a certain schema of Weltanschauung. Just as we understand that our Weltanschauung is determined by our original limitation, without being able to explain why we are limited in this particular way, why this is our Weltanschauung and not another one, the life and intellect of animals can only be a special, albeit incomprehensible, kind of original limitation. And only this kind of limitation would distinguish them from us. (1856–61, Abt. I, 3: 182)

The key point for us here is that human Weltanschauung is contrasted with animals' Weltanschauung. Kant's transcendental program still shines mightily through this anthropological argument. Schelling's deliberation of gravity and retarding force as two transcendental forces of nature reads: "For everything finite, there must be a limit of Weltanschauung. This original limitation is for the intellectual world what gravity is for the physical world—that which ties individuals to a certain system of things and assigns them their place in the universe. However, even within a certain system, the Weltanschauung is determined in respect to each individual object. Thus limitation is added to limitation" (ibid., 3: 265–66).

Besides giving a flavor of Schelling's sometimes rather opaque prose, both quotes show Schelling's preoccupation with a dialectical theory of limitation as the "condition for the possibility" of intuition (to use a catchphrase of *Transzendentalphilosophie*). In other words, Schelling continued Kant's transcendental epistemological project, but he followed it along the lines of a dialectical idealistic offshoot à la Fichte.

Schelling's *Einleitung zu dem Entwurf eines Systems der Naturphilosophie* (Introduction to the Draft of a System of Naturphilosophie) (1799) probably demonstrates most clearly that his concept of Weltanschauung was still very different from the modern one. There Schelling asserted that Weltanschauung is *nonconscious* human perception as performed by a mental capacity, thus again placing himself in the Kantian tradition. "The Intelligence is productive in two modes—that is, either blindly and unconsciously, or freely and consciously;—unconsciously productive in external intuition [*Weltanschauung*], consciously in the creation of an ideal world" (Schelling 1867 [1799], 193). Tom Davidson's appropriate rendition of Weltanschauung as "external intuition" in this English translation of Schelling's *Einleitung* indicates that Weltanschauung is still far removed from its more modern meaning.[19] Schelling's concept of Weltanschauung as unconscious intuition also appears somewhat related to Leibniz's "perceptions."

At the same time, in 1799, when Schelling, the supposed creator of the modern understanding of Weltanschauung, still argued along the older Kantian lines, the theologian Schleiermacher already advanced toward the modern usage. In his famous speech "Über das Wesen der Religion," he linked Weltanschauung with religion and knowledge of God. He explicitly used the term *Weltanschauung* itself and more frequently used the alternative forms of *Anschauen der Welt* or *Anschauen des Universums*.[20] Prophetically, he declared that Weltanschauung would be a key concept that would not go away: "Anschauen des Universums: please, get used to this term. It is the key to my whole speech; it is the most general and highest formula of religion" (Schleiermacher 1911, 4: 243). Schleiermacher considered contemplation of the infinite Universe the essence of religion: "Religion is the sense and taste for the Infinite" (ibid., 4: 242). Or, "[Religion] wants to behold the universe, it wants to listen reverently to it in its own representations and actions, and, in childlike passivity, it wants to let itself be gripped and fulfilled by its immediate influences" (ibid., 4: 240).[21] Using Spinoza as his guide, Schleiermacher fused the concepts of God and the universe (although he did not consider them totally identical) and, consequently, also tied the Anschauung of the world closely to the Anschauung of God. As we shall see, Einstein was also a follower of Spinoza, and he used to quote, from memory, Schleiermacher's praise of that philosopher (Pesic 1996, 195).

Owing considerably to Schleiermacher's teachings, Weltanschauung became a common term among theologians (Götze 1924, 48–49). One indication for Schleiermacher's central role in popularizing the concept of Weltanschauung is perhaps that he was occasionally—and erroneously—considered its creator. (See Dornseiff 1945–46, 1086.) At the other extreme, he was virtually banned from most standard histories of the word. A more balanced view can be found in Sandkühler (1990). Schleiermacher was certainly not the creator, but an important shaper and popularizer.

The few years after 1799 marked a widespread and dramatic transformation of the concept of Weltanschauung. As Götze (1924) described it, Romantic poets and writers in the Jena circle, who knew and appreciated the philosopher Schelling as an ally of the Romantic movement, immediately picked up Schelling's word *Weltanschauung*, and radically altered its meaning.[22] In a process of Romantic *Verinnerlichung* (internalization), Weltanschauung came to describe the creative, synthetic, and conscious activity of the spirit (*Geist*). One might surmise that some of these Romantics were more inspired by Schleiermacher's 1799 notion of Weltanschauung, which was already more modern, than by Schelling's writings. Although Schleiermacher lived in Berlin at the time, he was well connected to the Jena group. He corresponded with several of its leading members and came to Jena for visits. Jena's intellectual scene at the turn of the century may have been the cradle for the modern notion of Weltanschauung, but there were also outside influences, which in the case of Schleiermacher were underrated, but nonetheless considerable. Furthermore, Schleiermacher (who became a professor of theology at the Humboldtian flagship University of Berlin), was a renowned public speaker and became pivotal in transmitting and popularizing the concept among the German intellectual public.

As early as 1800, August W. Schlegel wrote in a sonnet, "Der Geist muß sich, um nicht der Welt zu fröhnen,—Zur Weltanschauung in sich selbst vertiefen.—Begreifend schafft er Kräfte, welche schliefen,—Die durch Bewußtsein sich als mündig krönen" (1846, 354).[23] ("So as not to be subject to the world, the spirit must be introspective to gain Weltanschauung. The spirit's insight creates forces that, once dormant, now proclaim their emancipation through consciousness.") In this notion of Weltanschauung, it is already the spirit (Geist) that brings about Weltanschauung in an introspective, reflective way.

From Novalis (Friedrich von Hardenberg), we have a fragment that is properly cryptic for a Romantic: "The world is the result of an infinite consensus, and our own inner plurality is the basis of Weltanschauung" (Novalis 1907, 204). The companion term *Weltbild* appeared in Novalis's novel *Heinrich von Ofterdingen*, where one of the characters proclaimed, "[T]he conscience appears in every earnest perfection, in every formed truth. Every predilection and skill that contemplation has transformed into a Weltbild becomes a manifestation, a transformation of conscience. All Bildung leads to that which we cannot call anything else but freedom" (Novalis

1960a, 331). Novalis wrote this part of the novel in the fall of 1800. It remained incomplete because of his death in 1801.

When Schelling used the word *Weltanschauung* again in 1802, he appeared to have already moved somewhat toward the modern usage:

> If we consider the union of philosophy and poetry even in its lowest synthesis, the pedagogic poem, it is necessary, because poems should be without any external purpose, that the pedagogic intention of the poem is sublimated and transformed into an absolute, so that the poem can appear to exist for its own sake. This however is only conceivable if the knowledge is already by itself poetic—knowledge as picture of the universe and in complete harmony with the universe, which is the most original and beautiful poetry. Dante's poem is a much more advanced interpenetration of scholarship and poetry [than a pedagogic poem is]. So much the more, its form must, even in its greater independence, fit the general type of Weltanschauung. (1856–61, Abt. I, 5: 157)

Here, *Weltanschauung* was connected to "knowledge as the picture of the universe," but the term continued to play a rather marginal and not entirely clear role in Schelling's thinking, until Schelling altogether ceased to use it after 1804 (Meier 1967, 84).

The Romantics, by contrast, were quick to cement the modern meaning of Weltanschauung. Jean Paul made a crucial connection between Weltanschauung and genius when he wrote that the passive genius is "different from the person of talent, who can perceive only parts of the world and objects of the world, but no world spirit, and [is] therefore similar to the genius whose first and last characteristic is a Anschauung des Universums. [Yet] for the passive geniuses, Welt-Anschauung is merely a continuation and adaptation of someone else's Weltanschauung of genius" (1963 [1804], 53).[24] Jean Paul also associated Weltanschauung with *Lebensanschauung*, even to the point of treating these two terms as synonyms. The heart of the genius, he wrote, contained a "new Welt- or Lebensanschauung" (Paul 1963 [1825], 64).

Very clear evidence of the modern meaning can be also found in 1807 in the work of Joseph von Görres, a major Romantic thinker from Heidelberg, another center of the movement: "[I]t is moving to see how it [the spirit]—groping, probing everything around, and winding in all directions—struggles for Weltanschauung" (1807, 13). Weltanschauung here again is the creative synthesis achieved by the human spirit. Among Görres's other references to Weltanschauung, the following clearly has an evolutionary flavor: "[T]he doings of the great multitude, of the community, have presented themselves to our contemplation: what Weltanschauung the community has gradually formed; how much it has appropriated from the stream of knowledge and experience that winds through the times" (ibid., 272). The same evolutionary thrust was even more explicit, as "higher Weltanschauung," in Görres's "Wachstum der Historie," which was completed in 1808: "What we said about the meaning of the present and the essence of the future must add to itself what it might lack in historical existence, lest it be considered an empty pipe dream. By proclaiming higher Weltanschauung and the divine lineage of higher ideas as the sanctum of the future, we go back along the thread which connects us with the origin, and these justifications must necessarily ground themselves historically in prior developments" (1926 [1808], 412).[25]

Hegel's preference for the word *Weltanschauung* over the competing synonym *Weltansicht*, which at the time was also frequently used, finally established the dominance of the former term.[26] Weltansicht soon became extinct. In his *Phänomenologie des Geistes* (Phenomenology of the Spirit), Hegel included a section on the "moralische Weltanschauung" (1952 [1807], 424–34). *Weltanschauung* was thus a term of moral philosophy and described "the relationship between the moral being-in-and-for-itself and the natural being-in-and-for-itself" (ibid., 425). Hegel's lectures on *Ästhetik* also used the term *Weltanschauung*. In his discussion of the epos, Hegel emphasized that the "embellishment and crafting of the epos lie not only in the particular content of a *concrete* action, but just as much in the *totality* of the Weltanschauung whose objective reality it endeavors to describe" (1985 [1835–38], 2: 450). There, Weltanschauung is "the original Weltanschauung of the peoples, that grand natural history of the spirit" (ibid., 2: 437).

Similarly, Hegel spoke of "the basic type of the Weltanschauung of that particular people" (ibid., 1: 315). This coupling of Weltanschauung with a people will be of paramount importance later on. Yet Weltanschauung, for Hegel, had also a strong developmental aspect that lay in the journey of the spirit toward self-consciousness. Within the development of the *Kunstgeist* (spirit of art), there was "the stage sequence of definite Weltanschauungen understood as the definite, but comprehensive consciousness of the natural, the human, and the divine."[27] In Hegel's philosophy, which viewed history as the process by which the rational spirit gained awareness of itself, the understanding of Weltanschauung as a comprehensive rational synthesis reached a high point.

In the course of a few decades, Weltanschauung had become the creative, synthetic, and conscious activity of the spirit (the Romantics, Fichte, and Hegel); had incorporated the dimension of morality (Hegel); and had even become the focal point of religion (Schleiermacher). Two different elements, the fault line between which was not quite visible at first, made up that comprehensive concept at mid-century. There was the universalist claim that passed in various metamorphoses from its origins in Kant's transcendentalism through the thinking of Fichte, Schelling, Schleiermacher, and Hegel. There was, on the other hand, the element of individuality and diversity that the Romantics introduced. The whole conceptual formation was formidable, perhaps too formidable. Almost instantly, erosive forces began chipping away at it.

The roots of Einstein's Weltanschauung. By the middle of the century, the meaning of Weltanschauung had stabilized, and the word was firmly established in the vocabulary of the gebildeten German. Nevertheless, the term increasingly vexed intellectuals who discovered and explored some basic problems with this concept—first of all, the problem of the plurality of possible Weltanschauungen. More and more philosophers, scholars, and intellectuals realized that the ambitious project of a comprehensive rational synthesis of the world was riddled with problems. In the following, we will survey how these dissatisfactions unfolded and unleashed the relativism inherent in the existence of several Weltanschauungen and how, in this process, some evolving variants of Weltanschauung radically altered the contents of the concept.

This point at the middle of the century is also a crucial juncture for understanding the roots of Einstein's own Weltanschauung. Its basic character was shaped by the meanings of Weltanschauung that had emerged in the early phase of the life of the concept. From those early formations, the notion of the scientific Weltanschauung originated—partly in a materialist mutation of Hegel's idealism—and became a separate branch among the varied offshoots of the concept. This was the branch that influenced Einstein. (And we shall discuss it in great detail later.) The philosophical and intellectual developments that ensued in the later part of the century outside of the scientific Weltanschauung were less important for Einstein. An indication for his detachment may be that, in his extraordinarily voluminous library, there was not a single publication by Dilthey, by Husserl, or by Heidegger—three main figures in the further evolution of the concept of Weltanschauung.[28]

Nonetheless, we continue to follow the conceptual development of Weltanschauung beyond the point at which it ceased to be influential on Einstein's own formation, because some strands of more recent notions of Weltanschauung became ascendant in the intellectual life of Kultur and then directly confronted Einstein in his middle and later years. We shall see that Einstein was so strongly committed to certain core ideas of that early scientific Weltanschauung—rigorous determinism and strict causality—that he refused to go along with most of his younger colleagues who, by contrast, found it much less repulsive to interpret certain results of quantum mechanics from a nondeterminist viewpoint. In a much less benign way, the National Socialist Weltanschauung, which became the official ideology in Germany between 1933 and 1945, directly attacked Einstein's physics, most other tenets of his Weltanschauung, as well as his ethnic background and forced him to leave the country.

Drift toward plurality. As we have seen, the Romantic predilection with individuality linked Weltanschauung with the individual genius—which, of course, might make each genius's Weltanschauung different from that of the next. This association of the exceptional individual with Weltanschauung prompted the genre of studying the Weltanschauung of famous persons.[29]

A first step toward linking Weltanschauung with particular human groups was taken in the study of languages. In this respect, Johann Gottfried Herder can be considered a forerunner. He had written, "How different is the world in which the Arab and the Greenlander, the soft Indian and the rock-hard Eskimo live! How different their Bildung, food, education, the first impressions they receive, their internal structure of perception! And on this rests the structure of thought and the expression of both: language" (Herder 1967 [1775], 302). Here language expresses perceptions and thoughts (i.e., Weltanschauung) that are shaped by environmental influences. Wilhelm von Humboldt (Alexander's brother) was one of the pioneers of a school of linguists who contended that a particular language shaped the Weltanschauung of the individual speaker and also, by extension, of a whole nation viewed as language-community.[30] According to Leo Weisgerber, "[Humboldt's] concept of the 'Weltansicht' emphasizes external conditions, which are shaped by the linguistic generative forces that operate in the *internal form of language*. In the Weltbild of the native tongue, these external conditions gain a lasting conscious existence in the existential category of reality" (1954, 2: 207). The linguistic interpretation of Weltanschauung became well known in wider circles and was reflected, for instance, in Mauthner's dictionary entry about Weltanschauung (1911, 579).

Going a step beyond this linguistic determination of the Weltanschauung, historians, students of culture, and other scholars attributed a Weltanschauung to all kinds of groups and epochs.[31] Writing in the early twentieth century, the psychologist and philosopher Richard Müller-Freienfels, the proponent of *Lebenspsychologie*, summarized this major shift from the unity of Weltanschauung to the plurality and diversity of Weltanschauungen: "In order to understand Weltanschauungen in their diversity, we must not consider them merely the results of a unified universal subjectivity, but we must seek to understand them from their individual particular forms" (1923 [1909], 5–6).

In the prevailing spirit of nationalism, the most prominent of these groups with a Weltanschauung were probably the *Nation* or the *Volk* (people).[32] In the wake of World War I, Max Wundt (1920) contrasted the Weltanschauung of individualism, which he saw dominating the Western nations, with German idealism. For the Romantics, the *Nation* (or the *Volk*) was a kind of supraindividual individuality. Although, of course, consisting of a group of persons, the Nation itself was considered a living, organic whole with its own unique individuality. This individuality of the Nation was grounded in deep nonrational forces and thus, the Romantics concluded, ultimately eluded rational analysis and comprehension. As Bernhard Giesen appropriately summarized, "Because the individual essences of nations were exempted from ordinary and mundane communication, only art could provide a way of approaching the charismatic core, and poets assumed the function of the priests of this sacralized nation" (1998, 245).

As a Weltanschauung was being attributed to a wide variety of ethnically, culturally, or historically defined groups, Weltanschauung bonded with Kultur, which, as we have shown, was also drifting toward plurality. The development from one Weltanschauung (as understood within transcendental or theologian frameworks) to several Weltanschauung*en* corresponded to that from a universal Kultur to several unique Kultur*en*. In their new versions, the two concepts drew closer; they were compatible and entered a close symbiosis: A particular Weltanschauung became the ultimate expression of a particular Kultur.

Along more individualistic lines, researchers tried to correlate Weltanschauung and individual characteristics, for instance, body-type. In *Körperbau und Charakter* (Body Type and Character), Ernst Kretschmer (1961 [1931]) postulated *Konstitutionstypen*, which combined body morphology and ways of thinking: Thin ("leptosom-schizothym") individuals, in this view, were prone to speculative and abstract thinking, fat ("pyknisch-cyclothym") persons, to empirical and concrete thinking. This distinction thus would indicate a clear difference in Weltanschauung. If we ask, slightly differently, which Konstitutionstypus is *interested* in consciously thinking about and expounding a Weltanschauung, it is again the leptosom-schizothym type. Hence, one would expect speculative philosophers to be skinny. In his study of portraits of geniuses, Kretschmer found this expectation corroborated (1961 [1931], 370–94; 1958 [1929]).[33]

Plurality as problem. The Romantic turn from the general to the particular, from *the* Weltanschauung to a plurality of Weltanschauungen, and the impossibility to establish *one* rational Weltanschauung posed an intense challenge to philosophers and other scholars—how to react to the apparent loss of a single objective way of looking at the world.

As already mentioned, the plurality of Weltanschauungen corresponds to the plurality of cultures, but the problem of relativism was more heatedly fought over in the arena of Weltanschauung than in that of culture, partly because culture was solidly attached to nations so that cultural pluralisms could be defined, and dealt with, as international rivalries. We already encountered the fever-pitch intensity of the defense of German Kultur against French civilisation in the World War I era. Weltanschauung, by contrast, attached to a much wider variety of groupings, in particular to social classes, so that there was a relatively less stable "us." In addition, it continued to carry, as an early Romantic legacy, the meaning of an individual's achievement of synthesis. Weltanschauung thus was, to some extent, subject to individual challenge and responsibility—a core element of the individual's Bildung—and was, in that respect, more volatile than the collective Kultur. This volatility clashed mightily with the roots of Weltanschauung in the Kantian program of Transzendentalphilosophie, which made it rather problematic to accept the existence of two or more Weltanschauungen. Perhaps most importantly, the emphatic assertion of its own comprehensive and absolute truth belonged to the very essence of a Weltanschauung. Thus, almost by definition, each Weltanschauung was extremely hostile to all other Weltanschauungen.

An early diagnosis of the loss of philosophical certainties was put forward by Rudolf Haym: After the collapse of Hegel's grand philosophical synthesis, he wrote, "At the moment, we experience a great and almost universal shipwreck of the spirit and of the belief in the spirit as such" (1857, 5). The plurality of Weltanschauungen and the dissolution of the traditional certainties in beliefs and values created a crescendo of laments that reached their highest volume in the wake of Germany's defeat in World War I and the ensuing revolution. In gripping words, the philosopher Karl Joel expressed what he and many others perceived as an overwhelming sense of chaos and loss of orientation in modern society:

> Our present time is without goal because it reaches for all too many goals, for which it can do no more than merely grope. . . . In the meantime, an abyss opened up, and life's strongest support and foundation were about to vanish in it. The abyss created a tension between ideal goals and reality so that the world of experience appeared to explode into chaos rather than to congeal into the cosmos of a Weltbild. In the horror of this time, humanity was frenzied by passion, hounded by anguish; it cried out for support, for firm moral standards, clear guiding stars, and waited for the redeeming word—but the spheres of the spirit remained silent. (1928, 1)

Already in 1911, Edmund Husserl had noted,

> The intellectual crisis of our time has indeed become insufferable. . . . It is the most radical life crisis from which we are suffering, a crisis that does not stop short of any aspect of our life. All of life means taking a position, all position-taking is subject to an "ought," a judgment of validity or invalidity according to pretended norms of absolute validity. As long as these norms remained unchallenged and were not threatened and derided by skepticism, there was only one question in life, how best to abide by these norms in practice. But what now, when each and every norm is challenged or empirically falsified and robbed of its ideal validity? (1911, 336)

In his essay "Die Zeit des Weltbildes" (The Time of the Weltbild, 1938), published in his book *Holzwege* (1957), Martin Heidegger described the plurality of Weltanschauungen as a characteristic feature of modernity.[34] He went as far as to argue that "Welt *als* Bild" ("world *as* picture") would be a more appropriate interpretation of

Weltbild than was the usual "picture of the world." Heidegger's notion emphasized that, in forming a Weltbild, humans construct reality as a picture—an act that in itself presupposes a particularly perspectivist way of looking at the world.[35] The very fact of having a "world picture" thus becomes a distinct characteristic of modern age; earlier epochs did not have this way of looking at the world. "Understood in an essential way, 'world picture' does not mean 'picture of the world' but, rather, the world grasped as picture. . . . Wherever we have a world picture, an essential decision occurs concerning beings as a whole. The being of beings is sought and found in the representedness of beings. Where, however, beings are *not* interpreted in this way, the world, too, cannot come into the picture—there can be no world picture" (2002, 67–68).

Dilthey's relativism. In the issue of Weltanschauung and Weltbild there culminated wider debates about facts and values, about objectivity and relativism, *Naturwissenschaft* and *Kulturwissenschaft.* Scholars such as Wilhelm Windelband, Heinrich Rickert, and Wilhelm Dilthey divided the academic disciplines into two large camps: the Kulturwissenschaften (or *Geisteswissenschaften*—humanities) and the Naturwissenschaften (sciences), which were seen as distinct in terms of both subject matter and general approach.[36] A major issue was to determine the position of the Kulturwissenschaften vis-à-vis human values. Contrary to philosophers and Enlightenment enthusiasts who thought that reason could not only understand the world but also determine how one should live, these scholars asserted that the realm of values was beyond Wissenschaft. Kulturwissenschaft could study values but not set them. In their conceptual framework, it consciously stayed valuefree and limited itself to objective conclusions about facts. This creed became the basis for Dilthey's study of Weltanschauung. His solution to the problem of the multiplicity of Weltanschauungen was to withdraw to historical and psychological surveys of the conflicting Weltanschauungen, while taking no position on their actual truth claims.[37] Before we explore Dilthey's work, we should note that his position was not the terminus but a way station on a pervasive movement toward relativism.

Fewer and fewer intellectuals were content with the stance implied by *Wertfreiheit* (freedom from value judgment); its asceticism was unappealing and would not satisfy people in search for meaning, as Max Weber (1967 [1917]) clearly saw. We will see how they filled the void left by reason.

Some scholars escalated relativism beyond Dilthey: They doubted whether the neat separation of facts and values was even possible. Relativism was seen at work also in the realm of facts—or "so-called facts," as they would tend to put it—in addition to the realm of *weltanschauliche* value judgments. Not only was reason, in the guise of Wissenschaft, no longer deemed able to create an all-encompassing rational vision of Weltanschauung; it could no longer achieve a rational Weltbild. Inescapable commitments to a Weltanschauung (no longer founded in reason) were conversely considered to bias statements of fact and thus to make an objective Wissenschaft impossible, even in the much more limited factual realm.[38] Georg Simmel, for instance, pronounced a severe "critique of historical realism, for which historiography constitutes a mirror image of past events 'how they actually were'" (1905, v). According to Simmel, historical "facts" are construed by the human subject. Rather vaguely, the ideal of historical truth "can only grow out of [historical truth] itself" (ibid., 169). Karl Mannheim and other sociologists of knowledge explored how human knowledge was pervaded by biases, including knowledge created by the Kulturwissenschaften themselves.[39]

We now return to Dilthey's landmark *Weltanschauungslehre*.[40] Its starting point was the mentioned fundamental dichotomy or dilemma—the existence of a number of mutually contradictory Weltanschauungen on the one hand and claims of each one of these Weltanschauungen to contain the absolute, objective truth on the other hand. "Can there be a solution to this dichotomy? If it was to be possible, it must be brought about by historical self-reflection. It must make these human ideals and Weltanschauungen themselves its objects" (Dilthey 1960, 7).[41] Convinced that the philosophers' quest for an overall metaphysical system could only end in skepticism, Dilthey retreated to the level of devising a historical-psychological theory of Weltanschauungen. He undertook to accomplish this by comparisons of art, religion, and philosophy. By replacing systematic philosophy with history and psychology, Dilthey took a relativist stance toward the validity of individual Weltanschauungen—arguing, "What is determined by historical circumstances is also relative in its value" (ibid., 6)—but at the same time preserved the possibility of their wissenschaftliche examination.

Dilthey's psychological foundation for Weltanschauungen posited three major mental functions of humans—thinking, feeling, and desiring. Correspondingly, every Weltanschauung had three main elements—Weltbild, *Lebenswürdigung* (evaluation of life), and *Lebensführung* (conduct of life)—although any one of these elements might be predominant in a particular Weltanschauung.[42] "All Weltanschauungen, if they undertake to provide a complete solution of the riddle of life, typically contain the same structure. In every instance, this structure is a context in which, on the basis of a Weltbild, the questions about the meaning of the world are decided. From this, in turn, the ideal, the highest good, and the highest precepts for leading one's life are deduced" (ibid., 82).

Among the Weltanschauungen, he distinguished three main and recurrent types, which he traced through the history of philosophy: *Naturalismus* (e.g., Democritus, Protagoras, Hume, Hobbes), *Idealismus der Freiheit* (e.g., Anaxagoras, Socrates, Kant, Fichte, Schiller), and *objektiver Idealismus* (e.g., Parmenides, Spinoza, Herder, Schelling, Hegel, Schopenhauer, Schleiermacher, Leibniz). Naturalism expresses a primarily cognitive attitude toward the world. It is characterized by reliance on the senses, by the acceptance of passions, and by manipulating nature through knowledge. Its epistemology is sensualist, its metaphysics mechanistic. Idealismus der Freiheit (subjective idealism) has an active, moral-driven attitude, with ideals that transcend reality. It emphasizes a priori categories of the mind and human freedom, especially the freedom of the will from any physical causality. Its prime epistemological category is consciousness. Its metaphysics centers on a God who controls the material world. Objective idealism focuses on the appreciative contemplation of the universal harmony of all things. Its epistemology is contemplative, esthetic, even artistic. Its metaphysics is a mysticist panentheism.

In the following, we will outline some responses to the historical-relativist plurality of Weltanschauungen à la Dilthey. The spectrum ranged from the celebration of diversity to the abandonment of reason, to attempts at defending the rational synthesis of Weltanschauung, or at protecting at least some of the core elements of that synthesis.

Diversity: Pieces of a puzzle. Karl Joel, from whom we heard earlier, rephrased Dilthey's central question in this way: "But if there is only *one* truth, where, then, does the multiplicity of Weltanschauungen come from? And reversely: if there are many truths, where, then, does the monopolistic truth claim in each Weltanschauung come from?" (1928, 2). He viewed Weltanschauungen as "functions of truth" and "intellectual styles," "which do not arbitrarily follow each other, but which respond to each other and complement each other out of the wholeness of life, which unfolds its unitary structure in different functions" (ibid., viii). Central to his exposition was the "polar opposition of connection and separation, may it be expressed in unity and differentiation, rule and independence, necessity and freedom, collectivization and individualization, synthesis and analysis, assimilation and distancing, monism and dualism, or other similar terms" (ibid). Joel's solution was to consider the different Weltanschauungen pieces of a puzzle that somehow—almost dialectically—are all part of the synthesis of the "wholeness of life," although it remained somewhat unclear what this wholeness might look like.

Theodor Litt appeared to echo this strategy when he suggested that one should regard the plurality of Weltanschauungen as a boon: "It seems to me that nothing can save us Germans but the love of our inner fate, a love that affirms this separation into a plurality of intellectual formations that not only stand side by side, but even oppose each other, because that love recognizes not only its anguish and painful dissatisfaction in this fate, but also its unsurpassable riches" (1930, 87). Litt thus turned the lack of unity into a celebration of diversity.

Lebensphilosophie—Existentialism. Existentialist philosophers and *Lebensphilosophen* took a somewhat different and more radical approach. Their starting point was also the plurality of Weltanschauungen. But rather than examining the historical and social conditions of Weltanschauungen in a scholarly and rational fashion (under suspension of their own truth claims), or trying to reclaim a rational basis for evaluating Weltanschauungen, they emphasized the necessity of nonrational choice. They filled the void left by Dilthey's relativism (or by any even more comprehensive relativism) by jettisoning the supremacy of reason—a idea so cherished by the Enlightenment as well as by the exponents of German idealism—and the creative synthesis of the Romantics' conscious spirit in favor of instinct, will, and life itself.[43]

From Schelling's understanding of Weltanschauung as nonconscious perception, a side stream led to Schopenhauer's usage of this concept. Schopenhauer valued instinct over conscious knowledge, arguing,

> All knowledge, everything of which one is abstractly conscious, everything upon which one reflects makes action unsure. Hence acting from instinct is much more unfailing and surer than our acting according to concepts. Hence everything in the visual arts and in music that has sprung from concepts is bad, and similarly in poetry: all this must come directly out of Phantasie (that is, sensuousness and intellect subject to the will). Even every good philosophical statement must first originate from an idea, that is, from a phantasm, then it is Weltanschauung; otherwise it is worth nothing. (1916 [1814], 155)

H. S. Chamberlain similarly dismissed the conscious spirit, usually a key component in the post-Kantian understanding of *Weltanschauung*: "Only in the relatively rare cases of higher education and well-trained reflection might a person be consciously aware of having a Weltanschauung. This, however, is no criterion for the richness or especially for the vigor of a Weltanschauung" (1917, 7).

An increasing fascination with lifestyles, rather than Weltanschauung types, was also expressed in scholarly work.[44] Still in the spirit of "geisteswissenschaftliche Psychologie," Eduard Spranger, one of Dilthey's students, developed a typology of lifestyles in a book titled *Lebensformen* (1914). Here, the emphasis clearly shifted from the world that is looked at to the person who looks at the world and to the person's preferences, values, and choices. Ways of looking at the world became indicators of persons' character or personality. Spranger distinguished six ideal types: the theoretical, economic, aesthetic, social, powerful, or religious person. For the theoretical person, "Only a single passion may be alive in him, the passion for objective understanding, this however in the most literal sense. For the theoretical persons of the purest intellectual type know only *one* passionate suffering: the suffering from the problem, from the question that produces an urge for explanation, context, theorizing. This is their metaphysical motivation: that they can despair over ignorance, that they can be euphoric over a merely theoretical discovery, even if it is an insight that kills them" (Spranger 1922 [1914], 111–12).[45] A few years later, Richard Müller-Freienfels (1923), in his *Persönlichkeit und Weltanschauung,* developed a very similar typology. His primary distinction was between the *subjektiven* (emotional) and the *objektiven* (rational) individuals.[46]

Max Scheler also recognized "the diversity of the chaotic life in these days," but he did not wish to surrender to relativism (1923a, vii). He replaced formal ethics with a material ethics that emphasized commitment to certain values: "It is always the standard-setting Weltanschauungsphilosophie that, as metaphysics and material theory of a value hierarchy, has to precede any theory of Weltanschauung" (ibid., 18). "Because the individual person of every single human being is immediately rooted in the eternal being and spirit, there is no universally valid, but only an individually valid 'material' Weltanschauung, which is at the same time historically determined in the extent of its perfection and adequacy. However, there is a strictly universal method according to which all human beings— whoever they may be—can find 'their' metaphysical truth" (Scheler 1929, 14). That method brings about metaphysical knowledge, directed toward God, which is not objective, but emanates from life itself (ibid., 12–13). The act of living itself solves the theoretical quandaries. Scheler's austere bottom line was: "Wissenschaft . . . has in essence no relevance for the creation and establishment of a Weltanschauung" (1923a, 8).

Karl Jaspers's (1919) work on the *Psychologie der Weltanschauungen* is largely a Diltheyesque scheme of different Weltanschauungen, in which Jaspers went to great taxonomic lengths.[47] Within his classification, he distinguished types of Weltbilder that corresponded to several basic aspects of human existence. The *sinnlich-räumliche* Weltbild refers to the immediate, empirical-sensory world; it has developed through mythical, historical, and mechanical stages. (The latter equates to what we call the scientific Weltbild.) The *seelisch-kulturelle* Weltbild, by contrast, refers to the world of values and to the realm of meaning. Finally, the *metaphysische* Weltbild aims at the totality and the absolute; its two main subtypes are the mythical-daemonical and the philosophical Weltbilder.[48] The following quote indicates that Jaspers had already embarked on a journey from reason to life—to experience and will—in his interpretation of Weltanschauung: "Any developed theory of the whole becomes a shell, robbed of the original experience of extreme situations, and it prevents the rise of the forces that restlessly search the meaning of existence in the future in self-determined experience, to replace them with the stillness of a comprehended, perfect and soul-satisfying world of perpetually present meaning" (1919, 225). Furthermore, "The book has meaning only for people . . . who experience life as a personal, irrational, and interminable responsibility" (ibid., v).[49]

The belief in action and power was one of the planks of the National Socialist Weltanschauung, which officially dominated Germany between 1933 and 1945.[50] Another plank of that Weltanschauung was the connection it made between Weltanschauung and race.[51] Already at the beginning of the nineteenth century, Joseph Görres had loosely linked the evolution of Weltanschauungen to human races (among whom the European race was deemed superior) (1926 [1808], 407). Toward the middle of the twentieth century, Gustav Wyneken discussed, in a relativistic mode, "Weltanschauungen der Rassen" (1940, 336–42). Whereas elements of tolerance for the different Weltanschauungen of different races were present in Wyneken's work, the National Socialists repudiated any such notions of tolerance and prized the Weltanschauung of the Nordic race over all others. The linkage between Weltanschauung and race—race understood in a biological, socio-Darwinian sense—pervaded National Socialist thought. According to National Socialist doctrine, race-based Weltanschauung expressed itself in many facets of life—and even in the realm of science, where Einstein's theory of relativity was denounced as a prime example of "Jewish physics."[52] The long process of Weltanschauung divesting itself of reason culminated in the National Socialist interpretation. We shall return to this issue in greater detail below.

Plurality as evolution. A nonrelativist way of dealing with the plurality of existing Weltanschauungen is to order them in a developmental scheme. Here, reason invokes the Enlightenment and idealist traditions to impose an evolutionary hierarchy on Weltanschauungen. As already mentioned, Hegel used Weltanschauung in a clearly evolutionary sense and spoke of the "stage sequence of definite Weltanschauungen" (1985, 1: 80; cf. 309–13)[53] Similar to Hegel, Ludwig Feuerbach, in *Das Wesen des Christentums* (The Essence of Christianity) (1909 [1841]), understood the term *Weltanschauung* in some kind of an evolutionary fashion. For Feuerbach, Weltanschauung was something objective that characterized the individual's rise above his or her mere subjectivity.

A comprehensive evolutionary scheme characterizes Marxism, which ranks among the most prominent examples of an antirelativist philosophy of Weltanschauung.[54] Its late dogmatic representation in the East German *Philosophische Wörterbuch* reviled "the notion of the impossibility of a Weltanschauung that is constructed according to wissenschaftliche principles, a notion that today is propagated in assorted variants by the imperialist philosophy. . . . For dialectical and historical materialism, there can be no contradiction between Wissenschaft and Weltanschauung, because the weltanschaulichen generalizations and conclusions are reached in accordance with methods inherent in the Wissenschaften and find objective corroboration in practical application" (Klaus and Buhr 1975, 1287, 1289).

While acknowledging that the various Weltanschauungen of historical epochs and social classes are determined by specific economic and social conditions, Marxism posits that its own Weltanschauung, which is the Weltanschauung of the proletarian class, is the only correct one, whereas the others are biased to various degrees. According to this view, the Weltanschauungen of all other classes are distorted by those classes' particular self-serving economic and political interests. The interest of the proletariat, however, is unique: It coincides with that of humankind at large. The Weltanschauung of the proletariat (i.e., historical and dialectical materialism) is based on this interest in universal emancipation and hence is the only one that is unbiased. In Marxism, partisanship for the communist Weltanschauung becomes a necessary precondition of true science. This, as Hollitscher put it, "results from the fact that the working class is the historically first and only class that is not subject to socially induced limits to knowledge. This is because the working class has an elementary interest in revolutionary-practical change and thus in the knowledge of reality" (1985, 242).

Along with other sociologists of knowledge, Karl Mannheim, who used "total ideology" as a synonym of Weltanschauung, radicalized the Marxist epistemological framework and questioned the Marxists' own claim to an objectively true Weltanschauung.[55] The Marxist Weltanschaung was shielded against this and other criticisms primarily by the clout of governments committed to upholding it.

Restoring unity. Reacting against Dilthey's historicist relativism, Edmund Husserl undertook the ambitious enterprise of reasserting a wissenschaftliche philosophy. He attempted to avoid relativism by retooling Kant's transcendental program. Where Kant had focused on the transcendental analysis of perception, Husserl turned to the transcendental analysis of concepts, which he called the transcendental phenomenology of *Wesensschau*. In a 1911 *Logos* article, Husserl took a decidedly negative view of the concept of Weltanschauung when he distinguished Weltanschauungsphilosophie and wissenschaftliche Philosophie.

The emergence of the new *"Weltanschauungsphilosophie"* is essentially determined by the transformation of Hegel's metaphysical philosophy of history into a skeptical historicism. This Weltanschauungsphilosophie appears to spread quickly especially in

these days and, by the way, claims to be nothing less than skeptical, with its commonly antinaturalistic and occasionally even
antihistoricist polemics. To Weltanschauungsphilosophie, especially, did the talk about the decline of the philosophical ambition
toward Wissenschaft apply, in as much as it appears no longer governed, at least in its whole goal and method, by that radical aspiration for wissenschaftliche doctrine, which has been the major
feature of modern philosophy up to Kant. (1911, 293)

Husserl criticized both naturalistic and historicist approaches.
His prime example for the naturalistic error was a psychology that
aspired to the methodological standards of the natural sciences. The
second target of Husserl's critique was historicism or, to be more
precise, a historicist skepticism that refrained from making any judgments about the validity of cultural phenomena and only examined
their historical development. He explicitly criticized the relativist
tendency of Dilthey's study of Weltanschauungen by reaffirming the
time-honored dichotomy of genesis and validity.

He further stated, "Naturalists and historicists are fighting about
the Weltanschauung, but both make an effort, from different sides,
to reinterpret ideas as facts, and to transform all reality, all life into
an incomprehensible idea-less jumble of 'facts.' The superstition of
the fact is common to all of them" (ibid., 336). Here Husserl joined
Simmel in the rejection of "facts," but he was, of course, arguing for
the higher truth of transcendental-phenomenological Wesensschau.

In his 1935 Vienna lecture about philosophy and the crisis of
European humanity, Husserl (1970 [1935]) took up again the
theme of the Logos article. For him, the essence of Europe was the
development, in ancient Greece, of a theoretical attitude: "Man
becomes gripped by the passion of a world-view and world-knowl-
edge that turns away from all practical interests and within the
closed sphere of its cognitive activity, in the times devoted to it,
strives for and achieves nothing but pure theoria" (ibid., 285). The
task of philosophy was "the function of free and universal theoretical reflection, which encompasses all ideals and the total ideal, i.e.,
the universe of all norms" (ibid., 289). However, Husserl noted that,
in modern times, the objectivist sciences, as experts of the "psy-

chophysical world-view," embarked on naive and misguided attempts to reduce everything, even the spiritual realm, to objective, physical facts (ibid., 294). According to Husserl, the way out of this crisis of reason was either "barbarity" or, preferably, the culmination of transcendental phenomenology. As we now know, transcendental phenomenology (ibid., 299) did not prevail.

Another, internally diverse, group of philosophers set out to consolidate a rational Weltanschauung in a more limited form. They considered the Diltheyesque turn to historicist relativism, in its sweeping range, an unnecessary capitulation and endeavored to sort out what they considered heterogeneous components within the comprehensive concept of Weltanschauung. They agreed that Weltanschauung contained parts that were beyond the wissenschaftliche purview, but they also emphasized that other parts lay squarely within the realm of Wissenschaft. As to the latter parts, they contended, it was possible to make objective judgments about one Weltanschauung being superior to another. These efforts, in effect, differentiated the Weltbild from the Weltanschauung (although sometimes different terminologies were used), and then defended a wissenschaftliche Weltbild. Inasmuch as this wissenschaftliche Weltbild was a scientific Weltbild, chapter 5 takes up its discussion in much greater detail.

Whereas Dilthey was primarily interested in Weltanschauungen, Heinrich Gomperz's (1905, 1908) *Weltanschauungslehre* was a systematic analysis of what, according to our definition, would be called Weltbild: "And thus we differentiate the Weltanschauungslehre [theory of Weltbild, in our terminology] from the Lebensauffassungslehre [theory of Weltanschauung, in our terminology] in a similar way as people have for a long time been used to distinguishing theoretical and practical philosophy" (Gomperz 1905, 4). Moreover, Gomperz's Weltanschauungslehre "does not wish to explain existing Weltanschauungen, but wants to found a Weltanschauung by itself; and as to the existing Weltanschauungen, it is not interested in showing their determinants, but in examining their truth. In a single word, it is not a descriptive and comparative, but a critical and dogmatic discipline" (ibid., 3–4). He postulated five Weltbilder (*animistisch*, *metaphysisch*, *ideologisch*, *kritizistisch*, and *pathempirisch*) and showed how they differed in their conceptions of a few basic concepts (substance, identity, relation, form). These Weltbilder were considered developmental stages. Whereas, for Dilthey, all three major Weltanschauungen had philosophers of the highest caliber as their representatives, Gomperz regarded what he called the pathempiric way of looking at the world as superior to all others.[56] A fairly elaborate evolutionary approach was also followed by Ernst Cassirer (1954 [1923–29]), who, in *Philosophie der symbolischen Formen* (Philosophy of Symbolic Forms), explored the differences between mythological and empirical-scientific Weltbilder.[57]

As we look back on the evolution of Weltanschauung, we realize how the concept slipped from the grip of reason, represented primarily by rationalist philosophers and allied scholars. It was an accelerating process that, using a term by Georg Lukacs (1954) in a somewhat different meaning, could be called the "destruction of reason." We shall see in chapter 5 that some scientists and theorists of science stood ready to inherit the project of a rational Weltanschauung from the philosophers, staking their own claim to a rational Weltanschauung that would be based on science—a scientific Weltanschauung. There it will become evident that this project encountered similar problems as those diagnosed for the philosophical Weltanschauung. But scientists also developed the notion of a scientific Weltbild, which became much more central than the scientific Weltanschauung and allowed them to move on safer and very productive ground. First, however, we complete our conceptual survey by examining the development of the term *Weltbild* in general.

Weltbild

Although the word *Weltbild* has a Latin precursor in *imago ideaque mundi* and has been documented as far back as the German of the early Middle Ages, it became more widely used only in the early nineteenth century (Grimm and Grimm 1955, 1552). An early instance of the modern use of Weltbild can be found in an already quoted passage of Novalis's novel *Heinrich von Ofterdingen*, written in 1800 (1960a, 331).

On the whole, there has been much less change in the meaning of Weltbild than in that of Weltanschauung. Only in very rare cases was Weltbild used in a very literal sense, as the image of the world provided by sensual perception. One of these instances can be found in Ludwig Büchner: "We too have seen that the material organization of the brain is the main determinant of mental development, but this development can only occur in the presence of external intuitions of objective reality. In the absence of the latter, there is no reflection of Weltbilder on the material plane of the brain, no matter how well prepared [the material plane of the brain] is" (1856, 185).

On the one hand, Weltbild often appeared as the poor relation of Weltanschauung: Individuals who were unable to reach the glorified height of a philosophical Weltanschauung were left with a mere Weltbild (Hildebrand 1910, 82; Petersen 1944, 372–73; and see below). But on the other hand, Weltbild did get used in a more scholarly meaning in a wide range of areas. Because the concept of Weltbild does not share the Weltanschauung's universalist root in the epistemology of human perception, it has always carried an element of pluralism. Scholars have commonly discussed a Weltbild by ways of comparing and contrasting it with a rival Weltbild or other Weltbilder.

Roughly parallel with the spread of Weltanschauung, the term *Weltbild* also entered a number of fields, especially theology, linguistics, and the historical and cultural disciplines. In these areas, Weltanschauung and Weltbild were often used as synonyms. One of the major and enduring questions for German intellectuals has been how the Christian Weltbild will fare in the face of alternative Weltbilder springing from the natural sciences. (See Hübner 1987.) Bernhard Bavink (1947), for instance, wrote about *Das Weltbild der heutigen Naturwissenschaften und seine Beziehungen zu Philosophie und Religion* (The Weltbild of Today's Sciences and its Relationships to Philosophy and Religion). Hans Rohrbach (1967) published a book titled *Naturwissenschaft, Weltbild, Glaube* (Science, Weltbild, Faith), with chapters such as "Does the Weltbild of the Bible Still Fit in Our Times?"[58]

The already mentioned linguistic school that contended that language determined Weltanschauung used the term *Weltbild* perhaps even more frequently. Leo Weisgerber (1953–54) issued a two-volume book titled *Vom Weltbild der deutschen Sprache* and said in another work, "The wonderful achievement of language [is], . . . that it enables humans to synthesize all their experiences in a Weltbild" (1929, 29).

In the realm of historical and cultural studies, scholars have examined the Weltbilder of epochs and peoples. Weltbild has also been used so refer to social groups (e.g., "Weltbild of courtly society," Petersen 1944, 378) or to great individuals. Friedrich Ranke, for instance, wrote about Wolfram von Eschenbach's Weltbild (1953, 29).[59] Within psychology, *Weltbild* has been used in the branch of developmental psychology, especially to describe the mental concepts children of different ages hold about the world. Wilhelm Hansen (1938), for example, wrote a book called *Die Entwicklung des kindlichen Weltbildes* (The Development of the Child's Weltbild). Other branches of psychology have, as already mentioned, tried to link Weltanschauung—and also Weltbild—to body type, or to race.[60] Julius Petersen mentioned the attempts by Ludwig F. Clauß and H. F. K. Günther (two major race theorists with National Socialist affinities) of describing "rassisch bestimmte [racially determined] Weltbilder" (1944, 376). Walter Groß, the director of the office for race policy of the NSDAP, advocated a "rassebestimmten Weltbild" (1936, 29). He emphasized that such a Weltbild is inherently relativistic because "it recognizes the racial determination and hence the racial subjectivity of value standards, with which alone peoples and individuals in this world can judge their own achievements and those of others. Thus the Weltbild protects us from the arrogance of a false objectivity of the liberal type, which so often used its own standards to evaluate foreign cultures, for which those standards could not be valid" (ibid.).

While Ernst Kretschmer and Karl Friedrich Schaer used scientific methods to correlate Weltanschauung with Konstitutionstypen and blood types (see above), Rudolf Kassner (1951 [1932], 1978 [1930]), the protagonist of the *physiognomische Weltbild*, declared science bankrupt and instead embraced intuitive knowledge. He posited the unity of character and personality with physiognomic features: "Face and idea form a whole and can be understood and perceived only as such" (1978 [1930], 303). His basic methodology was decidedly nonquantitative, but even more than that—nonlogical: "First we must state that physiognomy, as we understand it, is not a Wissenschaft, and cannot and should not be one. In its desire to explain things, every Wissenschaft operates with concepts. Physiognomy does not explain, but it interprets" (ibid., 377). It is based on the power of imagination. While excluding the quantitative approach from his physiognomy, Kassner instead subjected numbers and mathematics to the physiognomic approach, and consequently condemned Einstein's theory of relativity, along with communism: "In both [Soviet communism and Einstein's theory of relativity], something comes to the fore that we here would like to call a reverse or incorrect magic (*une magie a rebours*) of the number" (ibid., 486).[61]

Within philosophy, the issue of the Weltbild is much harder to avoid than that of the Weltanschauung. Because of the large role that value judgments and practical issues play in Weltanschauung, many contemporary philosophers tend to remove Weltanschauung from the realm of philosophy proper—to be decided by personal choice and to be examined by historians and social scientists. Weltbild, however, in its more restricted meaning, does not contain those elements and is therefore less vulnerable to relativist skepticism. It has remained a bona fide philosophical concern to examine basic questions underlying the Weltbilder, even though not all philosophers, of course, are expressly interested in this issue. Our study is going to focus particularly on the Weltbild that has come out of the sciences—the scientific Weltbild.[62] We will return to the scientific Weltbild, as well as to the scientific Weltanschauung, to depict how scientists adopted these two key concepts of Kultur—and we shall find Albert Einstein as one of the most devoted supporters of this cause.

Distinction between Weltanschauung and Weltbild

The two terms *Weltanschauung* and *Weltbild* are often used interchangably (e.g., Grimm and Grimm 1955, 1531–32), but a number of writers and scholars did make an effort to keep the two concepts apart. Two contrary opinions exist about the relationship between the two notions. The first posits that Weltanschauung is more objective and interindividualistic than Weltbild. The second asserts the exact opposite: that Weltanschauung is less objective and more individualistic than the Weltbild.

An adherent of the first—and older—school of opinion, Julius Petersen contrasted the Weltanschauung as "supraindividual totality, something whole and universal" with the Weltbild that contains only an individual's perspective (1944, 372). Thus, for Petersen, the Weltbild was much more subjective than the Weltanschauung: "There is more subjectivity in the notion of the 'Bild' than in that of the 'Anschauung'; the Bild is an artistic creation of the world, which in all its features is determined by the character of the creator" (ibid., 373). Here the grand philosophical claims of the concept of Weltanschauung, harking back to its youthful days, are still fully sustained. Rudolf Hildebrand had earlier expressed a similar distinction between Weltanschauung and Weltbild: "The philosopher with his potentially all-encompassing Weltanschauung works on this; and each individual does so as well with the Weltbild that, in the course of his life, he has put together from experiencing, and thinking about, his own piece of the world. For quite some time, this Weltbild has also come to be called Weltanschauung" (1910, 82). This passage hints at an additional factor that might have contributed to the way Weltanschauung and Weltbild were originally distinguished. The philosophical pedigree and usage of *Weltanschauung* gave the term a dignified, even grandiose, halo, and consequently Weltbild filled the need for the more pedestrian usages. However, as Weltanschauung became so popularized that all sorts of people claimed to have one, it inched closer to Weltbild in this respect. This, one may suspect, opened the door for the more recent differentiation.

That second differentiation, grounded in the relativist turn of Weltanschauung, currently dominates and appears in most dictionary definitions. A good example is *Meyers Enzyklopädisches Lexikon* (1979): "Weltanschauung: unified, pre-scientifically or philosophically formulated, . . . comprehensive concept of the world and humanity, which intends to influence action. Weltbild: synthesis of the findings of objectifiable knowledge to a comprehensive view of the world." Walter Brugger made this distinction with great clarity: "Weltanschauung says substantially more than Weltbild does; Weltbild is understood as the synthesis of scientific findings to a comprehensive wissenschaftlichen view. It hence remains purely theoretical and does not pose the last, metaphysical questions about existence and meaning of the world as a whole" (1976, 455).[63] In this distinction, Weltanschauung is the more subjective and individualistic term because it adds subjective value judgments to the Weltbild, which contains only objective facts about the world.

NOTES

1. The publisher, Querido Verlag of Amsterdam, used the title *Mein Weltbild* for the book at the suggestion of Einstein's son-in-law, Rudolph Kayser (Holton 1986, 315). By the way, the term *Weltbild* subsequently appeared in the titles of other writers' publications about Einstein, e.g., Kanitschneider's *Das Weltbild Albert Einsteins* (1988).

2. Paul Vogel's definition of the early Romantic ideal of Bildung can be applied more widely to the concept of Bildung in general; it illustrates how closely connected Bildung is to Weltanschauung: "According to [the early Romantics'] opinion, not he was gebildet who has acquired a large sum of individual pieces of knowledge that are more or less causally related, but only he who investigates the grand nexus encompassing all the life of humans and of Nature; he who is able to reduce the diversity of knowledge to unifying, universal principles; he who is capable of viewing knowledge as an organic unity; he who is thinker and poet at the same time; he who puts his philosophical insight into action" (1915, 5). For a pedagogic perspective on the ideals of Bildung, see Lüttge 1900.

3. In his study on *Freiheit und Form*, Cassirer set out to demonstrate the "Einheit des Prinzips" in the history of German thought (1961 [1916], xii). To Cassirer, the persistence of unifying schemes among German thinkers constituted this unity of principle. In an interesting metatheoretical twist, Cassirer's effort itself thus can be understood as a chapter in the very intellectual tradition it describes. Also see Lindsay 1915.

4. This discussion is not meant to postulate one specific American style that contrasts with "the" German style. Each country has contained a variety of styles and schools. It is intended only to illustrate differences in dominant trends.

5. Similarly, Wach noted "that the formation of a Weltanschauung is certainly not, as some erroneously assume, a purely intellectual process, but that it makes demands on attitude and character. If those demands are not met, the most penetrating and comprehensive intellectual achievements must remain pointless in this respect" (1930, 202–203).

6. *Trübners Deutsches Wörterbuch* 1957, 113; also Brugger 1976, 45; Hoffmeister 1955, 663.

7. Austeda 1989; *Brockhaus Enzyklopädie* 1994; Eisler 1930a; Fuchs and Raab 1972; Grimm and Grimm 1955; Hoffmeister 1955; Klaus and Buhr 1975; Kosing 1985; *Meyers Enzyklopädisches Lexikon* 1979; *Oxford English Dictionary* 1989; Schischkoff 1978; Prechtl and Burkard 1996; Sandkühler 1990. Exception: *Großer Brockhaus* 1981.

8. Auroux 1990; Blackburn 1994; Brugger 1976; Durozoi and Roussel 1987; *Enciclopedia Filosofica* 1957; Foulquié and Saint-Jean 1969; Honderich 1995; Legrand 1972; Mauthner 1911; Morfaux 1980; Mourral and Millet 1993; Runes 1983.

9. Götze 1924; Gombert 1901; 258–59; 1902, 156; 1906, 138; Kainz 1943, 237. The most exhaustive discussion of the concept (which, unfortunately, I came across only after much of the following was completed) can be found in the dissertation by Helmut G. Meier (1967). There is also an extensive bibliography on Weltanschauung—as well as on Weltbild and similar terms—on the Internet at http://www.muellerscience.com /SPEZIALITAETEN/Philosophie/Weltanschauung.htm and http://www .muellerscience.com/SPEZIALITAETEN/Philosophie/Weltanschauung _Literatur.htm.

10. See documentation in Grimm and Grimm 1955, 1532–33; and Götze 1924, 43, 45.

11. On Goethe's Weltanschauung, also see Carus 1948.

12. Auroux 1990; Durozoi and Roussel 1987; Foulquié and Saint-Jean 1969; Legrand 1972; Morfaux 1980; Mourral and Millet 1993.

13. We use the German word *Wissenschaft* because it has no precise English translation. The umbrella concept Wissenschaft comprises all branches of scholarship taught at a university, including sciences, arts, humanities, and professional fields.

14. Note, however, that, in the same volume of *Kosmos*, Humboldt also used *Weltanschauung* in its more modern meaning—for instance, when he wrote of the "history of Weltanschauung, that is, the gradual formation of the concept of the concert of forces within a totality of nature" (1845, xiii).

15. See Grimm and Grimm 1955, 1532; Eisler 1930b, 18–20; Ratke 1929, 28–29, 302.

16. This passage is from Kant's dissertation, *De mundi sensibilis atque intelligibilis forma et principiis.*

17. This stance differs substantially from Marxist approaches to Weltanschauung, which view developments at the level of Weltanschauung as entirely determined by socioeconomic developments. An example for such a deterministic approach is Franz Borkenau's *Der Übergang vom feudalen zum bürgerlichen Weltbild* (1971 [1943]).

18. For an opposing—and correct—view, see Meier 1967.

19. This translation, by the way, corresponds to the translation of *Anschauung* as "intuition" for Kant's works: "Intuition (Anschauung) is knowledge (Erkenntnis) which is in immediate relation to objects (sich auf Gegenstände unmittelbar bezieht)" (Smith 1962, 79).

20. "You see that that which is often meant to support the rejection of religion has in fact a higher value for it in the Weltanschauung, than does the scheme that first presents itself to us and lets itself be surveyed on a smaller scale" (Schleiermacher 1911, 4: 260).

21. Elsewhere Schleiermacher wrote, "Our knowledge of God is only complete with the Weltanschauung. As soon as there is a trace of the latter, the basic features of the former appear. To the extent that the Weltanschauung is defective, the idea of the deity remains mythical" (1839 [1811], 322).

22. Meier pointed out that Schelling's own use of the term ended abruptly in 1804—perhaps partly because of the drastic shift of meaning around the turn of the century (1967, 84–85).

23. The cited sonnet is actually undated in Böcking's Schlegel edition. Böcking only noted that the sonnet was written between 1798 and the spring of 1800 (in Schlegel 1846, 1: xv). Götze indicated 1800 as the year when Schlegel penned the sonnet (1924, 44). At the very least, this is a reasonable guess, given that Schelling's work to which the sonnet apparently refers was written in 1799.

24. On the development of the concept of genius in Germany, see Schmidt 1985. On the evolution of the somewhat related concept of individuality, see Willems 1995.

25. This German passage is extremely hard to understand (and to translate) because its style is fragmentary and nongrammatical.

26. See Gombert 1901, 258–59; 1902, 156; 1906, 138; Götze 1924, 45–46.

27. Hegel 1985, 1: 80; see 1: 498.

28. This information was gleaned from the catalog of Einstein's library that the Japan Broadcasting Corporation (NHK) produced.

29. See, for instance, the books on Lessing's Weltanschauung by Gent 1931, Klein 1931, Leisegang 1931, and Spicker 1883. On Goethe's Weltanschauung, see Michel 1920, Spranger 1949, and Steiner 1948[1897]. On Schopenhauer's Weltanschauung, see Cornill 1856; on Giordano Bruno's, see Kuhlenbeck 1899; on Pindar's, see Bippart 1848; on Dostoyevsky's, see Prager 1925; on Nietzsche's, see Katz 1892–93; on Gerhart Hauptmann's, see Herrmann 1926; and on Plato's, see Schneider 1898.

30. Humboldt still used the term *Weltansicht* that approached extinction.

31. For Weltanschauung and art, see Trahndorff 1827; for Weltanschauung and psychoanalysis, see Pfister 1928; for Weltanschauung and faith, see Gogarten 1937 and Stahl 1845; for Weltanschauung and relativity theory, see Driesch 1930. Other scholars, for instance, examined the Weltanschauung of the German Enlightenment (Wolff 1963), of the Romantic age (Joachimi 1905; Kluckhohn 1932), of primitive peoples (Frobenius 1898), of the Sumerians (Jeremias 1929), of the ancient Orient (Winckler 1905), of the Middle Ages (Schaller 1934), of the Catholics (Lippert 1927), of the Christians (Proksch 1968), of the Jews (Pick 1912), and of the modern human (Bürgel 1932). A survey of Weltanschauung from the ancient Greeks up to Hegel was provided by Schlunk (1921).

32. See Max Wundt's *Griechische Weltanschauung* (1910) and *Deutsche Weltanschauung: Grundzüge völkischen Denkens* (1926) (in which he also wrote of "völkische Weltanschauung") and Max Scheler's *Nation und Weltanschauung* (1923b). For the French Weltanschauung, see Groethuysen (1927–30).

33. Karl Friedrich Schaer (1941), in his *Charakter, Blutgruppe und Konstitution: Grundriß einer Gruppentypologie auf psychologisch-anthropologischer Grundlage* (Character, Blood Type, and Constitution: Outline of a Group Typology on a Psychological-Anthropological Basis), ventured to advance Kretschmer's Konstitutionstypologie by basing his own typology on a simple and objective criterion—blood type. From his research on Swiss civilians and soldiers, he concluded that character was indeed associated with blood types. In particular, an almost polar difference was found between blood types A and B. Blood type A individuals had a vital, natural personality in which experience and expression easily merged into one another. The carriers of blood type B were very different: "In contrast with group type A, however, experience and the expression of this experience do not spontaneously merge. Rather, a barrier is inserted between experience

and its expression, which is characteristic of its [group type B's] entire activity and inactivity. This is reflection, deliberation, often also the conscious processing of the stimulus of experience" (Schaer 1941, 26). In his summary, Schaer characterized the thinking of type B persons as "abstract, conceptual, partial to formal logic. Metaphysical, unearthly. . . . Romantic-rambling. Humanities. Weltanschauung [!]. Philosophy. Ethics. Metaphysics" (ibid., 103).

34. *Holzwege* literally means "paths in the woods" and has the figurative meaning of "paths that lead astray."

35. For another perspectivist approach to the Weltbild, see Nelson Goodman's *Ways of Worldmaking* (1978).

36. The main goal of the Kulturwissenschaften was "idiographic" (describing individual facts and events), as Windelband (1904 [1894]) put it, whereas the Naturwissenschaften were "nomothetic" (looking to establish general laws). The different goals went hand in hand with different methodologies. The study of Kultur, according to these scholars, required a distinct kulturwissenschaftliche methodology that differed from the scientific method. "Verstehen" (the hermeneutic understanding of meaning) became the centerpiece of the kulturwissenschaftliche method.

37. Many empirical social scientists and historians shared Dilthey's solution—among them, famously, Max Weber (1967). Dilthey's stance was related to a wider school, called *Historismus*, that abandoned value judgments concerning larger historical units, such as Weltanschauungen or, perhaps most typically, nations and *Völker*. The historian Leopold von Ranke's famous dictum, "every epoch is immediate to God," embodied this relativism that rejected a grand evolutionary scheme of human history (1971 [1854], 60). But at least Ranke still held onto epochs and collective subjects (implying a certain continuity). Nietzsche's (1984 [1874]) more radical and fabulously bleak approach to history drew even those continuities into question.

38. Hence these scholars doubted that it was at all possible to give a detached and objective account of the various Weltanschauungen, as Dilthey had claimed to provide.

39. Whereas Mannheim still exempted the sciences from his general presumption of bias, some more radical relativists questioned even the possibility of objective science, thus achieving a special kind of (re-) unification of all Wissenschaften.

40. The cited work (Dilthey 1960) was published posthumously as part of Dilthey's *Gesammelte Schriften*. Prior to the edition of the *Gesammelte*

Schriften, large segments of the work existed only as a handwritten manuscript, but a core part of Dilthey's Weltanschauungslehre was published in 1911 in a collection of various authors' essays on Weltanschauung (Dilthey 1911). This essay (Dilthey 1911) was the target of Husserl's (1911) critique in *Logos.* On Dilthey, see Makkreel 1975.

41. This point is echoed by Jaspers (1919, 123).

42. The implied distinction of Weltanschauung and Weltbild coincides precisely with our own.

43. Of course, the scholars we group under this heading differ from each other in various shades and degrees.

44. As mentioned earlier, Jean Paul was the pioneer of tying together Weltanschauung and Lebensanschauung.

45. The theoretische Mensch also came in a variety of subtypes: empirical vs. a priori–speculative, users of different types of categories, natural science vs. humanities, Analytiker vs. Synthetiker, specialists vs. encyclopedists (Spranger 1922 [1914], 125–27).

46. Among the latter category, there were *Sinnesmenschen* (sense-oriented persons), *Phantasiemenschen* (persons of creative imagination), and *Abstrakte* (abstract thinkers). Important further modifications of these three basic intellectual types were the dichotomies of *Speziellseher* (persons who focus on details) vs. *Generelldenker* (persons who think in general terms) and *Pluralisten* vs. *Simplifizisten.*

47. For other efforts in similar directions, see Hofmann 1914; Kern 1911; and Leisegang 1928.

48. The latter category is again subdivided into several types. First, there are the hostile twins of materialism (for which the *sinnlich-räumliche* Weltbild has become absolute) and spiritualism (for which the *seelisch-kulturelle* Weltbild has become absolute). Then, in the *rationalistische* (or *panlogistische*) Weltbild, the absolute is conceived as the spirit of all prior absolutes (Hegel). The Weltbild of *negative Theologie* recognizes the limits of rationality and speaks about the world in negative and paradoxical ways. Finally, the *mythisch-spekulative* Weltbild views the absolute either as the eternal essence of ideas, laws, and forms or as a singular historical process of creation that is extrasensual.

49. Also see Klages 1936.

50. On the National Socialist Weltanschauung in general, see Broszat 1960.

51. A practical version of the connection between biology and Weltanschauung was made by Fritz Dupré (1926), who, in his *Weltanschauung und Menschenzüchtung,* advocated the state-run breeding of humans.

52. By founding Wissenschaft on race, the National Socialists explicitly rejected the notion of wissenschaftliche objectivity. In his book *Wissenschaft, Weltanschauung, Hochschulreform* (1934), the National Socialist ideologist Ernst Krieck, for instance, argued for a partisan race-based Wissenschaft in the service of German Weltanschauung. Also see Krieck 1942 and Rust 1940.

53. The evolutionary tendency is also apparent: "Because humans think, neither common sense nor philosophy will ever cease to rise from, and out of, the empirical Weltanschauung to God" (Hegel 1843 [1812], 107).

54. In Eugen Dühring's (1875) *Cursus der Philosophie als streng wissenschaftlicher Weltanschauung und Lebensgestaltung* (A Course of Philosophy as a Strictly wissenschaftliche Weltanschauung and Way of Life), a rival Weltanschauung emerged within the wider socialist camp. It was fiercely attacked by Engels (1970 [1878]).

55. To some extent, Mannheim designated an alternate social group to the proletariat as the carrier of nonideological knowledge—the notorious *free-floating intellectuals*, mentioned above.

56. The pathempiric way of looking at substance is characterized by holistic sensations, *Gesamteindrucksgefühle*—sensations of the total impression (Gomperz 1905, 117). In the area of identity, the pathempiric way calls forth endopathic "feelings of the continuity of the I" (ibid., 156).

57. In another replay of the Kantian transcendental program, Aloys Wenzl's (1936) *Wissenschaft und Weltanschauung* attempted to rescue metaphysical thinking about Weltbild and Weltanschauung by protecting its rational core from relativism: "There is no complete human being, and there has never been one, who has not felt the need, on the basis of a 'Weltanschauung,' to address the ultimate questions about the essence, cause, purpose, and meaning of being and history in general, and of life and experience in particular. Here thinking and living meet; here philosophy, Wissenschaft, and practical life-decisions originate from a common root" (1936, 1). Yet, through a key differentiation between Weltbild and Weltanschauung (in our, not Wenzl's, terminology), Wenzl defended the possibility of a nonrelative Weltbild, while conceding the more subjective elements of Weltanschauung: "This book intends . . . to do positive philosophy, that is, it wants what true philosophy has wanted and always must want: to attempt to develop a unified Weltbild and a Weltanschauung" (ibid., vii). In Wenzl's view, Weltanschauung contains subjective and objective elements, and metaphysics has the task of examining the objective side of Weltanschauung.

58. Also see Sausgruber 1962.

59. On Stefan George's Weltbild, see Koch (1933).

60. As mentioned earlier, "Weltanschauungen der Rassen" were discussed by Gustav Wyneken (1940, 336–42).

61. "For [the physiognomist], the number is the fundamental expression of the break between phenomenon and essence, appearance and truth" (Kassner 1978 [1930], 496).

62. The scientific Weltbild is also a central issue for the philosophy of science.

63. Similarly, Müller-Freienfels wrote, "Unlike pure Wissenschaft, 'Weltanschauung' does not want to create a Weltbild that is detached from all subjectivity to the highest degree possible; but every artistic, religious, or philosophical 'Weltanschauung' takes a position toward the external world that arises from the needs of the I, wherein the subjective factor plays an essential role" (1923, 5). The *Brockhaus Enzyklopädie* (1994) made this distinction: "Whereas a Weltanschauung tends to conceive reality ideologically as a totality, the Weltbild is more realistic and aims at a representation of the individual domains of experience that is as comprehensive as possible."

4.

KULTUR AND SCIENCE

When World War I broke out, both the majority of the German Kulturträger who supported it and the dissenting minority (to whom Einstein belonged) argued about the war *in terms of* Kultur. Few things illustrate the overriding importance of Kultur in German thought more dramatically than the fact that, at this momentous turning point, the German intellectuals fought over what Kultur meant. The manifesto of ninety-three prominent Kulturträger in support of Germany was characteristically titled "Aufruf an die Kulturwelt" (Appeal to the World of Kultur) and included the remarkable words, "that we shall fight this fight to the end, as a nation of Kultur [*Kulturvolk*], to which the legacy of a Goethe, a Beethoven, a Kant is as holy as its hearth and its soil" (quoted in Fölsing 1993, 391). The countermanifesto titled "Aufruf an die Europäer" (Appeal to the Europeans) attacked the militarist fervor evident in the manifesto of the ninety-three as a lapse from Kultur:

"No passion can excuse such an attitude, which is an embarrassment to what the whole world up to now has understood under the term Kultur. If this attitude became common among the Gebildeten, it would be a disaster" (quoted ibid., 392).[1]

In a more sedate form—in the foreword to a scholarly book—the philosopher Ernst Cassirer expressed his confidence that German Bildung would overcome its troubles in the difficult times of World War I: "Even in these days [June 1916], German Bildung will not allow itself to be sidetracked from its original course, neither by its enemies' misconceptions and detractions, nor by a narrow-minded chauvinism" (1961 [1916], xvi). That original course, for Cassirer, led toward realizing Fichte's ideal of the "citizen of liberty."

The theme reappears almost twenty years later in Einstein's letter to his friend Solovine of September 29, 1932: "I hope you personally are doing well in this crazy world, in which militarism cannot be extirpated, owing to the hypocrisy of the 'Gebildeten'" (Einstein 1987, 74). The horror vision from 1914 that militaristic fervor would spread among the Gebildeten had become reality. For Einstein, militarism (in the original German, he used the uncommon and derisive word *Soldaterei*) constituted such a perversion of true Kultur and Bildung that those who hypocritically promoted it no longer deserved being called Gebildete—they were merely "Gebildete," in sarcastic quotation marks. Einstein's commitment was to the early cosmopolitan and universal notions of Kultur, as, for instance, expressed by Goethe. He detested what he perceived as the degeneration of Kultur to a nationalistic and aggressive project.

After having surveyed some key features of the German intellectual landscape in the nineteenth century, we now ask how science was positioned within that landscape. Recall that it is one of our main contentions that German science was shaped considerably by the larger cultural forces that surrounded it. Albert Einstein's physics will later serve as a prominent case example demonstrating the thorough influence of a cultural milieu on a scientific research program. Our thesis presupposes that different national host cultures make a difference in how science develops. To set the stage for our study of the relationship between German Kultur and German science, it will be necessary to address the issue of national scientific styles in a comparative way.

4.1. NATIONAL STYLES OF SCIENCE

Two kinds of minds—the strong and narrow mind and the weak and ample mind—were distinguished by the French scholar Pierre Duhem (1962 [1906]), building on a similar distinction by Blaise Pascal.[2] Although exemplars of these two basic types can be found everywhere, they are, according to Duhem, spread unevenly across the nations. The weak and ample mind predominates among the English, whereas the French (and other Continentals) usually possess a strong and narrow mind. The ample English mind relishes in a multitude of details and facts, but it is too weak to grasp abstract principles; the strong French mind, conversely, understands abstract principles with ease, but it is so narrow that details tend to overload and confuse it. Duhem thought furthermore that these general differences gave rise to distinct national styles of physics: The English physicists would emphasize illustrative models, whereas the French physicists would strive to build abstract systems of a few general and simple propositions.

According to Duhem, "Understanding a physical phenomenon is, therefore, for the physicists of the English school, the same thing as designing a model imitating the phenomenon; whence the nature of material things is to be understood by imagining a mechanism whose performance will represent and simulate the properties of bodies" (1962 [1906], 72). Moreover, "The English physicist does not . . . ask any metaphysics to furnish the elements with which he can design his mechanisms. He does not aim to know what the irreducible properties of the ultimate elements of matter are" (ibid., 74).

By contrast, the efforts of French, German, and other Continental physicists are directed toward "those majestic systems of nature claiming to bestow on physics the formal perfection of Euclid's geometry. These systems take as their foundations a certain number of very clear postulates, and try to erect a perfectly rigid and logical structure in which each experimental law is exactly lodged" (ibid., 80–81). And he further explained, "This unity of theory and this logical linkage among all the parts of a theory are such natural and necessary consequences of the idea that strength of mind imputes to a physical theory, that to disturb this unity or to break this linkage is to violate the principles of logic or to commit an absurdity, from this viewpoint" (ibid., 81).

During World War I, Duhem modified this scheme by including Germany as a separate category. His book *German Science*, largely based on four lectures given at Bordeaux in 1915, developed a tripartite system in which English science and German science now occupied the two opposite poles, and French science pursued a happy medium between these extremes. English thought, again, was considered intuitive, not given to rigorous reasoning or designing a systematic order, but very apt at understanding a multitude of concrete objects. By contrast, Duhem characterized German thought as mathematical, abstract, laborious, and deductive: "Endowed with a powerful geometrical intellect which allows him to deduce with extreme rigor, [the German] is deprived of common sense, of that subtlety of intellect which supplies the intuitive knowledge of truth" (Duhem 1991 [1915], 55). The French, in this new scheme, possess that subtlety of intellect; they have an intuitive mind and good sense (ibid., 75).

Duhem went on to criticize the German mind for not being grounded in real life. He also disparaged specific theories. Riemann's geometry was rejected because it "shocks common sense" (ibid., 91). The "new physics" emerging from Germany received nothing but scorn because it "did not recoil from contradicting common sense" (ibid., 104). "The assumptions of common sense do not hold if one admits the principle of relativity as that is conceived by Einstein, Max Abraham, Minkowski, or Laue" (ibid., 195). "That the principle of relativity disconcerts all the intuitions of common sense does not excite the distrust of German physicists. Quite the contrary, to accept it is, by that very fact, to throw over all the theories that speak of space, time, and motion, all the theories of mechanics and physics. Such a devastation possesses nothing displeasing to German thought" (ibid., 105–106). (Recalling the discussion about the lowly status of common sense in German philosophical thought, we surmise that many German Kulturträger might actually have agreed with this point made by Duhem—and might have taken it as a compliment.)

Ironically, while the Frenchman Duhem condemned Einstein's theories for being too German, some Germans condemned them for not being German enough. The "new physics" of relativity was denounced as Jewish physics by the followers of a self-described German physics (which Duhem, in turn, might very well have recognized as French physics). Two Nobel laureates, Philipp Lenard and Johannes Stark, championed this brand of "German" or "Aryan Physics" (Beyerchen 1977).[3] In the preface to the first volume of his *Deutsche Physik*, Lenard wrote: "'German physics?' people will ask. —I could also have said Aryan physics or physics of the people of the Nordic type, physics of the explorers of reality, of the truth seekers, physics of those who have founded natural research" (1936, ix). As its main opponent, German physics targeted Jewish physics. The "German physicists" rejected relativity theory in favor of the ether (Lenard) and quantum mechanics in favor of a different atomic theory (Stark). They also stressed the importance of experiment and observation and criticized the Jewish physics that, in their view, overemphasized theory and abstraction. In a similar vein, they contrasted Aryan pragmatism with Jewish dogmatism in physics. Furthermore, they discarded the notion of scientific objectivity and claimed that the race of a researcher would determine his or her physics, and they argued that science had to be socially relevant to the German people (Beyerchen 1977, 123–35).

Their ideas corresponded to the general and dominant view under National Socialism that science had no universal validity but was driven by, and subject to, particular Weltanschauungen, which, in turn, were rooted in the scientists' race. Ernst Krieck, one of the leading National Socialist propagandists of a German science, for instance, wrote, "Wissenschaft does not exist as a body of meaning that without doubt is uniform for all times and all peoples, but meaning and form of Wissenschaft themselves are subject to historical change, according to the questions it asks. Hence, Wissenschaft passively and actively participates in real life, and changes its meaning according to the particular situations and challenges with which it is confronted by the respective generations" (1934, 8). And he added, "All searching for the essence of wissenschaftliche truth will necessarily go astray, if Wissenschaft is not at the same time and ultimately seen and perceived within the larger framework of life itself, of the Weltanschauung, the Weltbild, the völkische character, the historical situation and task, the religious and fate-like basic decision of an era" (ibid., 16).

Deutsche Physik made few inroads in the German physics community. The conventional physicists went on the counterattack and reached the official recognition of relativity theory and quantum mechanics by the NS University Teachers League in 1940 (Beyerchen 1977, 176–79). Yet Deutsche Physik was only the tip of the iceberg. Heinrich Himmler, the leader of the SS, was also the patron of other, and sometimes entirely scurrile, efforts toward creating a "German science."[4] One of Himmler's main projects was to replace Christianity with his own brand of race-based secular religion—a neo-paganist revival of ancient Germanic practices and beliefs that had been suppressed by Christianity, with a strong admixture of ancestor worship (Ackermann 1970, 40–41). The central element of Himmler's religion was a glorified image of the Germanic past. To uncover this heritage, Himmler formed a number of research organizations, among them, first and foremost, *Das Ahnenerbe e. V.* (Ancestors' Heritage) (ibid., 43).[5]

As long as they supported, or at least were compatible with, the "correct" (i.e., National Socialist) Weltanschauung, even rather dubious theories were able to flourish under the regime. One of the diverse theories—many of them pseudoscientific or simply bizarre—that SS leader Himmler supported through the Ahnenerbe was the world ice theory, or *Welteislehre* (*WEL*, by common abbreviation).[6] A whole department of Ahnenerbe was dedicated to this theory, which had been thought up by the Austrian engineer and inventor Hanns Hörbiger.[7] The Welteislehre was published as *Hörbigers Glacial-Kosmogonie*, edited by amateur astronomer Philipp Fauth (1913), who later directed the Ahnenerbe Forschungsstätte für Astronomie (Kater 1974, 86). The following is Hörbiger's theory in a nutshell. Two types of celestial bodies exist: hot ones (suns) and ice-covered ones. Decelerated by the ether, planets and moons spiral down onto the bodies they orbit. In the past, several moons already crashed into the Earth (which explains certain catastrophes). When an ice body falls into a sun, superheated water vapor is explosively ejected again (sun spots!). The ejected vapor cools down and becomes cosmic ice. One part of this cosmic ice forms the Milky Way. Another part falls on Earth as hail. This general theoretical framework of the WEL claimed wide-ranging explanatory powers. It purported to explain basic facts of meteorology, geology, paleontology, and biology, as well as special events, such as the destruction of Atlantis (Cornwell 2003, 193–95; Nagel 1991, 22).

Whereas the controversy about Deutsche Physik was fought mainly *within* the established physics community in Germany—Lenard and Stark were physics professors of Nobel Prize distinction—the *Welteislehre*, created by an outsider, was disdained by virtually all physicists. It was one of the chief protagonists of Deutsche Physik, Philipp Lenard, who denounced the WEL in particularly caustic terms: "Should this (N.S.!) magazine be allowed to continue dumbing down the German people without restriction? It is irrelevant how much untruth or truth this 'theory' contains; in any case, it is *pure phantastic speculation*, an insult to all *knowledge* of nature." This was what Lenard wrote to the editor of the *Illustrierter Beobachter* after it had printed an article by Rudolf v. Elmayer-Vestenbrugg titled "Hanns Hörbiger, der Kopernikus des 20. Jahrhunderts."[8]

Although decisively rejected by the scientific community, the WEL became fairly popular among the general audience. Far more than one hundred books and monographs about WEL were published in Germany in the 1920s and 1930s—not even counting journal and magazine articles (Nagel 1991, 92). WEL supporters were also found in the highest reaches of the National Socialist regime. Next to Himmler, Adolf Hitler himself was pro-WEL (ibid., 67–68; see Picker 1965, 167, 298; Ackermann 1970, 46; Kater 1974, 51). Apparently, Göring and Reichsjugendführer Baldur von Schirach were also enthusiastic about WEL (Kater 1974, 51). Several aspects made the WEL attractive not only to the National Socialists but to the wider *völkisch* and anti-Semitic groups in the German populace. First, a "real German" created it—in contrast, for instance, to the theories of the "Jew" Einstein. Then, one can find some obvious correspondences of WEL with the ancient Germanic Edda saga of creation. And, perhaps most importantly of all, the WEL offered a comprehensive Weltbild, "a closed, ravishingly magnificent Weltbild as the scientific basis of a genuinely Nordic Weltanschauung," in the words of Hörbiger admirer Elmayer-Vestenbrugg (Nagel 1991, 62), that could serve as an antidote to the detested Jewish and materialistic Weltanschauungen that seemed to emerge from the sciences.

After those inchoate theoretical attempts at comprehending, and misguided practical efforts at promoting, national styles in science, this topic, perhaps understandably so, held little attraction for the research community. Few serious scholars would nowadays speak of a "German mind" or a "French mind" as if such reifications were real determinants of science.[9] However, scholarly interest in the national dimensions of science—more cogently defined—has recently grown again, and we now have a few excellent studies in this area. Jonathan Harwood (1993), for instance, examined the field of genetics; Pauline Mazumdar (1995) studied the field of immunology; and Anne Harrington (1996) demonstrated how holistic thinking pervaded various scientific disciplines in Germany.[10] The scientific development in a particular country certainly is shaped by a number of factors that operate at the national level, including the larger demographic, socioeconomic, and political forces, as well as, more specifically, by traditions and "schools" tying together subsequent generations of scientists and by the institutional structure of the national science system.[11] Within this mix, however, cultural values and ideas play a major role. In our case, this is particularly evident in the fact that, at the beginning of the nineteenth century, a major Kulturträger, Wilhelm von Humboldt, designed the German university to embody the core values of Kultur and Bildung. Our focus now is on the influences that the wider cultural milieu exerted on German science and physics in particular.

4.2. SCIENCE CLAIMS KULTUR

"Bildung durch Wissenschaft," the celebrated Humboldtian slogan, echoed countless times through German auditoria, whenever academic dignitaries had occasion to reflect on the mission of the university. Humboldt's motto tied Bildung, and through it Kultur, closely to Wissenschaft. In nineteenth-century Germany, Wissenschaft in general, together with its primary institutional base, the university, was almost universally recognized as an indispensable element of Bildung and Kultur. It is rather less obvious, however, how science (Natur-Wissenschaft) fit into this picture. Was science considered part of Kultur? Were scientists Kulturträger? These are the questions we now address. Because of the centrality of Kultur in German thought, and because of the prestige and power that membership in the group of Kulturträger entailed, these questions were of real importance for the scientists and for society.[12]

In the most general sense, science as a human activity and product is part of culture (as opposed to nature). Yet the widespread distinction of Kultur and Zivilisation called into question whether science belonged to Kultur. Our preceding survey of the history of the terms would suggest that science was typically considered a key ingredient of Zivilisation, and thus the opposite of Kultur. It was incumbent on the scientists to claim a place for science in Kultur.

Scientists who wanted to be recognized as Kulturträger had three main arguments to support their claim. First, as graduates of the Gymnasium, they had, of course, fulfilled a basic precondition of Kulturträger status. Second, they were doing Wissenschaft, after all. Wissenschaft was an umbrella concept that comprised all disciplines taught at the university.

Yet the second claim was perhaps not entirely solid. To be sure, according to Humboldt's vision, Wissenschaft was indeed a central (though not the only) element of Bildung and thus a qualification for being a Kulturträger, but Wissenschaft, insofar it was objective and universal, also pointed toward Zivilisation. The distinction of Kultur and Zivilisation reverberated, for instance, in the distinction between national Bildung and international Wissenschaft that the prominent Protestant theologian Adolf von Harnack made: "All Bildung has national characteristics; without them, it remains colorless and flat—but Wissenschaft is international; it does not tolerate border posts because any kind of limitation detracts from it" (1930, 617). Within the wissenschaftliche part of Bildung, it was the classics, and the humanities in general, that played the major role. In early nineteenth–century Germany, "Bildung had become so closely linked to classical philology that other interpretations of it were for a long time generally unacceptable" (Jungnickel and McCormmach 1986, 1: 4). The sciences, although increasingly influential in a modernizing society, were never considered an indispensable part of Bildung. Those who knew nothing of science still could pass as gebildet, but it was almost inconceivable to regard as gebildet anyone who did not know Latin.[13] Moreover, as we have seen, efforts were under way toward the end of the nineteenth century to establish an internal differentiation underneath the wide umbrella of Wissenschaften, along the lines of the Kultur-Zivilisation dichotomy, that is, between Kulturwissenschaften and Naturwissenschaften.

The previously mentioned two points that scientists could make to bolster their recognition as Kulturträger paled in significance compared with a third. It was paramount for scientists to declare science as a human endeavor explicitly part of Kultur (rather than accepting science's place within the less-valued Zivilisation). To fit science into Kultur, at least one of the concepts had to be modified from their dominant uses: Kultur, or science. Kultur was the main target in the nineteenth century. In that era, the scientists typically argued that science shapes Kultur—that science has *cultural benefits* and should be subsumed under a more inclusive concept of Kultur—while they retained the creed that scientific truths were independent of Kultur, which was why science was commonly counted under Zivilisation in the first place. In the twentieth century, the reverse argument—Kultur shapes science—gained force. Some scholars and scientists tried to make science an integral part of Kultur by "Kulturalizing" it, that is, by inviting the relativism, nationalism, and even irrationalism of Kultur into the domain of science and by downplaying its universal validity. This culminated in such developments as the National Socialist Weltanschauung and "Aryan physics."

But let's return to the first phase. The struggle for the place of science in Kultur was a struggle about the dichotomy of Kultur and Zivilisation, about the contents of Kultur, or to be more specific, about the role of reason in Kultur. The material (*zivilisatorische*) benefits of science for the German nation were impossible to overlook in the nineteenth century. Yet the German scientists who were eager to join the ranks of the Kulturträger found themselves in the somewhat paradoxical situation that it was not enough to point to the spectacular applied and technological achievements of science to gain recognition from the other Kulturträger. For, from that group's perspective, these achievements were suspect: At best, the Kulturträger regarded them as merely practical conveniences with no higher significance. At worst, they abhorred a mechanized, industrialized society so vehemently that they would rather have less than more technological progress altogether.[14]

The main battleground was the meaning of Kultur itself. Could science be established in Kultur? Many traditional Kulturträger were suspicious of what they considered encroachments by science from Zivilisation into Kultur, and especially of any new Weltanschauung or Weltbild that science might create. In the early twentieth century, the religious scholar Joachim Wach, for instance, wrote critically of "new attempts to carry out a—theoretically shallow—progressivist Weltbetrachtung on the basis of the empirical study of nature" by which he meant materialism (1930, 201).

By contrast, some scientists attempted to define science as directly relevant to Kultur. As early as 1829, at the Heidelberg meeting of the association of German scientists and physicians, Friedrich Tiedemann made such a suggestion. He celebrated the rapid scientific advances in many fields, acknowledging that their applications were dramatically improving the lot of humankind. Yet, Tiedemann added, science also had a positive impact on the *Veredelung des Geistes* (refinement of the spirit) and the *Gesetze der Vernunft und Sittlichkeit* (laws of reason and morality) (Schipperges 1976, 15). With a goal like Veredelung des Geistes, Tiedemann precisely claimed one of the central planks of Bildung and Kultur for science—of course, in a rationalistic, pro-Enlightenment meaning of Kultur.

In a 1877 lecture titled "Kulturgeschichte und Naturwissenschaft," Emil du Bois-Reymond exclaimed, "Science is the absolute organ of Kultur, and the history of science is the true history of humankind" (1912, 1: 596). In these words, science patently staked its claim as the central element of Kultur. Kultur, for du Bois-Reymond, was humanity's "slow rise from a partly animal state, its progress in arts and sciences, its growing mastery over nature, its daily increasing prosperity, its liberation from the fetters of superstition, in a word, its steady approach toward the goals . . . that make the human being human" (ibid.). In contrast to the dichotomy of Kultur and Zivilisation, Kultur here takes a more comprehensive meaning that includes elements of what is usually called Zivilisation. (This Enlightenment-type notion of Kultur thus harked back to earlier notions of the term; see chapter 2.)

Another noted protagonist of this view of the relationship between Kultur and Wissenschaft was Wilhelm Ostwald, who declared, "Hence all Kultur in every respect is based on Wissenschaft, which must be called both the highest flowering and the deepest root of Kultur" (1909, 170). One of the most prominent and outspoken scientists who argued for the cultural benefits of science was Rudolf Virchow. We will discuss his opinions within the framework of the Kulturkampf.

Kulturkampf

At a pragmatic level—joining forces against a common foe—science became an ally of the government and of wide segments of the Kulturträger when the Kulturkampf broke out. According to the 1885 Brockhaus' Conversations-Lexikon, the term is "a name for the struggle between the state and ultramontanism."[15] That struggle between the Prussian state and the Catholic Church occurred between 1872 and 1887. The word Kulturkampf itself was coined by Rudolf Virchow, one of the founders of the Fortschrittspartei (Progressive Party), in an

1873 speech (Tal 1975, 82). Both a famous pathologist and a prominent politician, Virchow was a prime representative of an alliance of science with Kultur and government against Catholicism.

One aspect of the Kulturkampf was a political power struggle between the Prussian government and the Catholic Church hierarchy. During the course of its enormous expansion, the originally Protestant Prussia had annexed large Catholic areas. Many state officials were wary of the Catholic Church as a centralized multinational organization ruled from abroad. They suspected Prussian Catholics of divided loyalties and saw the influence of the church on education and other areas of the social and political life as a threat to national sovereignty—especially after the Vatican Council reinforced the centralism of the church and proclaimed papal infallibility. But for our purposes, a second, more specifically cultural, aspect of the Kulturkampf is more important.

Notwithstanding notable variations among the diverse strands of German Protestantism, that denomination on the whole malleably accepted many of the major landmarks that arose in the German intellectual landscape in the aftermath of the Enlightenment, such as Bildung, Kultur, and Nation. In the Protestant academic classes, a *Kulturreligion* emerged, based on a mixture of the Bible and great German thinkers (Stern 1974, xxv; Timm 1990). In the Catholic Church, however, a greater measure of medieval ritualism and mysticism persisted. To many of those who wished to modernize the German nation, the Catholic Church appeared as an embarrassingly backward relic and a roadblock to progress.

In 1871, at the first meeting of the association of German scientists and physicians after the war between the German states and France and the subsequent German unification, Rudolf Virchow delivered a speech "On the mission of the sciences in the new national life of Germany": Science had two contributions to make to the new German Reich—furthering both its material and ideal values.[16] While it was fairly obvious what the first contribution meant, the second was more intriguing. Here it was the task of science, according to Virchow, to unite the nation spiritually by providing a shared basis of knowledge and a rational method of thinking. He contrasted Catholic dogmatism with the freedom of science and expressed his confidence "that we shall succeed, as knowledge advances, in also finding a motivation for higher moral aspirations and a source for the ever growing desire for truth, honesty, and dependability in our deeds" (Schipperges 1976, 47). In addition to its usual function, science thus was assigned a role in the national cause and in the moral advancement of the German nation.

During the next annual meeting in 1872, Virchow renewed his attacks on religion in general and the Catholic Church in particular. His speech titled "The Sciences and their Importance for the Moral Education of Humankind" accused religion of failure in the sphere of moral education. According to Virchow, it was time for science to provide a secure basis for morality:

> Gentlemen: We do not deny it; we acknowledge freely, openly, and gladly that only the introduction of the scientific method into medicine has broken the tradition, that two-thousand-year old tradition that has occupied all minds. . . . But if the sciences have given us their method, they should not go on casually to impose restrictions on our own field and to tell us: you are not permitted to tackle the problems of the mind and of conscience with our scientific method. Yes, gentlemen, this is what we demand. We even demand that all scientists participate in it. We demand that every one, from his own standpoint, contribute to developing morality as an empirical science according to the rules that general science has constituted. (ibid., 50)

Again, we find an Enlightenment view of Kultur grounded in rationalism and the belief in progress.

At the Eisenach convention, ten years later (1882), the location inspired Ernst Haeckel, the leading German exponent of Darwinism, to another attack on Catholicism and to a celebration of the advancing scientific Weltanschauung: "Just as, at this holy place, 360 years ago, Martin Luther's mighty hand tore the web of lies woven by the world-dominating papacy, and brought about a new era in the history of culture through his reform of the Church from top to bottom, so has Charles Darwin in our days destroyed the prevailing superstitions of mystical creation dogmas with equally over-whelming might, and has steered all the feelings, thoughts, and desires of humankind onto new, higher trajectories through his reform of evolutionary theory" (ibid., 56–57). Obviously, the common enemy was not quite enough to create a common outlook on Kultur among all the Kulturträger. When these scientists were talking about Kultur, they meant something very different from the dominant understanding of Kultur.

Even after the Kulturkampf was long concluded, some scientists proclaimed their own particular vision of Kultur. At the 1903 assembly in Kassel, chemist Albert Ladenburg's (1911a) speech "On the Influence of the Sciences on Weltanschauung" reiterated some of the anti-Christian themes and created quite a stir. (See Hoppe 1979, 142–45; Ladenburg 1911b.) For Ladenburg, the sciences had been the major driving force toward the *moral*, and not merely the tech-nological, progress of humankind: "I contend that almost all humane efforts during the past two centuries have come about through concepts that were formed on the basis of scientific discov-eries" (Ladenburg 1911a, 250). He celebrated the progress of liberty through the times—from the English Habeas Corpus Act to the American Declaration of Independence, to the Declaration of Human Rights in the French Revolution—and noted that "the scien-tific world view [*Auffassung der Welt*] leads to a spirit of tolerance, brotherhood, and the love of peace" (ibid., 252). All in all, a very pro-Enlightenment viewpoint.

Let us summarize the arguments made by these scientists: First, scientific and technological advances are important elements of Kultur and deserve a place in it. The concept of Kultur should be wide enough to include the sciences. Second, science also con-tributes to the spiritual progress of humanity and thus has a role even in the narrow view of Kultur. Within that view, the scientists clearly focused on the forward-looking face of the Janus-headed Kultur. Third, both the material and spiritual contributions of sci-ence are essential for the German *nation*. The last argument was the one the other Kulturträger were bound to like the best, given the widespread patriotic sentiment in this group. Therefore, many scientists tried hard to emphasize the national benefits, but their attempts did not surpass the level of superficially pleasing rhetoric—and one wonders if they could have. Their arguments remained rather weak because neither the application of the idea of unity nor the moral uplift provided by science nor the kulturgeschichtliche mission of science had any obvious connection with the German nation.[17] Such connections can be much more cogently made with humankind at large rather than with a particular nation. The most logical link between science and the German nation remained indeed the utterly pragmatic point that German science supported German economic and military power. And this was, of course, the very corner (defining science as merely an element of Zivilisation) from which the scientists tried to escape.

Du Bois-Reymond was one of the scientists who kept their distance to nationalistic rhetoric. His 1878 speech "On the National Sentiment" indeed warned against the excesses of nationalist fervor: "Wissenschaft is in essence cosmopolitan" (1912, 1: 670). "Wissenschaftlicher chauvinism, of which the German scholars have so far steered clear, is more odious than political chauvinism, insofar as one demands a higher moral standard from scholars than from the politically excited masses" (ibid., 1: 675).

Kulturpolitik—Kulturpropaganda

Whereas science joined forces with Kultur in the domestic Kulturkampf, it also joined Kultur, a few decades later, in the area of foreign policy, in the guise of *Kulturpolitik* and *Kulturpropaganda*.

Already before World War I, German politicians began to recognize external *Kulturpolitik* as an important facet of foreign policy.[18] In 1913, Chancellor Theobald von Bethmann-Hollweg noted the benefits of French "Kulturpropaganda" and British "Kulturpolitik" for the political and economic status of these countries and emphasized the need for a German Kulturpolitik. In a published letter to historian Karl Lamprecht, he wrote, "Like you, I am convinced of the importance, even the necessity, of a foreign Kulturpolitik. I do not overlook the political and economic benefits that France derives from this Kulturpropaganda, nor do I overlook the role that British Kulturpolitik plays in holding the British Empire together."[19] Foreign Kulturpolitik was seen as a vehicle to increase the understanding of German culture and language abroad, to spread sympathy for the German cause, to improve foreign relations, and to further Germany's political and economic standing in the world. Wissenschaft was to become a part of this Kulturpolitik.

After the military, political, and economic devastation in the wake of World War I, the importance of Kulturpolitik for German foreign policy increased. An immediate goal was to break through the isolation in which Germany had been placed by the victorious powers, an isolation that also extended to Kultur and science. German diplomats recognized Kulturpolitik as an innocuous way of reestablishing international relations and regaining sympathies abroad. The Akademie der Wissenschaften in Berlin concurred, stressing in particular the centrality of Wissenschaft in such a Kulturpolitik: "After the collapse of Germany, German Wissenschaft, first of all, is called upon and suited to reestablish the respect for German accomplishments abroad, and to revive the necessary relations with foreign countries" (quoted in Grundmann 1965, 3). Among the Wissenchaften, the sciences were particularly attractive to Kulturpolitik because they might seem less discredited than the humanities in the eyes of the former enemies.

In a letter to his friend Solovine, Einstein described himself "als Renommierbonze und Lockvogel" (as a bigwig and lure) for the Zionist cause (1987, 40). One might surmise that Einstein realized he served in a similar role for the German government. Einstein was most willing to play his part in Kulturpropaganda in the years immediately following World War I. In that period, his inclination to support the underdog and to help those who he felt were unjustly vilified made him an ally of German government policy. He was friendly with foreign minister Rathenau, who persuaded him, in 1922, to accept an invitation from the College de France, which was intended to reestablish connections between the French and German scholarly communities. During a trip to South America in 1925, when Einstein again served as a *Renommierbonze* for German Kulturpropaganda, he sarcastically noted in his travel diary, "What a droll bunch these Germans are. To them, I am a stinking flower, but they still keep pinning me to their lapel" (quoted in Fölsing 1993, 624).[20]

4.3. SCIENCE AS RELIGION

In 1999, the Kansas Board of Education dropped the teaching of evolution from the state's science curriculum, and this was only the latest of a series of similar anti-Darwinian initiatives throughout that century.[21] Thus, when one thinks about the relationship between religion and science in an American context, one of the first things that come to mind is probably the vigorous attack by fundamental Protestantism on the theory of evolution.[22]

The Kansas incident and other notorious clashes between religion and science have supported the commonly held view of science and religion as hostile opposites.[23] Indeed, conflicts between religion and science have been numerous and have often involved fundamental issues, extending beyond specific contradictions between individual religious doctrines and scientific theories to an epistemological disagreement over the methods of generating valid knowledge. Whereas religious knowledge is typically based on the revelation and interpretation of divine truths through authoritative channels (holy texts and/or a priestly hierarchy), or through immediate spiritual experience, the scientific method emphasizes demonstrable empirical evidence.

Nevertheless, there are also underlying communalities. In an abstract sociological framework, religion could be described as the societal function that produces meaning by addressing the deepest questions of human existence, for example, how the world was created, what happens after death, and so forth. To the extent that science also addresses these questions, it assumes a meaning-producing function, and it becomes a direct *competitor* of the organized religions in the narrower sense. Thus, it is not a dissimilarity but a similarity between science and religion that sometimes lies at the root of their conflict.

The position of science vis-à-vis religion has varied across time and space. In this section, we examine the relationship between science and religion in nineteenth-century Germany. We argue that science incorporated a particularly strong religious element in that era and that this could be understood, within the theoretical framework developed in this book, as science establishing itself on the Kultur side of the prevailing Kultur-Zivilisation dichotomy. Whereas earlier we had seen strategic alliances between science and Kultur in Kulturkampf and Kulturpropaganda, we now come to an essential and defining feature of nineteenth-century German science.

In a highly momentous speech, given in 1918 in honor of Max Planck's sixtieth birthday, Albert Einstein explained what science meant to him. He started that talk (of which we will read more later) with a reference to the religious metaphor of the "Tempel der Wissenschaft" (Einstein 1955a, 107). Einstein's choice of this religious imagery was not random or idiosyncratic. Rather, by employing the temple metaphor, he expressed his allegiance to a long and powerful tradition. We now elucidate that tradition by tracing back the roots of the religious self-image of nineteenth-century German science.

The annual assemblies of German scientists and physicians, which were founded by Lorenz Oken in 1822, had several purposes.[24] The most manifest function was, of course, the dissemination and discussion of scientific findings. In addition, the meetings gave the scientists the chance to get to know each other personally, to form networks, and to socialize with each other. As Myles Jackson (2003) documented, the assembled scientists and physicians particularly enjoyed singing songs together in a convivial atmosphere. Finally, the assemblies served as the prime forum for the scientists' collective self-definition and for discussions of the wider, weltanschauliche implications of science (Hoppe 1979, 141–45, Pfannenstiel 1958; Querner and Schipperges 1972; Schipperges 1976; Zevenhuizen 1937). A brief glance at the assemblies will show that they were rife with religious rhetoric and symbolism and that they anchored science firmly in the realm of religion.

At the 1826 congress in Dresden, for instance, poems expressly celebrated scientists as the "priests of nature" and mentioned "Isis's

newly founded altars" (Schipperges 1976, 11). The following meetings similarly produced a rich crop of such religious metaphors. The
scientists were likened to a priesthood, "priests of nature and of life"
at the Hamburg meeting in 1830 (ibid., 16). A poem dedicated to the
assembled scientists included the following lines (ibid.):

> Heilig ist der Gang der Zeiten,
> Sicher der Erfahrung Spur;
> Helle, süße Wahrheit quillet
> Aus der Urne der Natur.

> [Holy is the course of times
> Certain is the trace of evidence
> Bright, sweet truth flows
> From the urn of nature.]

Furthermore, the poem compared the scientists to angels:

> Ihr nahet Euch dem hehren Himmelsthrone;
> Euch blendet nicht das tausendfache Licht,
> Verklärt schaut Ihr des Höchsten Angesicht.
>
> . . .
>
> Ihr seid die Boten, die ein Gott gesendet,
> Durch die das große Werk hier wird vollendet
>
> . . .
>
> Es schmückt sich bräutlicher durch Euch die Erde,
> Euch ward, Herakles gleich, Unsterblichkeit
> Ihr führt den Menschen nun zum Tempel.

> [You approach the exalted throne of heaven
> You are not blinded by the thousand-fold light
> Transfigured you look into the face of the Highest.
>
> . . .

You are the messengers sent by a God,
Through whom the great project here will be completed

. . .

Through you the earth adorns itself more like a bride
Like Hercules, you achieved immortality
You now lead the humans to the temple.]

The temple of science is usually that of the Egyptian Goddess Isis. At the 1832 meeting in Vienna, a hymn proclaimed (ibid., 17):

Laßt uns vereint der Isis Tempel bauen
Der Göttin, welcher keine andre gleich
Die rätselhaft so nahe uns und ferne
Im Sandkorn thront wie dort im Flammensterne.

[Let us build the temple of Isis together
For the goddess who is unlike any other
Who is mysteriously near to us and far
residing in the grain of sand and in the flaming star.]

The 1834 meeting at Stuttgart reiterated the theme that, through their research, the scientist-priests will be able to lift the "veil of the Goddess" and glimpse at her true being, that is, at the true being of nature (ibid., 18). At these conventions, the Goddess Isis appeared not only in the speakers' rhetoric but also on the face of commemorative medals (Pfannenstiel 1958, 70–71).

The religious elements at those meetings permeated both form (the use of myth, icons, and allegory) and content (the definition of the fundamental nature of the scientific enterprise). Scientific activity was viewed as a religious service, with its ultimate goal of fashioning a Weltbild that would reveal the nature of the cosmos, and this led the scientists to define themselves as priesthood, the priesthood of Isis. The religious foundation made scientific research a meaningful activity *per se*. Science of this kind does not think of itself primarily as a tool to serve another purpose; it considers itself the ultimate purpose. In other words, such science justifies itself and does not require any outside justification (e.g., that it would advance technology, make life better, etc.).

At the 1865 assembly in Hannover, Rudolf Virchow, whom we encountered as an uncompromising critic of Catholicism, explicitly elevated Wissenschaft itself—implicitly meaning the sciences—to a religion: "for us, Wissenschaft has become religion" (Schipperges 1976, 40). At Wiesbaden in 1873, he similarly exclaimed, "We too have a faith: the faith in the progress in discovering the truth" (ibid., 50). Even toward the end of the century, scientists continued to use the religious metaphors of the "temple of truth" and the "righteous temple of Wissenschaft."[25]

The Monistic movement most fervently and radically expressed the impulse of making science a religion. Its leader, the distinguished German chemist Wilhelm Ostwald, who received a Nobel Prize in 1909, said at the beginning of the twentieth century, "Science, now and with immeasurable success takes the place of the divine" (quoted in Holton 1996, 12). But Ostwald was also keenly aware of an inherent tragic: "This is one of the laws to which almost all servants of the austere Goddess Wissenschaft are subject: that their life ends in sadness, and the more so, the more completely they had dedicated their life to her service" (1910, 401). The natural scientist thus came to the same conclusion that the philosopher and psychologist William James had reached in his *Varieties of Religious Experience*: "the purely naturalistic look at life, however enthusiastically it may begin, is sure to end in sadness" (2003 [1902], 122). Einstein's scientific life impressively illustrates these general statements. His last decades were filled with barren efforts to achieve the elusive unified field theory. Many lift the veil of Isis only to find another veil underneath. The progress of science so far resembles more an endless journey than the acquisition of final truths.

In sum, service of the Goddess Isis thus was a major metaphor by which science in nineteenth-century Germany understood itself. Scientists were the priesthood of Isis, the scientific Weltbild was the temple, and so forth. These ideas sounded much less obscure then than they do now. In the nineteenth and early twentieth centuries, a number of periodicals and scientific associations bore the name of Isis. Most significantly, Lorenz Oken founded not only the assemblies of German scientists and physicians but also the journal *Isis—oder Encyklopädische Zeitung* (in existence from 1817 through 1848), which published scientific and political contributions. The journal was headed by a picture of Isis sitting on her throne. She was flanked by Osiris and Anubis. Oken was certainly not the first to select the Egyptian goddess as the patron for a periodical. *Isis: Eine Monatsschrift von deutschen und schweizerischen Gelehrten* had existed from January 1805 through 1807. Among its more notable authors was Friedrich Schlegel (1806), who published poetry in it ("An Ida Brun").

The editorial preface to the first issue of that earlier *Isis* candidly stated, "Custom requires it. Walhalla and Mount Olympus have no more gods who were not yet obliged to give their names to the cover of some journal at its birth. Hence, instead of the employable choir of muses, of Mercuries, Minervas, Auroras, etc., the editors of this new journal selected Isis as its patron, and tie the exposition of its desires and goals to this name from the meaningful mythology of antiquity." It was fashionable for journals around that turn of the century to sport the name of an ancient deity in their title, and a practical reason for choosing Isis, as the editorial acknowledged, was simply that the pantheon of Greek and Roman antiquity was already somewhat overused. On the other hand, the choice of Isis was clearly not idiosyncratic. When Oken founded his own *Isis*, he did not even think it necessary to explain this choice of title in the first issue. The readers were apparently expected to be familiar with the mythology. In the 1830s, the Naturwissenschaftliche Gesellschaft Isis was established in Dresden. It issued *Sitzungsberichte und Abhandlungen* and *Denkschriften*, but, like Oken's *Isis*, these publications did not explain the institution's name.[26]

Whereas these earlier publications have long since sunk to obscurity, another journal of the same name is doing very well. Virtually all historians of science know the journal *Isis*, the preeminent journal of their field and the official organ of the History of Science Society. A somewhat smaller number, I presume, know the origin of the journal's name. The journal was of course named by George

Sarton, one of the founders of the discipline, who started the journal in 1912. (It first appeared in print in 1913.) But why did Sarton choose Isis as the name? An answer might begin by noting that Sarton's work on ancient science afforded him great familiarity with Egyptian mythology and hence with the goddess Isis (e.g., Sarton 1952a, 1959). When, in 1936, Sarton founded a companion journal to *Isis*, he called it *Osiris* (after the goddess's brother and husband). He also titled one of his books *Horus: A Guide to the History of Science* (1952b); Horus was another Egyptian god, the son of Isis and Osiris. However, it was not merely a personal predilection or eccentricity that led Sarton to name major publications after Egyptian mythological figures. Rather, he felt that he tapped into a cultural mainstream. An important clue is again the absence of any explanation of the name in the first issue of *Isis*. Apparently, the name was fairly obvious to the scholarly community Sarton addressed—or so he thought. About ten years later, in 1923, he did offer a belated explanation in the pages of *Isis*, acknowledging that the name choice was perhaps not as obvious as he had assumed:

> One more word. The name Isis has puzzled many persons. I chose it ten years ago because it was short, as I might have chosen Minerva, Athene, Hermes or Clio if those names had not been preempted. . . . The name Isis evokes in my mind the period of human civilization which is perhaps the most impressive of all,—its beginning. . . . However I discovered too late that this name evoked in other minds ideas of mysteries and occultism very remote from my own. I understand that some people have been attracted to *Isis* because of such ideas (they must have been very disappointed!) while others have been equally repelled. I heard of a prominent scientist declaring that a journal called Isis could but be superficial and that he would not even look at it. He did not realize that it would have been difficult to surpass his own superficiality! (*Isis* 6 [1923]: 39)

An example of the occultism Sarton referred to is the theosophical work by Madame Blavatsky (1972 [1877]) under the title *Isis Unveiled*.

To comprehend the relative popularity of an ancient Egyptian goddess during the nineteenth and early twentieth centuries in Europe, we have to take note of a long-standing trend among European intellectuals that mainly originated in the Renaissance. Pursuing their *ad fontes* (to the sources) motto, many humanists of that era, as well as the later Enlightenment protagonists and the neo-humanists of the eighteenth and nineteenth centuries, viewed themselves as some sort of pilgrims to the origins of Western culture in classic Rome and Greece. This "to the sources" mind-set could easily lead some of them even past classic antiquity, to the source of the sources. If ancient was good, more ancient was better—and thus the fascination with things Egyptian grew. A similar path led some of the more theologically inclined scholars from Christianity to the Hebrews, to Moses, and thus ultimately to Egypt (Assmann 1997).[27] But what meanings were associated with this Egyptian revival? A brief survey of ancient beliefs will help us understand better who Isis was and how she became linked to science.

Isis

Isis was a major goddess in the mythology and religion of ancient Egypt.[28] Because, in one of the ancient myths, she revived her slain husband (and brother) Osiris, she was primarily connected to *death* and *rebirth*. She was further revered for her prowess in magic, for her healing powers, and for her protection of the family. (As a beneficent mother figure, she was similar to the Christian Virgin Mary.) A central text of the Isis cult lists a long catalog of Isis's accomplishments, including the creation of the universe, of the floods of the Nile, of agriculture, and of the law (Merkelbach 1995, 115–18). The editorial in the inaugural issue of *Isis: Eine Monatsschrift von deutschen und schweizerischen Gelehrten*, whose starting lines we already quoted, gave a concise profile of the goddess:

> Isis was the great, general mother of things, the Io of the Greeks, the Cybele of the Phrygians. All peoples of the earth venerated her, and Plutarch therefore calls her the one with a thousand names. Her pic-

ture was the symbol of nature in whose womb the young arts swayed. Her many breasts indicated that she was the mother and nurturer of all. Her head adorned with a tower crown, with a cymbal or a sistrum in her hand, she sat on her throne among friendly, tamed lions. She had invented shipbuilding and had educated barbaric nations in the service of the deity and in the art of agriculture.

So may also these pages, which carry her name, be dedicated to nature, truth, the useful and the beautiful things that human arts yield.

Greek visitors, traders, and settlers who learned about Egyptian mythology tried to fit it into their own religious framework. The Greek pantheon contained a multitude of gods and goddesses with delineated fields of expertise and influence, and the Greeks attempted to match foreign gods with those they knew. In this, they followed a widespread practice in the ancient world (first observed in the glossaries of Sumerian and Akkadian words and concepts) of seeking correspondences and translatability between the various polytheistic systems (Assmann 1997, 44–47). Isis was, according to Herodotus's report, initially paired with Demeter (Merkelbach 1995, 51). This Greek goddess had brought the humans agriculture and laws, that is, civilization. Often, Isis was also identified with Io, but later, in Hellenistic times, with a whole variety of goddesses.[29] Behind these matching exercises, the conviction grew that all the polytheistic systems were mutually compatible because they all referred to the same underlying religious reality. Yet Hellenistic theological thought went even beyond believing that a basic master system ensured the mapping of polytheistic systems onto each other; it began to fuse elements of the master system itself into one concept. Isis's comprehensive range of competencies transcended Greek polytheism in the direction of monotheism, or *"cosmotheism"*— the term used by Jan Assmann, following F. H. Jacobi, to set its pantheistic gist apart from the monotheism of Judaism and Christianity (1997, 54).

The Isis cult, which was notorious for its licentious rites, spread in the Hellenistic world and then also in the Roman Empire (Hornung 2001). In the *Metamorphoses* (also known as *The Golden Ass*), written by Apuleius of Madaurus in the time of Marcus Aurelius, Isis appears in a devotee's dream and again claims that various peoples venerate her under various names, but that her true name is Isis. She describes herself as "the mother of the universe, the mistress of all the elements, the first offspring of time, . . . the single form that fuses all gods and goddesses" (Assmann 1997, 48). This passage clearly proclaims Isis as a cosmotheistic goddess of everything. Similar is an altar inscription at Capua, Italy, which had been published by Athanasius Kircher:

TIBI
UNA. QUAE.
ES. OMNIA.
DEA. ISIS

(For you—The one who—Is all—Goddess Isis)[30]

For us, the key question is how Isis was linked to science. Two strands form that linkage; they are closely interwoven, but nonetheless worth distinguishing. First, Isis was a goddess of healing, which might have connected her to the medical profession and biological knowledge (Merkelbach 1995, 199–201). She was at times portrayed with a snake (a parallel to Asclepius's serpent). Moreover, Isis—and Osiris—were gods of knowledge in general, not just of medical knowledge. During the time of the Roman Empire, a particular genre of popular philosophical works emerged in Egypt, according to some of which Osiris and Isis created a priestly class of prophets who were all-knowing and able to heal body and mind (ibid., 251). Plutarch, who around 120 CE wrote a book about Osiris and Isis from a Platonic perspective, also played a role in connecting Isis with science.[31] In his rather complicated scheme, it became Isis's domain to acquire knowledge about the universe. In Greek, the name *Isis* can be written *Eisis*, which, in a fabricated etymology, Plutarch related to the Greek word for knowing (*eidenai*). Furthermore, the Greek word for an Isis temple was *Eiseion*, which gave Plutarch another chance for a made-up etymological interpretation along the lines of "enter and you will know that which is" (ibid., 261–62). As an aside, another equally dubious etymology relating to Isis is found in Brewer's *Dictionary of Phrase and Fable* (1892), whose entry on Isis reports the speculation that the name of the city of Paris, France, may have derived from the Greek *Para Isidos* (near the temple of Isis).[32] The Hellenistic times reinforced Isis's link with wisdom. "As the embodiment of wisdom, as agent of cosmic order" (Kee 1980, 145), Isis showed close parallels with the Jewish *sophia*, and also with the *logos* in John's gospel (Conzelmann 1971; Kee 1980).[33]

The second strand linking Isis to science is the cosmotheistic interpretation of Isis as Nature. This aspect was to become dominant in the post-Renaissance revival of Isis and hence in the Isis rhetoric at the assemblies of the German scientists and physicians in the nineteenth century. We will return to this strand in more detail later.

With the triumph of Christianity, the goddess Isis faded into the background. Yet she was never completely forgotten, and she entered the stock of European mythology in medieval and later times, from which she occasionally popped up in literary products. Here are a few instances in which writers and poets referred to Isis, alongside other figures of antiquity, in their works. In his collection of profiles of famous women, written in 1361 and 1362 and titled *De mulieribus claris*, Giovanni Boccaccio included a section about Isis, "Queen and Goddess of Egypt." *Le Livre de la Cité des Dames*, a proto-feminist book written by the French court poet Christine de Pizan in 1405, hailed Isis as the goddess who introduced the art of planting to humankind (Ceres and Minerva were also mentioned as bringing similar benefits.)[34] In the late fifteenth century, an anonymous dream-vision allegory titled *The Assembly of Gods*—some allege John Lydgate to be the author—appeared in England. It mentioned Isis ("Ysys") together with Pan, the god of shepherds. At the end of the sixteenth century, Edmund Spenser's *Faerie Queene* (1590–96) included an episode in Isis's temple, and John Milton, in *Paradise Lost* (1667), named Isis, along with Osiris and Orus among the fallen angels.[35] In the late seventeenth century, at the German university of Greifswald, Brandan Heinrich Gebhardi initiated a program of comparing Jewish and classical myths that spawned a number of doctoral dissertations. Some of these studies also included Egyptian mythology. For instance, in *De Hecate* (Greifswald 1703), Joh. Friedrich Rahnaeus compared Isis to Eva (Gruppe 1921, 60).[36]

In the later part of the seventeenth century, Egyptian mythology was captured in the gravitational field of the emerging Enlightenment.[37] Among the pioneers of this development were John Spencer (1630–1693) and Ralph Cudworth (1617–1688). Whereas previously—especially since the *Corpus Hermeticum* appeared in the West at the beginning of the Renaissance—there had been a more magical and alchemist brand of Egyptian enthusiasts, the writings of Spencer, Cudworth, and others following their footsteps in the eighteenth century celebrated Egypt as the source of *cosmotheism*.[38] Cudworth characterized the Egyptian theology with the Greek slogan *hen kai pan* (One and All). Similar cosmotheistic leanings—postulating the coincidence of God and nature (*deus sive natura*)—were also evident in the work of Baruch Spinoza, Cudworth's and Spencer's contemporary. *Hen kai pan* is the Greek version of the *una qui es omnia* formula used by Isis devotees to praise the goddess. The seventeenth-century revival of cosmotheism thus brought the cosmotheistic link between Isis and science—Isis representing Nature and hence the object of science—to the foreground. Already in 1652, Athanasius Kircher had claimed that Isis's veil represented the mysteries cloaking Nature, and, since then, the veiled Isis became one of the customary *sujets* for illustrations of science books (Hadot 1982, 9).

While the appreciation of Egypt gained ground among the intellectual heavyweights around that time, it also spread among a wider audience through efforts by writers such as the Genuese Giovanni Paolo Marana (1696) and Abbé Jean Terrasson (1732). Marana wrote *Conversations of a Philosopher with a Solitary about Divers [sic] Matters Appertaining to Morals and Erudition*, in which ancient Egypt harbors both magic life-prolonging potions and true philosophy.[39] Terrasson's historical novel *Sethos* also portrayed ancient Egypt in a highly favorable light, and it succinctly stated the *ad fontes* theorem: "The Romans were an unciviliz'd people, till they learnt the sciences of the Greeks; as the Greeks themselves were, till they became acquainted with the knowledge of the Egyptians" (1732, 1: 63).

In December 1739, classical scholar Bernard de Montfaucon reported to the French Academie Royale des Inscriptions et Belles Lettres about Isis and other Egyptian gods ("Sur les Anciennes Divinités de l'Égypte," 1743). An inscription at the Isis temple of Sais, Egypt, became very famous for encapsulating the cosmotheistic interpretation of the goddess. In his *Kritik der Urteilskraft*, Kant called Isis Mother Nature and quoted that inscription: "Never perhaps has anything more dignified been said, or has a thought been expressed in a more dignified manner than in that inscription on the temple of Isis (Mother Nature): 'I am all that is, that was, and that will be, and no mortal has lifted my veil'" (1922 [1790], 391).[40]

Many Romantics found the Isis myth congenial (Hornung 2001). In the journal *Athenaeum*, for instance, Friedrich Schlegel wrote: "The demands and traces of a morality that would be more than the practical part of philosophy become louder and clearer. People even start talking of religion. It is time to rend the veil of Isis and lay open the secret. He who cannot bear the sight of the goddess may flee or perish" (1800, 4).[41] Similar to the Romantics, many of their friends, the *Naturphilosophen*, subscribed to some brand of cosmotheism and hence were sympathetic to the Isis mythology.[42] At that time, varied and often tight links existed between Naturphilosophie, "Romantic science," and the emerging empirical science. Lorenz Oken, who founded both the annual assemblies of German scientists and physicians and the journal *Isis*, was heavily influenced by Naturphilosophie. He dedicated his *Lehrbuch der Naturphilosophie* to "His friends Schelling and Steffens"(1831 [1809–11]). The book had a conciliatory purpose: "Through this work, I hope to reconcile the opponents with Naturphilosophie. Opposing schools of thought must exist for science to advance, but the ill will between these schools need not be" (ibid., v). Here, in the person of Oken, we find a first continuity that bound the evolving empirical science to its predecessors. (We shall explore that tie more thoroughly when we examine "Romantic science.") The strands connecting the two successive epochs of science formed a crucial part of the pathway that allowed cosmotheistic ideas, and especially the quest for Weltanschauung and Weltbild, to enter and to shape the subsequent eras of science in Germany.

Two concepts were particularly prominent in the scientists' rhetoric revolving around Isis. First, the metaphor of "building the temple" could be heard frequently at their conventions (e.g., in Stuttgart 1834; Schipperges 1976, 18)—and Einstein, too, as already mentioned, prominently spoke of the "Tempel der Wissenschaft" (1955a, 107). For the scientists, the metaphor denoted the project of constructing a complete scientific model of the world, that is, of generating a scientific Weltbild—which was the ultimate goal of the scientist-priests' service. For an explanation of this usage, one might consider that, in ancient Egypt, the temple was often considered a model of the world (Bonnet 1952, 787).

Second, there were recurrent references to lifting the veil of Isis, the "Schleier der Göttin" (Schipperges 1976, 18). As mentioned, the unveiling of Isis, which symbolized the research mission of the scientist-priests, was also a favorite image for the illustration of scientific books.[43] Even in the twentieth century, at the Nobel Prize festivities, the laureates' feats were occasionally described as lifting the veil of Isis, for instance, in the presentation speeches for the 1921 and 1930 chemistry prizes, both delivered by H. G. Söderbaum, the chairman of the Nobel Committee for Chemistry of the Royal Swedish Academy of Sciences.[44]

We should emphasize that, in this cosmotheistic interpretation, it would not be precise to call Isis the goddess of science, in the way Hephaistos was the god of metallurgy, and so forth. Rather, Isis was the goddess of *nature*, and thus the *object* of scientists' efforts and desires. Their noble quest was to gain knowledge of nature by "lifting Isis's veil." At the Hannover assembly in 1865, Johann Jakob Nöggerath, a leading functionary of the scientists' association, exclaimed, "We are not kings, but priests, joined in a republic, of the exalted queen, whom we have often seen half unveiled, and with whom we desire to unite ourselves intellectually" (ibid., 41).[45] It is not hard to see how such rhetoric could be interpreted as a variant of the gendered imagery of science that has been particularly emphasized by feminist science scholars. Those scholars have claimed that modern science is "male" and that this maleness expressed itself in the desire to dominate—and, in the most extreme view, even rape—"female" nature.[46] Things might be somewhat more complicated, however, than a simple analogy between the male/female and science/nature dichotomies. Whereas nature has indeed consistently been considered female by those viewing it in gender terms, the status of science has been less clear. "Woman is," according to Londa Schiebinger, "the dominant image of science throughout the seventeenth century and deep into the eighteenth"—and this image persisted to more recent times, as Einstein's Nobel Prize medal illustrates (1989, 119).[47] Like all Nobel Prize medals awarded for physics or chemistry after 1902, it displays two female figures—Scientia lifting the veil from Natura's head.[48] In that typical imagery, thus, the activity of science was portrayed as an interaction between two females (science and nature). Curiously, the collective priesthood of science, though consisting almost exclusively of male individuals, was metaphorically portrayed as a female. Be this as it may. We now focus on a more sociological aspect of the "veil of Isis" metaphor—the secret.

Science and Secret

Secrets are central to many religions; one might even say that few religions can do entirely without them. On the other hand, we have seen that the revelation of nature's secrets—by lifting the veil of Isis—was designated the noblest task of science in speech after speech at the scientists' conventions. The notion of the secret, thus, appears in both religion and science, yet it plays different roles. It should be instructive to compare more closely the different ways in which the secret functions in religion and in the sciences.

Our starting point is the classic sociological analysis of the secret and the secret society by Georg Simmel. He observed that "the strongly emphasized exclusion of all outsiders [from the secret] makes for a correspondingly strong feeling of possession. For many individuals, property does not fully gain its significance with mere ownership, but only with the consciousness that others must do without it" (1964 [1908], 332). According to to Simmel, and this is one of his key points, a secret is an actively maintained asymmetry of knowledge; it is a "purposive hiding and masking, that aggressive defense, so to speak, against the third person" (ibid., 330).[49]

In his sociological analysis of group membership, Simmel noted in passing that there were different kinds of nonmembership, an idea whose usefulness was recognized and explicated by Robert K. Merton in his essay "Continuities in the Theory of Reference Groups and Social Structure" (1968, 342–51). The sociologically crucial distinction within the larger category of nonmembers is that between those who do not fulfill the criteria of membership and those who do not belong to the group even though they fulfill the criteria. It appears particularly useful for our discussion of the secret to develop a somewhat analogous distinction in the area of nonknowledge (or ignorance). To be comprehensive, one would have to set up a whole taxonomy of nonknowledge, but it may suffice here to focus on only the two types of nonknowledge most relevant for our purpose: (1) knowledge that nobody knows and (2) knowledge that some know and others do not and to which access is purposefully restricted.[50] The second variety is the secret in Simmel's understanding—the secret requires an actively maintained gradient of knowledge.

One secret that by definition is common to all secret societies is the secret of their existence, but, in addition, some of them are also vessels of a secret content. In that case, they are a "special type of secret societies whose substance is a secret doctrine, some theoretical, mystical, or religious knowledge" (Simmel 1964 [1908], 355). Among the social features of the secret society, Simmel discusses the growth of ritual and hierarchy (various stages, degrees, or ranks, often defined by differential knowledge of the secrets). Secret societies often turn into what is known as total institutions that envelop the whole person. The secret society "quite characteristically claims to a greater extent the whole individual, connects its members in more of their totality, and mutually obligates them more closely, than does an open society of identical content. Through the symbolism of the ritual, which excites a whole range of vaguely delimited feelings beyond all particular, rational interests, the secret society synthesizes those interests into a total claim upon the individuals" (ibid., 360). Through emotional and motivational engagement, ritual provides a stronger bond than rational interest alone would provide.[51]

At the time when the German scientists and physicians relished religious rhetoric at their conventions, an alternative science existed that was much more thoroughly and utterly religious—joining together Isis mythology, cosmotheistic beliefs, the definition of non-knowledge as secret, and the organizational form of secret societies. This was Freemasonry.

The Masonic Connection

Masonic orders formed a vital institutional base for Enlightenment thought. "In the discourse of the Enlightenment, [ancient cosmotheism] was reconstructed as an international and intercultural mystery religion in the fashion of Freemasonry" (Assmann 1997, 8). The secret orders of Freemasons were interested in Egyptian mythology, in Isis and other members of that pantheon, as fountainheads of cosmotheism. The following passage, written by Ignaz von Born, a leading figure of the Austrian Enlightenment and also a Mason, in the *Journal für Freymaurer*, exemplifies the Masons' cosmotheistic interpretation of the Isis cult: "The knowledge of nature is the ultimate purpose of our application. We worship this progenitor, nourisher, and preserver of all creation in the image of Isis. Only he who knows the whole extent of her power and force will be able to uncover her veil without punishment" (1784, 22). It might appear somewhat ironic that the Masonic paragons of the Enlightenment operated "in the dark" of secret societies. Whereas, at the beginning, a pragmatic reason—needing to hide from hostile and oppressive regimes—may have made secrecy necessary, we realize from Simmel's discussion of the secret that the social dynamic of secret societies tends to perpetuate secrecy even once the environment has become tolerant to bringing the secrets out into the open.[52]

The *Allgemeines Handbuch der Freimauerei* (Lenning 1900–1901, 1: 492–93) noted that many French Freemasons were inclined to believe that the Egyptian mysteries, among them the cult of Isis, were the origins of masonry—although the *Handbuch* itself expressed doubts about the validity of that view. This work was the third, "completely revised" edition of C. Lenning's *Encyclopädie der Freimauerei* (1822–28), which had been somewhat less skeptical of the Isis cult's significance.[53] That earlier *Encyclopädie* also had an entry about the Table of Isis, which was missing in the third edition. Lenning described the Table of Isis as a copper plate, which supposedly recorded the deepest secrets and holiest rites of the ancient Egyptians (1822–28, 2: 109). The Table of Isis made its way into the Egyptian Museum of Turin, Italy. (A picture of the table is reproduced in Hornung 2001, 85.)

An important link between Freemasonry and mainstream Kultur was Karl Leonhard Reinhold (1757–1825). He was a professor at Jena, a hotbed of German poets and thinkers, where he befriended Schiller and was well connected to other major Kulturträger (Assmann 1997, 116). His major claim to fame rested on a commentary on Kant's philosophy, but he also wrote about Hebrew mysteries in the *Journal für Freymaurer*. Reinhold equated Isis and Jehovah. His thinking influenced Schiller's ballad *Das verschleierte Bild zu Sais* (The Veiled Image at Sais) and his work on *Die Sendung Moses* (Moses's Mission), which, in turn, influenced Freud's writing on Moses (Assmann 1997, 117; 1999). Goethe's sympathies for hermetic traditions and cosmotheistic thinking are well documented (Hornung 2001; Zimmermann 1969–79). Egyptian mythology and Masonic symbolism entered also other works of Kultur, such as Mozart's *Magic Flute*.

The Masons were also interested in science, but that interest was related to older esoteric traditions, consciously shrouded in secrecy, with knowledge reserved only for the elite few. In the Masons' beliefs and practices, the religious underpinnings of their science were patent. Their cosmotheism was past oriented. The highest state of knowledge was deemed to have existed in the past in ancient Egypt.[54] History since that Golden Age has, from that perspective, been a history of loss or concealment: the ancient knowledge was either lost from view (often in some catastrophe) or purposefully hidden, or some combination of these events occurred.[55] In the first case, the quest for truth becomes an effort to regain a level of knowledge that had already been attained in the past. In the second case, it becomes an initiation process. In that latter scenario, nonknowledge clearly has the form of the secret—knowledge that is already known to some, but access to which is granted only to those considered worthy enough. While those initiates partake of the secret, the secret is preserved vis-à-vis the uninitiated. One typical reason for maintaining secrecy is that the secret knowledge might afford its carriers certain advantages that would vanish if the knowledge became generally accessible. Another reason is that the secret might be considered so compromising, subversive, or abhorrent by prevailing standards that its carriers would face disadvantages or even persecution if found out. William Warburton (1698–1779), bishop of Gloucester, England, who studied the role of secrecy in religion, gave yet another reason for why the secrets had to be kept secret. Having examined form and content of the ancient mystery cults, and the function of hieroglyphic writing, he asserted that the greater mysteries had to be kept secret to protect the political order because only few would be able to handle the truth (Assmann 1997, 97–101).

Modern science, by contrast, is future oriented. Most of its practitioners believe in scientific progress, in the more or less steady accumulation of scientific knowledge.[56] What is currently unknown (nonknowledge of the first type!) might soon be known, thanks to the discoveries that flow from the scientists' empirical or theoretical research. The present state of knowledge is deemed superior to past states, and future states are expected to be better still. This discovery-based view of expanding knowledge is different from the revelation of, or the initiation into, preexisting ancient truths. The scientific community perceives itself as laboring at the frontier of human ignorance, with the goal of expanding the realm of human knowledge into those *previously unknown* areas.[57] The underlying attitude is captured in the metaphor that most present-day scientists may be dwarfs compared with the past scientific giants, but, standing on their shoulders, they can look farther.[58] Einstein's general theory of relativity, a work of astounding genius, is now routinely taught to beginning physics students; the research frontier has moved on.

Once knowledge is gained in an area of previous nonknowledge, the ethos of science, as described by Robert Merton, requires that the new knowledge be made public (1968, 606–15). Among the various domains of modern society, science is probably the one least tolerant of secrets (nonknowledge of the second kind).[59] Whereas for most religions as well as in the Masonic beliefs, the secret is a fundamental premise, modern science relegates it to an ethical flaw. To put it somewhat hyperbolically: For secret societies, the betrayal of secrets is the violation that most seriously undermines them; for science, it is the keeping of secrets. Although the reality of practiced science may sometimes fall short of this ethos of nonsecrecy (Mitroff 1974; Mulkay 1976), the difference is still palpable when one compares science with other societal domains. For instance, the tensions about whether or not research results can be published, which frequently occur when scientific research is conducted within business or military frameworks, illustrate that secrets are a much larger part of these two areas than of science.

Having drawn a clear-cut distinction, in the interest of conceptual clarity, between Freemasonry and modern science, we must add that these two were deeply entangled historically: Modern science had a substantial Masonic root.[60] The second Masonic degree provided the main justification for a turn toward empirical science. As one of the Masons' ceremonial texts explains, "In the Second Degree, we are admitted to participate in the mysteries of human science, and to trace the goodness and majesty of the Creator, by minutely analysing His works" (Knight and Lomas 1998, 9–10). This sentence from the Masonic rite hints at a type of argument that can be employed to help bridge the conceptual differences between the different types nonknowledge: If the secret is lost irretrievably, or if God never communicates the secret, then, for all practical purposes, the secret transforms into nonknowledge of the first kind, and the quest for the secret, into research. To the extent that the obtained research results were confined within the institutional framework of Masonic secret societies, however, shifting the conceptual starting point of knowledge production from nonknowledge of the second kind (i.e., the secret) to nonknowledge of the first kind did not affect the status of the end product: Once knowledge was wrested from the domain of nonknowledge (of the first kind) by research, it instantly transmuted into a secret to be shared only by the initiates. Here lay a fundamental difference between the Masonic and the modern ways of doing science.

The driving force behind creating this science-oriented second Masonic degree was, as Knight and Lomas speculate, none other than Francis Bacon, supposedly a Mason himself (1998, 332). We find more evidence for the Masonic root of modern science when we look at the Royal Society of London for the Improvement of Natural Knowledge, chartered in 1662. Almost all early members were Freemasons.

The scientists who attended the German assemblies in the nineteenth century practiced open, modern science in line with the ethos described by Merton, but, at the same time, they retained elements of religion. The Isis rhetoric at the scientists' conventions expressed a quasi-religious desire for a deep understanding of the nature of the cosmos, but it did so at an entirely metaphorical level. Isis appeared as an abstract symbol, not as the actual Egyptian goddess. Although some scientists doubtlessly belonged to Masonic Lodges, the assemblies as a whole did not seriously contemplate practicing any Egyptian mystery cult. Nonetheless, the scientists were very familiar with the Isis symbolism, knew its cosmotheistic implications (hen kai pan), and wholeheartedly subscribed to it—while also embracing the modern ethos of science. Lifting Isis's veil remained the goal, but it was now to be achieved by scientific, empirical methods. For the scientists, the unity of the cosmos was not a revelation but the ultimate hypothesis of their empirical research program.

The realm of nonknowledge was metaphorically defined as the secret of the goddess, or the secret of nature. According to Simmel's definition, speaking of the secret of nature makes little sense, unless nature is personalized as somebody—perhaps some god or goddess—who knows the secret.[61] The talk of the secrets of the goddess or of nature implied a metaphorical switch between the types of nonknowledge. In actuality, the "secret" was nonknowledge of the first type, and it was something to be overcome. Whereas the Masons desired to partake in a secret (which was to remain a secret to the uninitiated), the scientists desired to dispel it (and replace it by universally accessible knowledge).

The religious self-image of science projected at the assemblies was more than just empty feel-good rhetoric; it had important effects. Within the cultural landscape of meaning, the religious rhetoric anchored science in the domain of the sacred—which, in turn, constituted an integral part of Kultur.[62] Placing science in Kultur was meant to forestall any attempts to reduce it to the mere practical usefulness of an element of Zivilisation. At the social level, it cemented the role of the scientists as Kulturträger and bestowed upon them the prestige and social benefits of that elevated status. And in terms of science itself, it shaped the collective research program and outlook of the scientific community. The religious metaphor symbolized and reinforced a unifying principle of scientific activity. For the German scientists, the point was not to accumulate heaps of unrelated knowledge, an amorphously multiplying mass of scientific "factoids," of short-range theories, and of practically useful information, but to understand the underlying basic principles that govern all of nature. In short, the religious element manifested itself in resisting specialization, downplaying practical applicability, and emphasizing the "ultimate" questions.[63]

Different Traditions

Cosmotheism and the cult of Isis were not limited to Germany, as the preceding section has shown. This seems to cast some doubt on our claim that the religious foundation of science was a German peculiarity. Might that blend of science and cosmotheism not rather be an essentially pan-European or Western—to include America— phenomenon? By briefly surveying the traditions of nineteenth-century science elsewhere, we intend to show that German science indeed differed markedly—though certainly not diametrically—from its counterparts in other countries. We focus on Britain and the United States. In both countries, organizations analogous to the Assembly of German Scientists and Physicians came into being in the nineteenth century: the British Association for the Advancement of Science (founded in 1831) and the American Association for the Advancement of Science (founded in 1848).[64] Hence, we can compare the way those organizations related to religion with what went on at the German Assemblies.

The religious Isis-temple-veil metaphors were certainly known also in Britain. Erasmus Darwin's poem *The Temple of Nature* had as its frontispiece a scene, created by Henry Fuseli, of the unveiling of a female statue with three breasts (Isis multimammaria, similar to the Artemis of Ephesos).[65] The priestess who is pulling back the veil averts her eyes; a female initiate who kneels before the statue is clearly shocked. (By the way, the poet's grandson Charles was to become one of the all-time great unveilers of Isis.) However, the predominant tone among the members of the British Association for the Advancement of Science differed strikingly from that of its German model and counterpart. As Jack Morrell and Arnold Thackray (1981, 1984) pointed out, most of the leading members of the British Association, whom they called gentlemen of science, were deeply embedded in the Church of England. Many of them were ordained and held positions in the church. That these people would publicly associate with the goddess Isis was highly unlikely. Instead, they usually supported the convenient and optimistic "natural theology," whose central tenet was that scientific research would ultimately reinforce Christian faith. Natural theology "permitted the English parson to botanize and geologize knowing that his activity was in accord with his religion" (Basalla, Coleman, and Kargon 1970, 18). Although some critical voices did accuse the British Association of pantheist leanings, the British mainstream never came close to a German-style science-religion. For instance, when the Reverend William Harcourt, in his presidential address of 1839, talked about the "sacred building" and the "temple of truth," it was meant in the spirit of natural theology that emphasized the compatibility of research and Christianity (quoted in Morrell and Thackray 1981, 243).

This belief was tested in a few controversies about religion. The first revolved around contradictions between Scripture-based accounts of the history of the earth and geological data. In the face of this clash, natural theology was augmented by the idea that Scripture and nature were separate pointers to God and that the words of the Bible should be looked upon not as statements of fact but as highly figurative or poetic accounts of religious and moral truths (ibid., 244). In general, the British Association pursued a pragmatic strategy—avoiding the topic of religion as much as possible. A similar problem was posed, and similar pragmatic accommodations were sought, when Darwin's theory of evolution challenged the Biblical story of creation in the latter half of the nineteenth century.

John Tyndall's presidential address at Belfast in 1874 provided the most noteworthy deviation from the pragmatic course (reprinted in Basalla, Coleman, and Kargon 1970, 441–78). Some denounced his address "Science and Religion" as blasphemous. Many others simply thought it bad form for an association president to use his official address to bring up such a highly controversial issue. In fact, Tyndall merely advocated the freedom of science from religious interference in its own scientific realm, while acknowledging the value of religion's contributions in the nonscientific realm: "And grotesque in relation to scientific culture as many of the religions of the world have been and are—dangerous, nay, destructive, to the dearest privileges of freemen as some of them undoubtedly have been, and would, if they could, be again—it will be wise to recognize them as the forms of a force, mischievous, if permitted to intrude on the region of objective *knowledge*, over which it holds no command, but capable of adding in the region of *poetry* and *emotion*, inward completeness and dignity to man" (ibid., 474). This was somewhat similar to du Bois-Reymond's famous effort, two years earlier, at the self-limitation of the scientific Weltanschauung ("ignorabimus").[66]

While the British scientists showed relatively little inclination to present themselves as priests of science at their meetings, one finds some elements of self-ironic humor—which behooves the gentleman. The following is a passage from an ode for the British Association's 1860 meeting at Oxford, composed by a "new life-member" (The British Association 1860, 13):

The Lectures call—their interest so divided,
"Twixt this and that you ponder undecided.
"I must hear this about the pole explained,—
"And there the battle on the rock's maintained,—
"Here the new bottled motion that propels
"So fast that it the telegraph excels—
"There's tunnel-boring (nothing now can stop one),
"Cuts stone like wads and shoots them *a la pop-gun;*
"But universal progress here's disclosed,
"All will be happy, and the doubters posed;
"Figures are facts, and figures go to prove
"The truth, if we the obstacle remove;—
"I can't miss this, that does before one lay—
"The value of the land at Hudson's Bay;
"Monopoly being broken and undone,
"A virgin land our Canada has won,
"Will send home corn and tallow, coal, besides
"A fleet of timber, all exchanged for hides:—
"But here is one I really must not miss, for
"'T is how to tame a Jackal by a whisper:—
"But waiting idle we shall lose the day,"
So in despair you enter section J.

The British Association was also keen on the technological applications that could be garnered from scientific advances. Quite
frequently, the annual meetings were convened in one of the boom-
towns of the Industrial Revolution, like Liverpool (1837), Newcastle
(1838), Birmingham (1839), Glasgow (1840), Manchester (1842),
among others, and the association stressed "the symbiotic relation
between esoteric research and material progress, and the union of
natural knowledge with technical advance" (Morrell and Thackray
1981, 158). This aspect was to prevail more and more. In the course
of the nineteenth century, British science (just like science in other
highly developed nations) underwent a deep transformation—it
became professionalized (Mendelsohn 1964a). A growing proportion of the scientists came from the lower middle classes and even
from the working classes, and they needed and expected to be paid
for their work. In the twilight of the gentlemanly ethos of science,

British scientists, as a group, were swiftly and inexorably pulled into the forcefield of technological and industrial progress. The Gentlemen of Science transformed into applied scientists in the service of the Industrial Revolution. The Priests of Science, by contrast, succeeded in making the German universities their institutional rampart and temple, where their steadfast pursuit of the most basic and profound questions of the scientific Weltanschauung was admired by society and underwritten by the government. Thus they protected their priestly ethos, even while they accommodated the more pedestrian demands of the new age to the extent necessary. After all, "priest" could become a full-time job, but "gentleman," almost by definition, could not.

For an account of the nature of American science in the early nineteenth century, we can rely on the insights of a brilliant French observer, Alexis de Tocqueville. His acclaimed analysis of American democracy was remarkably perceptive also in its assessment of science under democratic conditions. Ever since Robert K. Merton published his classic essay titled "A Note on Science and Democracy" in 1942, it has become fairly customary to assert that science flourishes best in a democracy rather than under dictatorial regimes (specifically, in National Socialist Germany and later in the Soviet Block), and there is much in the track records of the respective science systems to support this view. But what is of interest here is that science in democratic America, according to the early French observer, has taken a peculiar *shape*.

Tocqueville devoted one chapter of his magnum opus, *Democracy in America*, to this issue. The chapter was titled "Why the Americans Are More Addicted to Practical Than to Theoretical Science" (Tocqueville 1862 [1840], 2: 47–55). "In America, the purely practical part of science is admirably understood, and careful attention is paid to the theoretical portion, which is immediately requisite to application. On this head, the American always display a clear, free, original, and inventive power of mind. But hardly anyone in the United States devotes himself to the essentially theoretical and abstract portion of human knowledge" (ibid., 2: 48). The cause for this, according to Tocqueville, lies in the restlessness and fast pace of democratic society that is not generally suited to the contemplative lifestyle necessary for the pursuit of pure science:

> The world is not led by long or learned demonstrations: a rapid glance at particular incidents, the daily study of the fleeting passions of the multitude, the accidents of the moment and the art of turning them to account, decide all its affairs. . . . To minds thus predisposed, every new method which leads by a shorter road to wealth, every machine which spares labor, every instrument which diminishes the cost of production, every discovery which facilitates pleasures or augments them, seems to be the grandest effort of the human intellect. (ibid., 2: 50, 52)[67]

From a Tocquevillian perspective, thus, nineteenth-century American science could be described as utilitarian, with the prime goal of providing concrete benefits for the people—the very antithesis to the quasi-religious German concept of science, with its ultimate goal of creating a Weltanschauung.[68]

This stark dichotomy is an exaggeration because German and American science did not develop in isolation from each other (many leading nineteenth-century American scientists having received some of their training at German universities), and because neither one of them was monolithic. The view of science as religion also filtered into the United States, although it never found a substantial number of adherents there. For instance, the American publisher Paul Carus gave a presentation titled "Science, a Religious Revelation" to the World's Parliament of Religions, which was held on the occasion of the World Exhibition of 1893 in Chicago (Holton 1996, 10). In it, he advocated the replacement of conventional religions with a science-based "religion of truth." Carus was strongly influenced by German philosophical-scientific thought. An immigrant from Germany, he was a dedicated follower of monism, and he was best known for making Ernst Mach's works available to the English-speaking public.

The American Association for the Advancement of Science (AAAS) was founded in Philadelphia in 1848.[69] The major initiative came from American geologists and naturalists. In 1840, a precursor organization, the American Society of Geologists, had been formed, and, in 1842, "and Naturalists" was added to the society's name to indicate its broadened membership base. A very pragmatic reason lay behind creating a national society of geologists and naturalists. This was an age of geological and other surveys of the vast and relatively unknown American landscape, and the geologists and allied scientists felt the need for a national association that could provide some coordination of the state-level surveys and a forum where scientists would be able to learn from each others' undertakings. "For those geologists and naturalists, 'advancement' meant acquiring better intellectual and material resources for their work" (Kohlstedt 1999, 12).

Reflecting the overall democratic condition of the country, the issues of popular access to science and of the dissemination of scientific interest and knowledge among the people were particularly important to the American association. A vigorous opposition against "pseudoscience" went hand in hand with the association's public advocacy for science. When the association was founded, no scientific qualifications were needed for membership (contrary to the practice, for example, of the British association). Turf disputes occurred frequently between science enthusiasts and professional scientists. For instance, attempts in the 1850s to create an "associate membership" were hotly contested and failed, but, in 1874, the category of "fellow" was introduced, recognizing an elite group of professional scientists. Later the even more distinguished "honorary fellow" came into being.

One of the rare instances when the topic of religion was broached at the American association was Simon Newcomb's 1878 presidential address "The Course of Nature," which addressed the relationship between science and Christian religion (ibid., 27). Science and religion, in Newcomb's view, were distinct and complementary. He argued that "science concerns itself only with phenomena and the relations which connect them, and would not take account of any questions which do not in some way admit of being brought to the test of observation" (quoted ibid.). This position was somewhat reminiscent of Tyndall's address to the British association four years earlier, and it did spark some lively debates. However, AAAS "did not become the ongoing forum for a discussion that moved largely into the periodical literature and chapels on college campuses" (ibid., 28). Furthermore, on the issue of religion, the AAAS leaders "gave latitude to speakers to express their opinions from atheism to deeply held Christian beliefs, but authorized publication only of scientific papers and plenary addresses that would not embarrass or discredit the association" (ibid., 45).

In sum, although the Priests of Isis and the Gentlemen of Science were not entirely different from each other, and easily recognized each other as fellow members of the same overarching scientific community, distinct emphases were obvious. The Priests took religion much more seriously than the Gentlemen did, even though—or, perhaps more precisely, because—the gentlemen, but not the priests, held offices in established churches. The American scientists were less interested even than their British colleagues in any religious aspects of science. They were more prone to view their association in an entirely pragmatic light, as a support structure for their scientific activities. The democratic environment encouraged a uniquely American emphasis on the dissemination of science among the people, and on organizational openness vis-à-vis the countervailing trend of science developing into a hierarchical profession.

NOTES

1. When the "Aufruf an die Europäer" talks about "what *the whole world* up to now understood to be 'Kultur,'" it reveals a naive ethnocentric generalization. From the preceding discussion, it should be clear that the concept of Kultur, as it is used here, is specifically German.

2. In a more general way, national styles of thought were also discussed by John Theodore Merz (1965) two years earlier.

3. The attempts at Deutsche Physik had parallel efforts in mathematics. Fritz Kubach, one of Philipp Lenard's students, was a proponent of "germanische Mathematik" (Kater 1974, 50), and similarly Theodor Vahlen and Ludwig Bieberbach, who were influential in the Prussian Academy of Sciences, promoted a "deutsche Mathematik" (Walker 1995, 86–87, 97; also Cornwell 2003, 198–203). Bieberbach exemplified the type of the academic

opportunist. He had shown no signs of National Socialist sympathies before 1933, but, after Hitler had come to power, he soon emerged as a chief supporter of Deutsche Mathematik (Walker 1995, 85–87).

4. On science under National Socialism in general, see Cornwell 2003.

5. Some of the leading National Socialists were "reactionary modernists" who combined an enthusiasm for modern technology with reactionary sociopolitical viewpoints (Herf 1984).

6. Other Ahnenerbe activities included research on divining rods; mineralogical studies, especially methods for gold prospecting; cave research; and anthropological studies on races and race mixing (Kater 1974, 87). Plans existed for an Ahnenerbe division devoted to the study of the "sogenannten Geheimwissenschaften" (so-called occult sciences) (ibid., 87–88). To combat food shortages, Himmler instigated the production of mead, an old Germanic drink based on fermented honey (ibid., 215–16). Sea algae were used to make bread. Most concentration camps had rabbit-breeding stations (*Angorakaninchen*) to supply the army with woolen underwear. Ahnenerbe scientists also bred hardy strains of cereals, plants for vegetable oil, and a plant that would combat cancer. Having a special interest in horses, Himmler directed his researchers to breed a winter-hardy horse suited for the future settlement of the East (from Equus Przewalski and Equus Gmelini) (ibid., 217). Horses furthermore should serve as sources of meat and milk. The Ahnenerbe also studied Tibetan dogs and developed a strong general focus on this region. A Tibet expedition took place in 1939. A planned Caucasus expedition was canceled because German troops had to withdraw from the area (ibid., 214–15). Himmler was not impressed by the military potential of nuclear energy, but in 1944 supported the development of a "Wunderwaffe" called "Strahlengerät" with several advertised benefits: It could kill or paralyze living beings but also heal them, and moreover could serve as a remote sensing device for oil (ibid., 219). The Ahnenerbe also searched for gold in German rivers, making use of the divining rod division that had initially concentrated on finding water.

7. Hanns Hörbiger was the father of the well-known actors Paul and Attila.

8. Lenard's outburst was reported to Himmler by Ahnenerbe scientist Scultetus on February 10, 1937 (reprinted in Nagel 1991, 126). In this letter, Scultetus huffed about the "verkalkten wissenschaftlichen Bonzen [senile scientific bigwigs]" of the anti-WEL persuasion (ibid., 127).

9. At the individual level, the assumption of different cognitive styles in science may be less problematic. The idea that some scientists are primarily

interested in synthesis and the "big picture," whereas others relish in splitting up phenomena into ever more minute details, has had a long tradidtion. In his *Kritik der reinen Vernunft*, for instance, Kant distinguished two cognitive styles, when he discussed the "very different thinking styles of scientists, some of whom (who are supremely speculative) are adverse to dissimilarity and always aim at the unity of a category, others of whom (supremely empirical minds) incessantly seek to split nature into so much diversity that one might almost have to abandon the hope that one would be able to evaluate natural phenomena according to general principles" (1925 [1781], 403).

10. Also see Abir-Am 1993.

11. For the latter aspect, see, for instance, Ben-David 1984, 1991; Farrar 1975; Merton 1973; Turner 1971; Zloczower 1981.

12. The relationship of Kultur and Wissenschaft has been an enduring preoccupation among German intellectuals; see *Die Stellung der Wissenschaft in der modernen Kultur* (1984).

13. As a side note, Matthew Arnold also contrasted culture and science. For him, the source of science is curiosity, the scientific passion, "a desire after the things of the mind simply for their own sakes and for the pleasure of seeing them as the are" (1994 [1869], 30). Culture, by contrast, is "properly described not as having its origin in curiosity, but as having its origin in the love of perfection; it is a *study of perfection*. It moves by the force, not merely or primarily of the scientific passion for pure knowledge, but also of the moral and social passion for doing good" (ibid., 31).

14. An even more radical antiscience section of the Kulturträger rejected science not only for what were considered its deleterious applications but wholesale, out of principle, for what was considered its flawed epistemology. An exponent of New Romanticism, Arthur Bonus, for instance, emphatically separated science from true reality: Scientific facts were for the plebeian herd; they only impeded the elite's quest for the higher reality of vision and will (Mosse 1964, 65).

15. Ultramontanism (from the Latin for "beyond the mountains") means Catholicism because, from the German perspective, the highest Catholic authority, the pope, resides on the other side of the Alps mountains.

16. The official name, after 1890, was Gesellschaft Deutscher Naturforscher und Ärzte.

17. The facile analogy between the unification of scientific theories and the political unification of Germany is rather arbitrary. If anything, the unification of the *world* would have been the more appropriate counterpart to scientific unification in this analogy.

18. It is important to distinguish, within the concept of Kulturpolitik, the internal and external spheres. (For a summary of the meaning of the concept, see *Lexikon zur Geschichte und Politik im 20. Jahrhundert* 1971.) Internal Kulturpolitik means the governmental support of the arts and sciences, Bildung, and Kultur in Germany. Internal Kulturpolitik in the Wilhelminian Empire was primarily carried out by the states and the communities.

19. *Vossische Zeitung*, December 12, 1913, quoted in Grundmann 1965, 1.

20. Einstein's trips were carefully monitored. The document collection by Kirsten and Treder contains reports to the German foreign office from German diplomats in The Hague, Oslo, Copenhagen, Paris, Buenos Aires, Tokyo, Madrid, Montevideo, Rio de Janeiro, Chicago, New York, and Vienna (1979, 1: 225–40). When it was rumored in 1920 that Einstein planned to leave Germany, German diplomats in Britain grew alarmed. A letter from London to the German foreign office read: "Especially at this moment, Professor Einstein is a cultural factor [*Kulturfaktor*] of the first order for Germany because Einstein's name is known in the widest circles. We should not drive away from Germany such a man, with whom we can carry out true 'Kulturpropaganda'" (ibid., 1: 207).

21. The decision was reversed in 2001.

22. In Europe, by contrast, the Christian churches have become more resigned, and even accepting, vis-à-vis scientific theories. Unlike their American co-religionists, European Protestants have made no serious effort to supplant the theory of evolution with the teaching of "creationism" based on a literal interpretation of the Bible. Furthermore, the Catholic resistance against the heliocentric theory of astronomy was, in fact, abandoned long before pope John Paul II officially rehabilitated Galileo in 1992. The same pope also pronounced Darwin's theory compatible with the Catholic faith.

23. The view of science and religion being contradictory was also dominant in the scholarly community, until Merton's (1970 [1938]) groundbreaking study demonstrated the contribution of Puritanism to the rise of modern science in seventeenth-century England.

24. On Oken, see Ecker 1883. The model for Oken's creation was the Gesellschaft Helvetischer Naturforscher (Society of Swiss Natural Researchers) in Geneva (Jackson 2003, 124). A regional precursor of the assembly of German scientists and physicians was the Vaterländische Gesellschaft der Ärzte und Naturforscher Schwabens (Patriotic Society of Swabian Physicians and Natural Researchers), which had existed from 1801 to 1805 in Southwestern areas of Germany (ibid.).

25. Julius Baumgärtner, Baden-Baden 1879, quoted in Schipperges 1976, 54; Adolf Kussmaul, Straßburg 1885, quoted ibid., 58.

26. Scientific associations in Meissen (founded in 1845) and Bautzen (founded in 1846) also carried the name of Isis (Wilson 1952, n. 13). An apparently short-lived radical journal that was published in London in 1832 was titled *The Isis*. It also contained suffragist writings, perhaps signifying a certain affinity between feminism and the Goddess Isis. In 1839, a literary magazine called *The Isis* started to appear in Oxford at the university (Wilson 1952). A student magazine of the same name was issued at Oxford in 1892, yet its name was taken from the Oxford stretch of the river Thames, which in that locale is also known as Isis (Billen and Skipworth 1984, 11). I also found a 1947 publication by "Isis, Gesellschaft für biologische Aquarien- und Terrarienkunde" of Munich.

27. Also see Syndram (1990, 54–61). He pointed out that ancient Egypt was celebrated as the original Golden Age of humankind by scholars such as Bossuet, de Goguet, and the Comte de Caylus.

28. Among useful sources for more detailed information about Isis are Harless 1830, Merkelbach 1995, Roscher 1890–94, and Solmsen 1979. A substantial amount of our knowledge about Isis and Egyptian mythology can be traced back to Herodotus and Plutarch. For a comprehensive account of Greek scholars who wrote about Egypt, see Hornung 2001, 19–25.

29. A hymn to Isis, written by Isidoros of Narmuthis in the first century, enumerated her many names, claiming that different worshippers might venerate different goddesses in name but that, in reality, they all worshipped Isis. Her Egyptian name was given as "Thiuius" ("the only one") (Merkelbach 1995, 95). A similar text was found on a papyrus from Oxyrhynchos, Egypt (Assmann 1997, 50).

30. See Assmann for the quoted form (1997, 88). For the full text, see Merkelbach (1995, 98) and Assmann (1999, 27, 60n33).

31. Platonists and Gnostics typically rejected the material world in favor of the realm of ideas, and it might therefore seem unlikely that any interest in science would be compatible with that perspective. However, we should point out that the path from Platonic views to the—seemingly opposite—pantheistic view that the world is God is less contorted than one might think. Both notions share a crucial common element: They both motivate a quest for transcending the superficial and deficient knowledge that rules our everyday life in order to gain true knowledge—of what lies beyond this world, or of what is the essence of this world. Hence, science, if defined in a pantheistic mode, does not appear utterly irreconcilable with

a Platonic and Gnostic mind-set. Historically, the Neoplatonism of the Italian Renaissance is sometimes considered to have in some ways even prepared the emergence of modern science. (For a critical view of this thesis, see Vickers 1984b).

32. That Isis temple, according to the dictionary, was to become the church of Ste. Geneviève (the present-day Panthéon). Brewer's dictionary further reports that a statue of Isis was preserved in the church of St. Germain des Près, until Cardinal Briçonnet (who died in 1514) destroyed it, when he observed it to be the focus of veneration.

33. See Kee 1980; Kloppenburg 1982; Warner 1985, 177–80.

34. Although Chaucer's *Legend of Good Women* does not mention Isis, there may be parallels in this work with *Le Livre de la Cité des Dames* (Chance 1995, 42–43).

35. Book V, canto vii; Spenser 1978 [1590–96], 799–804. Book I, line 478; Milton 1993 [1667], 139.

36. A copy of Rahneus's work, mentioned by Gruppe, could not be located.

37. On the relationship between the Enlightenment, esoteric knowledge, and secret societies, see Neugebauer-Wölk 1999.

38. The hermetic tradition traces itself back to the—apocryphal—Hermes Trimegistus, a sage (or, according to some, even a god) who was the greatest teacher of knowledge and wisdom in ancient Egypt (Hornung 2001). In fact, the body of hermetic works was written between the second and fourth centuries CE (Churton 1997, 45). The *Corpus Hermeticum*, as it became known, was preserved in Byzantium and brought to Italy after the fall of that city. At Cosimo de Medici's direction, Marsilio Ficino, the head of the Florentine Academy, translated the work into Latin (1463).

39. This work by Marana was mentioned by Hazard (1963 [1935], 15); I could not find a copy.

40. This Kantian passage contains an explicit reference to Segner's (1746) *Einleitung in die Naturlehre.* The vignette there shows three half-naked fat boys or *putti* (the scientists) and a cloaked woman (Isis), the first boy pondering, the second boy measuring the woman's footprints, and the third boy lifting the woman's cloak. The inscription at the Isis temple of Sais came to encapsulate Egyptian wisdom. Beethoven, for example, had it standing framed on his desk (Hornung 2001, 134).

41. On this and other contemporary periodicals, see Horn 1994.

42. Some of the Romantics, for whom the self, the *Ich*, was the center of everything, had always seen an ulterior purpose in the whole project of

unveiling Isis—self-knowledge. As Novalis (1960b, 110) wrote in a fragment belonging to *Die Lehrlinge zu Sais,*

"Einem gelang es—er hob den Schleier der Göttin zu Sais—Aber was sah er? Er sah—Wunder des Wunders—Sich Selbst." (One succeeded—In lifting the veil of the Goddess at Sais—But what did he see? He saw— Wonder of wonders—Himself.)

Achieving a comprehensive Weltanschauung, in the eyes of the more radical *Naturphilosophen,* meant that Nature would become conscious of itself. Everything would be different in a thus awakened Universe: All divisions would be healed; all polarities, synthesized; subject and object, Self and Nature, merged.

43. For instance, on the frontispiece of Alexander von Humboldt's German edition of his *Ideen zu einer Geographie der Pflanzen,* Apollo could be seen lifting the veil of Isis (Bruhns 1872, 1: 199).

44. See http://www.nobel.se/chemistry/laureates/1921/press.html and http://www.nobel.se/chemistry/laureates/1930/press.html.

45. On the widespread metaphor of unveiling the secrets of nature (understood as a female entity), and its sexual overtones, see Merchant 1982. There is also a long-standing metaphor of the "naked truth"— Horace's "nuda veritas" (Warner 1985, 294–328). A different usage of the metaphor appeared in Linda Shepherd's 1993 *Lifting the Veil: The Feminine Face of Science.* Here it means reclaiming the feminine aspects of science that, in the author's view, have been hidden and denied in the currently predominant way of scientific practice. Similarly, Carolyn Merchant (1982) argued for introducing a feminist perspective into the history of science in an article titled "Isis' Consciousness Raised." The title of *Women of Academe: Outsiders in the Sacred Grove,* by Nadya Aisenberg and Mona Harrington (1988), evokes another religious metaphor in the context of a study of women teaching at colleges.

46. See, for instance, Merchant 1980. David Noble (1992) argued that the "maleness" of the contemporary science culture was rooted in Christian clerical traditions.

47. See picture in Schiebinger 1989, 150.

48. The following is the description of the medals for physics and chemistry on the Nobel Web site: "The medal of The Royal Swedish Academy of Sciences represents Nature in the forms of a goddess resembling Isis, emerging from the clouds and holding in her arms a cornucopia. The veil which covers her cold and austere face is held up by the Genius of Science" (http://www.nobel.se/nobel/medals/physics-chemistry.html).

49. On the secret, also see Voigts 1998 and other contributions in Assmann and Assmann 1997, 1998.

50. For a more comprehensive discussion of the secret in science, see Eamon 1994. In his taxonomy of secrets, he distinguished social, epistemological, and epistemic secrets of nature (ibid., 11). The first corresponds to our understanding of type-two nonknowledge (the secret as an actively maintained gradient of knowledge). The other two provide a further differentiation of type-one nonknowledge. The epistemological secret is defined as nonknowledge that is unknowable in principle, whereas the epistemic secret is defined as historically contingent, i.e., current nonknowledge that can be replaced by knowledge. Another important distinction in this area is that between known nonknowledge (one is aware of one's ignorance about certain things) and unknown nonknowledge (one does not even realize that one does not know certain things). (See Weingart 2001.)

51. Interestingly, even rationalist opponents of religion sometimes chose quasi-religious forms. John Toland's *Christianity Not Mysterious* (1696), for instance, became notorious for lashing out venomously at the Christian religion and its doctrines and was ordered burned by the Irish Parliament. Toland propagated the conviction "that Reason is the only Foundation of all Certitude; and that nothing reveal'd, whether as to its Manner or Existence, is more exempted from its Disquisitions, than the ordinary Phenomena of Nature. Wherefore, we likewise maintain, according to the Title of this Discourse, that there is nothing in the Gospel contrary to Reason, nor above it; and that no Christian Doctrine can be properly call'd a Mystery" (1696, 6). Yet when the rabble-rouser and militant opponent of religion founded the *Socratic Society* to promote his rationalist principles, it was replete with its own hymns and rituals (Hazard 1963 [1935], 264–65). This curious combination of mystical forms and rationalist ideals—a religious antireligion, one might say—was far from unique. It may have inspired the London Grand Lodge of Freemasons opening in 1717 and the first French Lodge, which was founded in 1725 (ibid., 265). The French Revolution even spawned a short-lived public cult of Reason. In 1793, Jacques Louis David's *Fontaine de la Régénération* was dedicated on the ruins of the former Bastille. The water of this fountain streamed from the breasts of Isis (Hornung 2001, 133, with picture). Finally, one might note that the Royal Society, the pioneer scientific association established in 1662, owed much of its inspiration to Rosicrucian ideals, according to Churton (1997, 158). The Rosicrucians were a secret mystical movement that began in the early seventeenth century.

52. Also see Hardtwig 1989.

53. Initially, the Freemasons were not interested in Egypt, but, from the second half of the eighteenth century onward, they increasingly integrated elements of Egyptian esoterics into their teachings and rituals (Hornung 2001, 121–25). A major force in this development was the Count Cagliostro, a mesmerizing figure whom many denounced as a charlatan and who died in a papal prison in 1795. Apart from the flashy Cagliostro episode, serious scholars like the previously mentioned Ignaz von Born and Karl Leonhard Reinhold solidly grounded Freemasonry in Egyptian lore. Another work in this tradition is Johann Gottfried Bremer's *Die symbolische Weisheit der Aegypter aus den verborgensten Denkmälern des Alterthums* (The Symbolic Wisdom of the Egyptians from the Most Hidden Monuments of Antiquity) (1793).

54. Many Masons subscribed to the theory that Hermes Trimegistus, the ancient Egyptian sage, was the fountainhead of the most profound knowledge.

55. A central piece of Masonic lore, for instance, is the murder of the master builder Hiram Abif, who died rather than betray some extremely profound secrets. These events are reenacted in the ceremony to attain the third Masonic degree (Knight and Lomas 1998).

56. There has been some disagreement about whether there are epistemological secrets of nature, as defined by Eamon (1996, 11), i.e., things that are unknowable in principle, or whether all the secrets of nature are epistemic, i.e., knowable in principle. For instance, in his famous *"ignorabimus"* speech (which we will discuss in more detail later), Emil du Bois-Reymond took the position that there are things that will forever, and in principle, remain unknown, and this statement drew heated criticisms from within the scientific community.

57. Within the nonknowledge of the first type, which science intends to dispel, one might further distinguish between manifest nonknowledge and pseudoknowledge (i.e., faulty existing knowledge—for instance, popular "everyday knowledge," or even what some social scientists might call "ideologies"). In the second case, science calls into question knowledge that almost everybody in a given culture routinely possesses, and it reveals that this supposed knowledge does not satisfy rational or empirical criteria (Oevermann 2003). It thus first generates the area of nonknowledge that it then seeks to replace with (scientifically more adequate) knowledge. In the case of manifest nonknowledge, by contrast, science addresses questions about which, by a wide consensus, no valid prior knowledge had existed (e.g., the structure of the subatomic world). On the whole, the natural sci-

ences may be more likely than the social sciences to deal with evident non-knowledge, but each branch of science has been addressing both of these two subforms of nonknowledge of the first type. It goes almost without saying that the instances in which science tackles pseudoknowledge are particularly prone to creating cultural controversies.

58. On this metaphor, see Merton 1965. The typical outlook of the scientific community is akin to the position of the moderns in the famous "querelle des anciens et des modernes" in the late seventeenth and early eighteenth centuries in France. (The moderns asserted the superiority of contemporary over classical literature, whereas the ancients took the opposite view.)

59. According to Merton, the ethos of science, and particularly the norm of "communism," implies "the imperative for communication of findings. Secrecy is the antithesis of this norm; full and open communication its enactment" (1968, 611).

60. The relationship of modern science and Freemasonry is connected to a much larger—and controversial—question in the history of science: What role did various occult traditions, such as alchemy, magic, numerology, etc., play in the emergence of modern science? See, for instance, Pagel 1982; Vickers 1984a; Yates 2002.

61. Indeed, the popular usage of "secrets of nature" has typically remained close to Simmel's explication of the term. The expression does not simply denote nonknowledge of the first type, but it often retains the flavor of ownership—the subtext that somebody, or some entity, or God, possesses this secret knowledge. Eamon (1994) explicated in great detail the rich tradition of speaking of the "secrets of nature."

62. Emile Durkheim introduced the classic dichotomy of the sacred and the profane. When, in 1811, the philosopher Fichte gave his inaugural speech as the first rector of the newly founded University of Berlin, he called the institution of the university "the most holy thing that humankind possesses" (1812 [1811], 6). A science positioned in the realm of the sacred (and hence of Kultur) was a legitimate and worthy member of the "most holy thing."

63. As the nineteenth century went on, the religious rhetoric in German science weakened, although it did not disappear completely. Jochen Hörisch (1998) perceptively traced a crucial semantic shift in how German intellectuals in the nineteenth and early twentieth centuries spoke of the unknown. What was earlier called Geheimnis (secret) was increasingly being called Rätsel (riddle, puzzle). In 1880, Emil du Bois-Reymond

gave a lecture titled *Die sieben Welträtsel* (1912, 2: 65–98), and nineteen years later, Ernst Haeckel's *Die Welträthsel* was issued. This is not merely a switch of synonyms but signifies the migration of the concept of the unknown from the realm of the sacred to the realm of the profane. Whereas the *secret* metaphorically implies a higher power and inspires awe, the *puzzle* is an intellectual challenge and fun. To partake of a secret, one has to be initiated; to solve a riddle, one has to be smart.

64. In Britain, there existed also the much older Royal Society (established in 1662), which in the early nineteenth century, however, was not in the best of shapes (Mendelsohn 1964a). On this organization, see Stimson 1948.

65. The frontispiece is reprinted in Assmann 1999, 46. It is from the 1809 Edinburgh edition. The edition to which I had access (Darwin 1804) did not have that frontispiece. The personification of philosophy in the Vatican, created by Renaissance artist Raphael, displays the "breasts of Nature." According to Hornung, Raphael's work was based on a multibreasted variant of Isis that existed since late antiquity (2001, 91). Pierre Hadot explained the iconographic syncretism between the multibreasted Artemis of Ephesos (who also was a goddess of nature) and Isis (1982, 4–5). He further noted that the multibreasted aspect of Artemis might have been a misinterpretation of ornamental chains and necklaces (ibid., 5).

66. In his address, Tyndall, who had received his doctorate at the University of Marburg, Germany, also explicitly referred to du Bois-Reymond.

67. Allan Bloom's more recent assessment of the humanities in the United States echoed Tocqueville's fundamental insight: "The democratic concentration on the useful, on the solution of what are believed by the populace at large to be the most pressing problems, makes theoretical distance seem not only useless but immoral" (1987, 250).

68. His diagnosis prompted Tocqueville to make a case for the government funding of basic science: "possessing education and freedom, men living in democratic ages cannot fail to improve the industrial part of science; . . . henceforward all the efforts of the constituted authorities ought to be directed to support the highest branches of learning, and to foster the nobler passion for science itself" (1862 [1840], 2: 54).

69. For a history of the early period of AAAS, see Kohlstedt 1999.

5.

WELTANSCHAUUNG AND WELTBILD IN THE SCIENCES

In his essay "Die Zeit des Weltbildes" (The Time of the Weltbild), Martin Heidegger considered the wissenschaftliche Weltbild a defining feature of modernity: "Now if Wissenschaft as research is an essential phenomenon of the modern age, that which constitutes the metaphysical ground of research must, earlier and long beforehand, determine the essence of that age generally" (1957, 80–81). In the preceding section, we have seen how German scientists tried to establish science within Kultur, for instance, by attaching religious imagery to it. A general hallmark of religion is that it addresses the "ultimate questions." Those questions are also the focus of Weltanschauung and Weltbild—concepts that owe their revered position in Kultur largely to addressing these very issues. Hence, at the very core of the German scientists' efforts to afford science entry into Kultur was the issue of a scientific Weltanschauung and Weltbild. Proponents and detractors alike believed that it was the claim of generating a scientific Weltbild that made science a serious contender in the realm of Kultur. Among the scientists, Albert Einstein excelled in his devotion to the project of a scientific Weltbild; he solemnly declared it "the supreme task" of the physicist to arrive at the fundamental physical laws that would constitute a comprehensive Weltbild (1954, 221).

The notion of the scientific Weltbild implies not only an inclusive explanation of the world but also responses to a number of questions that are fundamental in several dimensions—epistemological (what is the basis of scientific knowledge?), methodological (how is scientific research conducted?), ontological (what is the status of the scientific Weltbild vis-à-vis the world and vis-à-vis other Weltbilder?), and social (what ramifications does the scientific approach have for the social world?), even ethical and religious (what does the scientific Weltbild mean for morality and religion?). These questions have become important issues in the philosophy of science.[1]

5.1. HISTORY OF SCIENTIFIC WELTAN-SCHAUUNG AND WELTBILD

How did German scientists adopt the concepts Weltanschauung and Weltbild in their science? Before setting out on our historical survey to answer this question, we should note that it would have required more of an explanation if German scientists had rejected the concepts. For, as we have shown, they were so central to the intellectual climate that it appeared almost inevitable that they had to be used or at least reacted to. Especially those who sought acceptance as Kulturträger had to take a position on Weltanschauung and Weltbild.

Romantic Science

In his profiles of scientific geniuses, Wilhelm Ostwald (1910) introduced the dichotomy of Classical and Romantic scientists. For Ostwald, these terms denoted differences in a person's style of doing science—the thorough, deep, and retiring Classical scientist on the one side and the quick, charismatic, and flamboyant Romantic scientist on the other. Yet not only individual scientists but whole epochs have been classified according to similar schemes. A long-term perspective on the history of the sciences is conducive to discerning oscillations or cycles in collective scientific styles. In his study of science and culture in the nineteenth century, Stephen Brush (1978), for instance, distinguished three comprehensive trends that shaped the sciences (as well as the wider cultural domains): Romanticism in the early part of the century, realism in the middle, and Neoromanticism at the end. In a much more extended usage, the Classic/Romantic dichotomy—often termed Apollonian/Dionysian—has been harnessed to interpret fundamental changes in the fabric of Western civilization.[2] W. T. Jones (1973), in a similar vein, delineated four broad "syndromes" (the Medieval, Renaissance, Enlightenment, and Romantic syndromes) that each covered the whole range of sciences, arts, philosophy, and similar products of the human mind.[3] And drawing a parallel between the contemporary situation and what went on in the late eighteenth and early nineteenth centuries, Arthur I. Miller (1996) identified today's postmodernism as "Postmodernism II" that had been preceded by "Postmodernism I," which for Miller was Romanticism.[4]

These oscillations are, of course, real, but they should not be exaggerated to the point of imagining a radical break in the middle of the nineteenth century, a "turn from an idealistic and speculative outlook on life to a realistic and materialistic attitude toward both nature and history," as W. J. Bossenbrook put it (1961, 335). Although such a view is aided by some prominent scientists' vitriolic condemnations of Naturphilosophie, we emphasize continuity over discontinuity.[5] We will argue that important elements passed on from the earlier, Romantic epoch of science to the later empirically oriented period. In particular, the shared quest for a Weltanschauung or Weltbild and, with it, commonly held basic inclinations toward synthesis, unification, and wholism constituted a major similarity. Although the later, empirically oriented scientists vehemently rejected the Romantic scientists' speculative approaches in favor of systematic empirical research, they accepted the Romantics' chief goal as their own. (That enduring emphasis on the Weltbild, in turn, formed the bridge linking science to Kultur and the scientists to the other Kulturträger.) Moreover, many close connections in terms of teacher-student and friendship relationships linked the successive cohorts of German scientists. Because, in those respects, Romantic science was an important forerunner of the later science, we start our historical survey with that era. An additional reason for devoting a few pages to Romantic science is that Einstein's life and work, in particular, reveals some salient continuities to Romanticism. (We will later take up the question to what extent he could be considered a Romantic.)

In the German lands, at the turn of the eighteenth and nineteenth centuries, science had not yet clearly differentiated from other human pursuits. It was still closely related to philosophy (especially to the influential school of thought called *Naturphilosophie*) as well as to other forms of intellectual and artistic activity. Most Romantics tried to preserve that state of affairs and emphatically rejected the boundaries between science and art, between poetry and philosophy, boundaries that have nowadays become commonplace. The outlook of those Romantics thus cannot be captured in any sharp dichotomy of science versus nonscience. They typically felt they could be scholars, artists, philosophers, and scientists, all in one. Those whom we shall call Romantic scientists espoused a particular kind of science that was based on Naturphilosophie. However, whereas the *Naturphilosophen* remained entirely speculative, the Romantic scientists complemented speculation with some actual research. And whereas some of the Romantic poets rejected science wholesale, the Romantic scientists engaged in science—their special brand of science.[6]

Kant's philosophy contained the seed that sprouted into Naturphilosophie and the allied views of the Romantic era. L. Pearce Williams (1973) held that Kant's work shaped the outlook of a group of scholars and intellectuals whom we collectively call the Naturphilosophen. They, of course, gave Kant's philosophy their own spin. "But to the *Naturphilosophen* what counted was the reinstatement of the divinity of human reason. And, more importantly, the reinstatement of the human reason as capable of understanding the whole of creation, as a whole" (Williams 1973, 7). Whereas Kant had an immediate influence on some *naturphilosophische* scientists, such as Oersted, there were also important intermediaries in the chain from Kant to Naturphilosopie.

One of them was Fichte's extreme idealism. Whereas Kant still thought that a world "an sich" existed, although we as humans could become aware of it only through our mental framework of perception, Fichte's starting point was the absolute ego (the "I") that posits itself, but then also posits the nonego. For him, there is no Kantian thing "an sich"; the world is nothing but the Ego's field of operation created by the Ego itself. In this philosophy, the well-known oppositions of subject and object, of the knower and the known, of spirit and nature are only superficial—everything originates in the absolute ego. In a similar vein, Hegel's philosophy centered on the spirit. The whole of history was conceptualized as a process of self-actualization during which the spirit is gaining awareness of itself. Whereas Schelling began as a follower of Fichte's philosophy, he rejected the extreme emphasis on the spirit in Fichte's and also in Hegel's thinking. Rather than considering nature a mere epiphenomenon of the spirit, Schelling believed that nature and spirit had formed an original whole, that nature and spirit then split, and that it was the destiny of world history to accomplish the reunification of nature and spirit. This strong teleological motif of the spirit's or nature's journey to self-consciousness is a hallmark of *naturphilosophischen* thought (Hennemann 1967–68).

For Schelling, the chief purpose of the sciences was to elucidate the unity of nature, which would serve the even grander purpose of demonstrating the unity of nature and spirit—that is, to bring the universe to consciousness. His framework of Naturphilosophie also gave the sciences a decidedly nonempirical bent. Not all Naturphilosophen, however, were content with inward contemplation and speculation; some followed a more empirical approach. The Dane Hans Christian Oersted perhaps best exemplifies a "Romantic scientist" whose belief in Naturphilosophie aided significant scientific contributions. (See Snelders 1990.) Oersted was personally acquainted with several leading exponents of German philosophy, among them Schleiermacher, Baader, Fichte, and Schelling, and possessed a thorough grasp of Naturphilosophie. In his view, which clearly reflected this philosophical influence, Divine reason constructed nature. Human reason was the reflection of Divine reason, and, therefore, human reason was able, by its own devices, to comprehend nature. However, Oersted also believed that human reason was only a *dim* reflection of Divine reason. Hence, he argued, we should not naively accept what our mind tells us about nature, but we need empirical experiments to double-check our insights.

What was the essence of Romanticism? As so often when one deals with a social or intellectual movement, it would be a gross oversimplification to put all its exponents into a single neat category.[7] Nonetheless, all the Romantic thinkers shared some characteristics. What united them more than anything was a common enemy. They all opposed the mechanistic and rationalistic Weltanschauung whose principal representative, in their eyes, was Isaac Newton. William Blake, for instance, waged fierce battles against what he considered the unholy trinity of Bacon, Newton, and Locke (Ault 1975). Even though there were considerable differences between Newton and Descartes, Blake rejected them both as "co-conspirators against the imaginative vision of reality" (ibid., 22). He raged against the tyranny of reason and denounced the scientific method as a straightjacket that constricted spirituality and genius, writing, in his 1827 Laocoön engraving, "Science is the Tree of Death. Art is the Tree of Life." In his *The Marriage of Heaven and Hell* (1790), he condensed, in one famous sentence, how the Romantics felt about life: "The road of excess leads to the palace of wisdom" (Blake 1975 [1790], plate 7).[8]

Mechanical was a dirty word for the Romantics. They abhorred the mechanical world-clock or world-machine that Newtonian physics implied (Harrington 1996).[9] Opposing the mechanical notion of that "all-powerful, blind, lonesome machine," the Romantics emphasized organic and dynamic schemes that gave room to meaning, intuition, and spirituality (Paul 1963 [1804], 96).[10] As Bernhard Giesen perceptively put it, "In a Romantic context, nature referred to infinity, continuity, and uniqueness and was contrasted with the artificial dissecting activities of modern society" (1998, 244).

According to Hegel, "The true content of romantic art is absolute inwardness, and its corresponding form is spiritual subjectivity with its grasp of its independence and freedom. The inherently infinite and absolutely universal content is the absolute negation of everything particular, the simple unity with itself which has dissipated all external relations, all processes of nature and their periodicity of birth, passing away, and rebirth, all the restrictedness in spiritual existence, and dissolved all particular gods into a pure and infinite self-identity."[11] In counterdistinction to the world-machine, Schelling proposed the world-soul, which became a key concept for many of the Naturphilosophen, as well as for Romantic scientists, such as Oersted (Snelders 1970). The Romantics' fundamental belief about nature was that "[t]he world is a living entity. . . . [E]verything is alive, everything is connected in cause and effect, there is nothing dead in the world. The cosmos is an organism, and each part of it is its representation, and carries the traits of the cosmos" (Huch 1920, 2: 46).

At the political level, the Romantic rejection of the mechanistic worldview led to a deep disdain of the modern state—for which France was the detested prototype: "In French government, centralization had been victorious for centuries—the bleak, life-negating principle of mechanicism, which Romanticism combated in all areas. Just as the Romantics opposed Newton as the one who replaced life with mechanicism in physics and astronomy, so they opposed France as the state that wanted to replace the manifold creation of organic form, and dynamic interaction with arbitrary calculation and construction" (ibid., 300). In the minds of Romantics like Joseph Görres, the French Revolution and a mechanistic view of nature were twin evils: "Struggle without goal, waste without purpose, striving without end, that would be the fate of humankind! The world only a putrid drop of water in which millions of microorganisms aimlessly oscillate around their tiny axes; a disgusting image of an eternal soul-less life, and of a pointless mobility" (quoted in Eberle and Stammen 1989, 210). The Romantics instead favored an enchanted universe and, in terms of social organization, a *Volk*—the organic community of a people that, through the chain of generations, is anchored in the past.

At the level of lifestyle, the Romantics riled against mediocrity and conventionalism, against the pedestrian world of the shopkeepers' and the public officials' common sense—in short, against "philistinism." In this aspect, Einstein ("the rebel") was clearly of one mind with those earlier rebels. The Romantics also championed genius, individualism, and authenticity beyond all external limitations. Crane Brinton beautifully described the Romantic in *The Encyclopedia of Philosophy*: "Sensitive, emotional, preferring color to form, the exotic to the familiar, eager for novelty, for adventure, above all for the vicarious adventure of fantasy, reveling in disorder and uncertainty, insistent on the uniqueness of the individual to the point of making a virtue of eccentricity, the typical Romantic will hold that he cannot be typical, for the very concept of 'typical' suggests the work of the pigeonholing intellect he scorns" (Edwards 1967). The Romantics were part of a generational rebellion that espoused new social mores as well as new ideas. Thus they had a greater resemblance to Mannheim's free-floating intellectuals than to the staid "philistine" Bildungsbürger. On an 1819 visit to Tübingen, Jacob Berzelius, the eminent Swedish chemist and enemy of the Naturphilosophen, was appalled by the "barbarian and shabby look" of the slovenly students he encountered there (quoted in Lindroth 1992, 23). The scandalized Berzelius even drew a sketch of a typical Tübingen student—long-haired, mustachioed, and pipe-smoking—which survived in the archives of the Royal Swedish Academy of Sciences (reproduced ibid.). The root cause of this wretchedness, according to Berzelius, was Naturphilosophie, or "a trusting and unthinking devotion to the ideas of those whose unintelligibility has gained them a reputation for profundity" (quoted in Lindroth 1992, 22).

At a sophisticated level of philosophical reflection, the Romantics rejected the rationalism of the French Enlightenment with its strong undercurrent of mechanical materialism. Baron d'Holbach (1984 [1770]) had presented a materialistic *System of Nature*, and the group of *ideologues* (Condillac, Destutt de Tracy, Cabanis, *inter alios*) had endeavored to understand the human mind in materialistic terms. The Romantics instead related back to the vitalistic tradition of the alchemists, of Paracelsus, Helmont, and the mystic Jacob Boehme, which, in turn, can be traced to the ancient Gnostics (Mason 1962, 349–54).[12] The scholars in that tradition cherished analogy; they believed that man was a microcosm, a miniature copy of the macrocosm.[13] The Romantics, too, made liberal use of analogy in their reasoning, and they wholeheartedly embraced that mystical idea of the correspondence between macrocosm and microcosm. To them, the universe appeared as a gigantic organism—in analogy to the human or other organisms.[14] In connection with this central Romantic tenet, many Naturphilosophen and their allies regarded the polarity of opposites as the key principle that drove the dynamics of the universe.

The young Goethe, although certainly not a hard-line Romantic, participated in the Romantic overture, the "Sturm und Drang" movement, with its antirationalistic, anti-Enlightenment overtones. Goethe also carried on a protracted battle against Newton's theory of optics. (See Sepper 1990.) Whereas his rejection of Newtonian optics looked rather Quixotic to most scientists, his scientific approach paid off when he discovered the intermaxillary bone in humans. It is rather astounding that this most famed of German poets repeatedly asserted that he made his greatest contributions in the field of science.[15] For instance, in 1829, only a few years before he died, Goethe told his confidant, Eckermann, "I make no claims at all for what I have achieved as poet. Fine poets were my contemporaries, even finer ones lived before me, and there will be others after me. But that I alone in my century know what is right in the difficult science of colour, for that I give myself some credit, and thus I have a consciousness of superiority to many" (quoted ibid., 189). He insisted on dissolving the boundary between subject and object that mechanistic physics à la Newton had erected and on bringing the "I" back into science. Goethe's science was chiefly that of the poet-philosopher, and it still moved in the neighborhood of the Naturphilosophen.

Goethe's strong interest in science was not an idiosyncratic predilection. He was the best known, but not the only, poet-scientist. Ludwig Achim von Arnim, for instance, both wrote Romantic literature and conducted research in physics (Snelders 1970). Under his real name, Friedrich von Hardenberg, the Romantic poet Novalis held a day job as an official in the Saxon salt mine administration, after having received a science education at the Freiberg mining school (Hansen 1992). Novalis believed in the unity of natural forces, and thought, like Schelling, that oxidation played the central role (Wetzels 1971). But he also rejected Schelling's deductive and speculative flights of fancy as "scientific depravity" (Novalis 1960c, 668). Novalis embarked on compiling a comprehensive encyclopedia, which remained unfinished, like several other of his projects, owing to his untimely death. Novalis's goal was to transcend the boundaries between science and poetry and to create a "realen, wissenschaftlichen Poesie" (Novalis 1976 [1798], 252). Such a synthesis was also intended in Jakob Grimm's and Joseph Görres's concept of "Wissenschaftspoesie" (Frühwald 1991, 294). And Friedrich Schlegel wanted to create a new mythology—based on the myths of the Orient, the philosophy of Spinoza, and physics (Wetzels 1971).

The Romantic scientists could hardly ignore the complete triumph of Newtonian physics in the core area of mechanics. Yet perhaps, they hoped, more enigmatic forces, such as magnetism, electricity, chemical reactions, or hypnosis, might provide an opportunity for rejecting Newton's bland world-machine and creating a reenchanted, organic Weltanschauung.[16] With great enthusiasm, therefore, the Romantic scientists turned to studying these fields. Similarly, the hidden treasures beneath the surface of the earth held an enormous attraction for the Romantics, some of whom, like Novalis, studied at the famous Freiberg mining school. Suitably mysterious, geomancy became a favorite Romantic method of prospecting.

Physiognomy was another field that attracted the Romantics. Its central premise—that external features indicate a person's inner traits—was entirely convincing to them, given their creed that everything is connected. This creed also set them on a quest for the origin. The Romantics believed in an original people in Central Asia, an original language, and an original religion. Like Goethe, many Romantics postulated an archetypal plant and an *Urkraft*. Furthermore, they tended to believe in one illness as the root of all illnesses and consequently also in the possibility of a cure-all—for which

Mesmer's magnetism was viewed as a promising candidate. In the Romantic worldview, medicine was connected with religion, as was science. Franz Baader, for instance, wrote of "the vibrant connection of religion and physics," and Karl Windischmann proclaimed that physics had to lead to God (quoted in Huch 1920, 2: 71, 70). In this respect, the Romantic scientists found themselves—perhaps unexpectedly, or unwittingly in some cases—in the company of their nemesis Newton.

The naturphilosophische speculation in which biologists, such as Lorenz Oken, Richard Owen, and Johannes Müller, engaged, led them to several important scientific discoveries.[17] Lorenz Oken, whom we encountered as the founder of a political and scientific journal called *Isis* and of the annual meetings of the German scientists and physicians, did extensive empirical work in comparative zoology, anatomy, and physiology and also studied "animal magnetism," or hypnosis. (See Pross 1991.) He crossed swords with Goethe over the publication of *Isis* as well as in a scientific priority battle. In terms of his Weltanschauung, he was a kind of cosmotheist who believed that all natural laws were "congealed" thoughts of God. Like other Romantics, Oken revered mathematics. He stated that "Naturphilosophie is only science if it is mathematizable, i.e., if it can be equated to mathematics" (quoted in Snelders 1970, 209). Yet, characteristically, the Romantic view of mathematics heavily emphasized a mysticism of numbers (Huch 1920, 2: 79–85).

The naturphilosophische biologist Carl Friedrich Kielmeyer proposed the law of embryological recapitulation, according to which an embryo, as it develops, passes through stages resembling a developmental progression of species (from relative primitive ones to successively more advanced ones in the hierarchy of life). To Romantic eyes, this law was immediately plausible because it reaffirmed that all life came from a common source, and it furthermore alluded to a certain microcosm-macrocosm analogy. The physiologist Johannes Müller is often counted among the Romantic scientists, but his exact relationship with Naturphilosophie is controversial. (See Lenoir 1980.) Some of his students, like du Bois-Reymond and Helmholtz, formed the influential school of biophysics that domi-

nated the later part of the century. Although these scientists explicitly rejected Naturphilosophie, the program of the German bio-
physicists bore a strong resemblance to its ancestor (Culotta 1974).
Moreover, as Steven Turner (1977) argued, Helmholtz's empiricist
psychology still betrayed an intellectual debt to Fichte.

Johann Wilhelm Ritter, the foremost physicist of the Romantic
circle, conducted path-breaking experiments about galvanism, or
"animal electricity," and about the relationship between chemical
reactions and electricity. He had a clear interest in experimentation,
but he was also given to speculation. When he learned of Herschel's
discovery of infrared light, he concluded that, by virtue of analogy,
there must also be invisible rays on the other side of the spectrum.
Thus he discovered ultraviolet light. He also believed that chemical
forces, electricity, magnetism, and heat were variations of one basic
force—light (Snelders 1970). After Ritter moved from Jena to
Munich in 1805, he risked his scientific reputation on experiments
and séances involving divining rods and pendulums, which fewer
and fewer scientists would take seriously. For Ritter and his friends,
human divining faculties opened an exciting avenue toward the ultimate Romantic goal of a communion between human beings and
nature. Here was the magic that foreboded a re-enchanted universe.
In true Romantic fashion, Ritter followed his genius without any
philistine regard for a secure career, material rewards, or social
respectability. The victim of a bad marriage and hard liquor, he died
young and poor (Huch 1920, 2: 151, 291–92).

Hans Christian Oersted discovered the link between electricity
and magnetism in 1820. For major parts of his career, the Romantic
preoccupation with finding a unified Weltanschauung had oriented
his research program toward the connection between the seemingly
distinct forces of electricity and magnetism (Gower 1973). Julius
Robert Mayer, whom Helmholtz considered a speculative theoreti-
cian, not a rigorous scientist, was one of the co-discoverers of what
we now call the first law of thermodynamics. His work was influenced by Naturphilosophie, or, as Ostwald called it, by an "atavistic
trait of a religious belief in miracles" (Ostwald 1910, 87, 73). In
Mayer's case, the central plank of the naturphilosophische creed that

all natural forces were merely different manifestations of the same thing again guided a major discovery.[18]

Romantic science was a "strange but stimulating mixture of fact and fantasy" (Wetzels 1971, 52). The Romantic thinkers relished wild and fantastic ideas and often disdained careful empirical work as pedestrian. They produced a lot of pompous and opaque nonsense that drew the ridicule of the more empirically oriented scientists of the following generation. But they also produced some first-rate scientific breakthroughs.

By the 1820s and 1830s, Naturphilosophie was losing ground, and by mid-century, empirical scientists had firmly established themselves in the universities. As the Romantic scientists faded away, their attempted synthesis of science, poetry, and philosophy gave way to a more sober differentiation. Yet there was no radical break. A closer examination reveals more continuities, continuities that are easily overlooked amid the more noticeable antagonistic battle cries. The harsh rejections of Naturphilosophie by Liebig and du Bois-Reymond are well known. The chemical pioneer Justus Liebig's bitter experiences with Naturphilosophie led him to his much-quoted indictment of this philosophy in the *Annalen der Chemie* of 1840: The activities of the Naturphilosophen were "the pestilence, the black death of the century" (quoted in Paul 1984, 8). As a student at the University of Erlangen, Liebig was initially attracted to the lectures of the dean of Naturphilosophie, Schelling himself, yet, after a couple of years, his scientific mind was turned off by the Naturphilosoph's fantastic flights of speculation. His denunciation of Naturphilosophie, as expressed about twenty years later, could have hardly been more hostile:

> I did part of my studies at a university where the greatest philosopher and metaphysicist of the century seduced the studying youth into admiration and imitation. Who could at that time protect himself from being contaminated? I too have lived through that period, which was so rich in words and ideas and so poor in true knowledge and solid studies; it cost me two precious years of my life. I cannot describe the horror and the disgust when I awoke from this dream and regained consciousness. How many of the

brightest and most talented did I see perish in this swindle; how many laments about totally failed lives did I have to listen to later! (Kohut 1905, 17)

Emil du Bois-Reymond's (1912, 2: 495) condemnation of "the mental illness of the false Naturphilosophie" was equally blatant. Alexander von Humboldt ironically spoke of a "bal en masque der tollsten Naturphilosophen" (Bruhns 1872, 1: 230). That scholar, who was a major figure in the transition from Naturphilosophie to a more empirical science, ardently rejected the speculative approach of Naturphilosophie and complained about "[a] chemistry in which you didn't get your hands wet. The diamond is a pebble that has achieved consciousness. Granite is ether."[19] The ranks of the opponents of Naturphilosophie also included people like the mathematician Carl Friedrich Gauss, the physicist Christian Pfaff, and the chemist Leopold Gmelin (Snelders 1970, 214).

Dieter Wittich further argued that this antipathy led the non-Romantic scientists to spurn the project of developing the synthetic totality of a scientific Weltbild altogether and, instead, to direct their energies toward concrete, empirical research: "As to them, one could not at all speak of any somehow comprehensive materialistic Weltbild" (1971, xiii). However, it would be an oversimplification to characterize these early natural scientists as pure empiricists with no interest in larger philosophical questions. A detailed study of early nineteenth-century physics textbooks showed that they did include discussions of the scientific method, epistemology, and philosophical worldviews (Jungnickel and McCormmach 1986, 1: 23–33). As Christa Jungnickel and Russell McCormmach pointed out, these textbooks distanced themselves from Naturphilosophie, not from philosophical questions generally: "Except for matters of emphasis, the physicists' quarrel was not with philosophy as a whole but only with a part of it, Naturphilosophie, which they often did not even dignify with the name of philosophy" (ibid., 1: 27). Furthermore, other thorough and careful investigations by historians have revealed a measure of compatibility between Romanticism and empirical science.[20] The antagonists quarreled about what divided them, and they had little cause to discuss the fundamental assumptions they all had in common. What the Romantic scientists shared with the later cohorts was, first of all, the goal of unity, of a unified Weltanschauung, or at least Weltbild, as our look at post-Romantic science will show. (See Culotta 1974, 4–5.) This goal, in its most general form, is not a Romantic invention; it is a recurring dream of humankind, going back at least to the ancient Greeks (Berlin 1979, 1992). But the Romantic era, that dream powerfully asserted itself. And it was this dream that inspired Albert Einstein.

The work of the Büchner brothers represented two different rejections of Naturphilosophie, both of which at the same time still clung to its original goal of a comprehensive Weltanschauung. Ludwig Büchner (whom we shall discuss in greater detail) was one of the exponents of a radically materialistic Weltanschauung, a project that carried on the original Romantic intention of the grand synthesis, albeit with an entirely antithetical content. As Frederick Gregory observed, "The protestations against Naturphilosophie had a very hollow ring about them, for Büchner's estimation of Darwin, determined as it was by his belief in the progress of history, was born from his unconscious sympathy for the best of German Romanticism" (1977, 188). Ludwig's brother Georg was a poet (who also held a doctorate in comparative anatomy). Although Georg's work was virtually ignored in the nineteenth century, he achieved considerable popularity in the twentieth. John Reddick's (1990, 1994) analysis of this enigmatic writer suggested that a vision of wholeness shone through his fragmented work. Yet for Georg Büchner, the goal of wholeness was unattainable; his was a wholeness that had been lost or never achieved. His work thus reflected the severe distress caused by the absence of the wholeness deeply desired and by the wounds inflicted by the elusiveness of the quest for a Weltanschauung.

Scientific Weltanschauung

The search for a scientific Weltanschauung is one variant of the search for a Weltanschauung, which we discussed earlier in more general terms. This pursuit was particularly attractive to the philosophically inclined, whereas scientists typically emphasized the more modest goal of a scientific Weltbild. The quest for a scientific Weltanschauung, I suggest, was eventually doomed, just as the quest for a philosophical Weltanschauung was, and essentially for the same reason: hubris-induced internal collapse (see chapter 3); but the quest for a scientific Weltbild proved more worthwhile and fruitful. The latter legitimized research into the deep questions of science, and it gave the scientific community the opportunity to create an amazing flowering of basic science.

We have already encountered one major expression of a scientific Weltanschauung: cosmotheism, or service of the Goddess Isis, as the early assemblies of the German scientists metaphorically put it. Another more philosophical but related root of the scientific Weltanschauung was the philosophical materialism that descended from idealist German philosophy—both Feuerbach and Marx were students of Hegel's. From this philosophical source sprang Marxism, which became the most popular and influential version of a self-proclaimed scientific Weltanschauung, with a strong emphasis on politics. It seems unnecessary, in the framework of this study, to elaborate this well-known Weltanschauung further. Around the same time when the Marxist Weltanschauung took shape, there were similar, albeit much less successful, attempts (which placed their focus more narrowly on science and less on economics and politics). We shall have a closer look at one of them now.

Ludwig Büchner. Though Ludwig Büchner's influence paled in comparison to that of Marxism, he was a widely popular writer. Most importantly for us, a direct line leads from him to Albert Einstein, who as a young teenager was exposed to Büchner's ideas of a scientific Weltanschauung and was greatly impressed by them. As already noted, he later recalled that he, "at about thirteen years of age, read Büchner's 'Kraft und Stoff' with enthusiasm" (Seelig 1954, 14).

The book *Kraft und Stoff*, Büchner's major work, which he wrote in the middle of the nineteenth century, propagated the "naturalistische" and the "materialistische" Weltanschauung: "The current official Wissenschaft and Weltanschauung—supported by the old forces of habit, tradition, ignorance, lethargy, and domination—

may still remain in control for a while. Yet, by necessity, the time will come when they will have to undergo a thorough transformation toward freedom, positivism, and sound natural truth. This will also be the day that will bring not only intellectual and moral, but also political and social liberation to humanity!" (Büchner 1867, 263, 264, xc).[21] Originating in science, Büchner's scheme also encompassed moral, political, and social spheres; it was a Weltanschauung, not merely a Weltbild.

Büchner's scientific Weltanschauung called for the full application of the findings of the natural sciences. The main tenet was "that macroscopic and microscopic existence in all aspects of its creation, life, and demise obeys only mechanical laws or laws that lie in the things themselves" (1872, xiv). Büchner drew his optimism about the ascendancy of this Weltanschauung from its foundation in reality: "The ultimate victory of this real-philosophical knowledge over its opponents appears to us to be beyond doubt. The force of its proofs consists of facts, not of unintelligible and meaningless phrases" (ibid.). Büchner was against abstract speculation and in favor of empirical research. His main targets were idealistic philosophy and theology, and he particularly reviled the German brand of speculative philosophy for using opaque and incomprehensible language in an attempt to affect the false aura of being "deep."

To introduce his chapter "Kraft und Stoff," Büchner quoted the noted scientist Emil du Bois-Reymond (see below): "If one goes to the bottom of things, one soon realizes that neither forces nor matter exist. Both are abstractions, from different perspectives, of things how they are. They complement each other and presuppose each other" (1856, 1). This became the centerpiece and motto of Büchner's Weltanschauung: "No force without matter—no matter without force!" (ibid., 2). Büchner emphasized that matter was the ultimate source of all being and that nature works according to eternal and universal laws—there are no miracles. He explicitly placed this naturalistic Weltbild in a genealogical line that included Alexander von Humboldt, about whom we will also hear more (ibid., 34).

Although Büchner explicitly disavowed any claim of presenting an exhaustive, systematic Weltanschauung, *Kraft und Stoff* covered many important areas that together formed a fairly comprehensive materialist philosophy. For instance, he supported the concept of natural evolution and the sensualist theory of thinking, and he rejected teleological causation in nature, the concepts of life force and free will, and the ideas of God and of a personal afterlife. In Büchner's work, the term *Weltanschauung* had shed all its connotations of idealistic philosophy or of the Romantic emphasis on the spirit—in fact, was openly hostile to such elements—and aligned itself with materialist philosophy. This scientific reintepretation of Weltanschauung was not without adherents in the scientific community. At the 1865 meeting of the German association of scientists and physicians, for instance, the biologist Emil Adolf Rossmässler called for a "natürliche Weltanschauung" on the basis of science to replace the dominant supernatural Weltanschauung (Schipperges 1976, 41).[22]

Büchner's contemporaries Jakob Moleschott and Karl Vogt promulgated a scientific Weltanschauung similar to his. From a Marxist perspective, they all were denounced as exponents of "Vulgärmaterialismus" (vulgar materialism), a Weltanschauung contrasted with the supposedly more advanced dialectic materialism that around the same time was developed by Marx and Engels (Wittich 1971). Vulgärmaterialismus, in this view, was the ideology of a disgruntled and declining Kleinbürgertum, which was defeated in the revolution of 1848 and was going to dissolve in the inexorable progress of societal development. Because of this base in a vanishing class, Büchner's brand of materialism was deemed unable to arrive at the fully correct Weltanschauung that, according to Marxist doctrine, could only arise from a proletarian base in the form of Marxism, or the later Marxism-Leninism (ibid., lxviii). This view of the social base of vulgar materialism is questionable if one looks at the social origins of its main creators. Vogt came from a scholarly family—his father was a professor of medicine—and he himself was a professor of zoology (ibid., xv). Moleschott was the son of a physician, and became a professor of physiology (ibid., xxii). A physician's son, Büchner also became a physician. This is hard-core Bildungsbürgertum (medical wing), not Kleinbürgertum. It should not be entirely surprising that a scientific concern with biology triggered some materialistic conclusions. Furthermore, many of these ideas (in moderated form) were promulgated by people at the top of the academic establishment (Virchow, du Bois-Reymond, et al.).

If one asks from which social strata the mass following of Vulgär-materialismus came, the Marxist analysis again seems dubious. Smitten with the idea that capitalist society would polarize into the two antagonistic classes of capitalists and proletarians, Marxists tended to overlook some actually rising intermediate strata—among them, importantly, engineers, technicians, and other workers in the science and technology fields. Many in that constituency, whose societal position was built on science and technology, identified with the optimistic and forward-looking message of Vulgärmaterialismus. The Einstein family, as we shall see, was a case in point. They sympathized with this Weltanschauung, and young Albert was indeed exposed to Büchner and other authors of that ilk at a young age.

Einstein's rejection of religious dogmatism was couched in similar terms as Büchner's. In 1630, Galileo Galilei's (1967 [1630]) classic *Dialogue Concerning the Two Chief World Systems* ("sistemi del mondo") had contrasted the Ptolemaic and Copernican theories of astronomy, or, one might say, astronomical Weltbilder. Einstein's foreword to an English edition of the *Dialogue* emphasized the wide-reaching implications of Galileo's argument for the new world system: "In advocating and fighting for the Copernican theory Galileo was not only motivated by a striving to simplify the representation of celestial motions. His aim was to substitute for a petrified and barren system of ideas the unbiased and strenuous quest for a deeper and more consistent comprehension of the physical and astronomical facts. . . . The *leitmotif* which I recognize in Galileo's work is the passionate fight against any kind of dogma based on authority" (in Galilei 1967, xi, xvii). This *leitmotif*, which was, of course, Einstein's own, was also that of Büchner and his comrades: "Perhaps only the sciences will succeed in redeeming humankind from the unnatural shackles of that cold and heartless dogmatism, into which the Christian religion was perverted, and in restoring the correct view of nature to humankind" (Büchner 1872, xxxii).

In the later part of the nineteenth and in the early twentieth centuries, a number of prominent scientists promoted the cause of a comprehensive scientific Weltanschauung.[23] Perhaps the most important institutional base of such a Weltanschauung was the Monistenbund around Haeckel and Ostwald.[24] Another group of scientists who not only subscribed to the goal of a scientific Weltbild but also stubbornly clung to the even more comprehensive one of a scientific Weltanschauung (although, in order to pull this off, the concept received a certain twist) became known under the name "Vienna Circle." That circle of scholars began meeting in the early twentieth century and flourished in the 1920s and early 1930s. Ernst Mach was the group's main intellectual ancestor. Because Einstein drew vital inspirations, in his youth, from Mach's philosophy, and because he later maintained a certain affinity to the scholars of the Vienna Circle, we shall now briefly describe the group's outlook.[25]

Ernst Mach and the Vienna Circle. Mach and the Viennese philosopher-scientists who considered themselves his disciples pursued the quest for the scientific Weltanschauung in the shape of a program for the unity of science, in which the emphasis had shifted from ontology to methodology. One major purpose of the unity program was to establish the same scientific method (based on logic and rigorous standards of empirical evidence) in all fields of scholarship.[26] This program was directed against various attempts of establishing two or more types of Wissenschaft (e.g., by the followers of Windelband, Rickert, and Dilthey; see the previous discussion). It was in accordance with Mach's dictum, "I only seek to adopt in physics a point of view that need not be changed the moment our glance is carried over into the domain of another science; for ultimately, all must form one whole" (quoted in Culotta 1974, 29). A public *Aufruf* (appeal), issued in 1911 or 1912, asked scientists and scholars to join a newly emerging *Gesellschaft für positivistische Philosophie*. The ultimate goal of this initiative was, according to the appeal, "a comprehensive Weltanschauung based on factual material" on which all wissenschaftliche disciplines should cooperate (Holton 1993a, 12–14). An impressive array of scholars signed this Aufruf, among them Ernst Mach, Sigmund Freud—and Albert Einstein.

A second major plank of the unity program was, at the theoretical level, reductionism. The goal was to minimize the number of general scientific laws—a goal with a long tradition in science. Following Descartes, Jean le Rond d'Alembert eloquently described the role of reduction in the scientific comprehension of nature: "Indeed, the more one reduces the number of principles of a science, the more one gives them scope, and since the object of a science is necessarily fixed, the principles applied to that object will be so much the more fertile as they are fewer in number. . . . The universe, if we may be permitted to say so, would only be one fact and one great truth for whoever knew how to embrace it from a single point of view" (1995 [1751], 22, 29). Moritz Schlick, a major figure of the Vienna Circle, shared this general attitude when he wrote, "It is obvious that in the progress of knowledge, the number of concepts necessary for a description of nature will become increasingly reduced; so that what is denoted by the term 'world-picture' will become more and more unified. The world will become a '*Uni*-verse'" (1949, 18). The last phrase poignantly expressed the reductionist drive.

A third major purpose of the unity program was to combat obscurantism wherever it would show itself—be it in astrology, metaphysics, religion, art, or political and social ideologies. Here the Vienna Circle clearly entered the realm of the scientific Weltanschauung. A pamphlet issued in 1929 by the Verein Ernst Mach (the association providing the organizational structure to the Vienna Circle) under the title *Wissenschaftliche Weltauffassung: Der Wiener Kreis* identified metaphysics and religion as the major opponents of the scientific Weltanschauung. "Many assert that metaphysical and theologising thought is again on the increase today, not only in life but also in science. . . . But likewise the opposite spirit of enlightenment and *anti-metaphysical factual research* is growing stronger today, in that it is becoming conscious of its existence and task. . . . In the research work of all branches of empirical science this *spirit of a scientific conception of the world* is alive."[27]

Although the scholars associated with the Vienna Circle obviously addressed philosophical questions (see Gustav Bergmann's [1967] *The Metaphysics of Logical Positivism*), they were often reluctant to identify themselves as philosophers because, in their view, the philosophy of their times was severely tainted by metaphysical and speculative elements. This attitude goes back to their intellectual ancestor Ernst Mach, a professor of philosophy at Vienna University who never tired of proclaiming that he was not really a philosopher but a natural scientist (Moszkowski 1922, 171).[28] Nonetheless, the scholars of the Vienna Circle did tackle metaphysical tasks in the sense of probing the foundations of scientific thought and of the scientific method, which formed the central element of their Weltanschauung. The previously mentioned 1929 pamphlet by the Verein Ernst Mach further proclaimed, "The scientific world conception is characterised not so much by theses of its own, but rather by its basic attitude, its points of view and direction of research. The goal ahead is *unified science.* . . . Everything is accessible to man. . . . The task of philosophical work lies in this clarification of problems and assertions, not in the propounding of special 'philosophical' pronouncements."[29]

Ernst Mach (1923) grounded the scientific Weltbild (the core of the scientific Weltanschauung) pragmatically in the requirements of humans' interaction with the enviroment: It is superior, because it is more efficient than other Weltbilder. "To interact somehow with our environment, we need a Weltbild; and we do Wissenschaft to achieve this in an economical fashion" (Mach 1923, 394). "The goal of the scientific economy is, as much as possible, a complete, comprehensive, unified, and steady Weltbild, which is no longer exposed to significant disturbance by new events—a Weltbild of the highest stability possible" (ibid., 366).

For Mach, the scientific Weltanschauung was not yet complete. Echoing Alexander von Humboldt's view of the scientific world picture as a work in progress and as the ultimate goal of scientific research (see below), Mach stressed,

Science does not claim to be a ready-made Weltanschauung, but it consciously works toward a future Weltanschauung. The highest philosophy of the natural scientist is to tolerate an incomplete Weltanschauung, and to prefer it to a seemingly complete, but insufficient one. . . . We must let reason and experience develop freely in areas where they alone are decisive. Only then will we—

slowly and incrementally, but surely—approach the ideal of a unified Weltanschauung to the benefit of humankind, one hopes. This
is the only ideal consistent with the economy of a healthy mind.
(1889, 437–38)

Moritz Schlick used the term "naturwissenschaftliche[s] Weltbild," although he felt that the term *Weltbild*—alluding to a "picture"
that can be seen—was too concrete to fit the entirely abstract scientific Weltbild (1925, 269). He therefore preferred *Weltbegriff*. For
him, the Weltbild was "only a system of signs . . . , which we assign
to qualities and complexes of qualities, whose totality and interrelationships are the universe" (ibid.). He distinguished three realms:
reality itself, quantitative scientific concepts of reality, which in their
totality form the "physikalischen Weltbegriff" (ibid., 270), and the
representation of the more concrete concepts of the second realm in
human consciousness. The third realm, consisting of psychological
phenomena, is part of the first; as such it can be encompassed by the
physical Weltbegriff. "Physics is the system of exact concepts that our
knowledge assigns to everything real. *Everything* real, because,
according to our hypothesis, the *whole world* can be described by this
system of concepts. Nature is everything; everything real is natural.
Spirit, consciousness, is not in opposition to nature; rather, it is a
segment of the totality of nature" (ibid., 271).

Although the Vienna Circle is commonly identified with hard-
core empiricism, the pages of the journal *Erkenntnis* reveal arguments that are more complex. Kasimir Ajdukiewicz (1934, 1935),
for instance, argued for a "radical conventionalism": Empirical data,
in his view, cannot unambiguously determine the Weltbild because
they do not derive purely from experience. Rather, they depend, to
some degree, on the choice of the conceptual apparatus for representing them. Those who want to evaluate a Weltbild, he contended,
can do so only on the basis of their accepting a conceptual apparatus
themselves. While this appears to lead straight into the relativist
quagmire, Ajdukiewicz devises an evolutionary escape route—based
on the assumption that, in the long run, science is going to get better
and closer to the real world.

The epistemologist therefore is not suited for the role of an impartial judge in the struggle between two world-perspectives over the claim to truth. Hence, he should not barge in to assume this role. Rather, he should set himself another task. He should devote his attention to the changes that actually occur in the conceptual apparatus of Wissenschaft and in the corresponding world-perspectives, and he should seek to determine what the motives are that generate these changes. Perhaps it is possible to conceptualize these changes in world-perspectives as a process that advances toward a final stage, which is coming about as if someone consciously wanted to achieve this final stage through that change in world-perspectives. (Ajdukiewicz 1935, 30)

This application of evolutionary principles to the development of the scientific Weltbild has obvious similarities with Karl Popper's views on the subject.

In accordance with the unity program, Philipp Frank emphasized that the scientific Weltanschauung had no room for insoluble problems: "For a purely scientific Weltauffassung, the notion of an 'eternally' insoluble problem is nonsense. All questions that science can pose in a meaningful way, i.e., as questions about real experiences, are solvable in principle by scientific methods, as especially Ludwig Wittgenstein and Rudolf Carnap have emphasized" (1932, 275).[30] In his *Tractatus logico-philosophicus*, Wittgenstein had earlier reduced the Weltbild to the totality of thoughts that are true—thereby excluding all statements whose truth cannot be scientifically established: "The totality of existing states of affairs is the world. . . . The totality of true thoughts is a picture of the world" (1974 [1921], 8,11). In sum, the Vienna Circle certainly attempted to go beyond the scientific Weltbild and rescue a more comprehensive, scientifically tenable Weltanschauung. This scientific Weltanschauung, however, came at the cost of defining what others might consider substantial parts of human existence away as meaningless.

Scientific Weltbild

Many practicing scientists do not spend a great deal of time thinking explicitly about a scientific Weltbild; they rather think in terms of theories, hypotheses, and data. Typically, even theories are much narrower than the grand synthesis of a Weltbild, let alone Weltanschauung.[31] Nonetheless, some practicing or trained scientists—often the most outstanding scientific leaders—have joined the debate about the most fundamental issues. The majority of these scientists have directed their energies toward a scientific Weltbild, rather than toward a scientific Weltanschauung.[32] That scientific Weltbild has been rooted in the tradition of wissenschaftliche synthesis and unification. The *goal* of the scientific Weltbild thus is very similar to that early Romantic goal of the grand synthesis of the spirit, yet the *means* by which this synthesis is to be achieved are restricted to scientific ones. The scientists entering this debate have typically focused on defining the limits of the scientific Weltbild. This demarcation is one of the two overarching themes that will become apparent in the following discussion of the scientists' efforts. The other theme concerns the struggles within individual scientific disciplines—we will focus on physics—between rival Weltbilder. Importantly, the coexistence of rival theories would be perceived as a state of crisis most urgently against the backdrop of the Weltbild imperative. Those who subscribe to the goal of creating a comprehensive Weltbild would be most bothered by the multiplicity of contradictory approaches.[33]

We will now see how the concept of the scientific Weltbild evolved—the research imperative that Einstein grew up with, made his own, and fulfilled to an extraordinarily high degree. By the middle of the nineteenth century, the notion of the scientific Weltbild had appeared. Alexander von Humboldt was one of its major protagonists.[34] His role in linking the earlier Romantic science to the later epochs of science was crucial indeed.[35] In his 1828 opening address to the assembly of German scientists and physicians, Humboldt emphasized, next to some political rhetoric about German national unity, the unity of the sciences. "In the true and deep sentiment of the unity of nature, the founders of this association have intimately united all branches of physical knowledge (of the descriptive, quantifying, and experimental kinds)" (Pfannenstiel 1958, 68). Scientific synthesis, a unified scientific Weltbild, was Humboldt's goal, not an assortment of specialized findings. It was this goal of a scientific Weltbild that made Humboldt's writings immensely popular among the general public (Waschkies 1990). His *Kosmos* is considered the first scientific bestseller in Germany (Beck in Humboldt 1978, v), and among its many readers was young Albert Einstein (Winteler-Einstein 1986, lxii).

Humboldt corresponded with Goethe (Bratranek 1876) and even dedicated the German edition of his *Ideen zu einer Geographie der Pflanzen* (the original edition was in French) to the poet, who reciprocated by drawing an illustration to this work and dedicating it to Humboldt (Bruhns 1872, 1: 199). Although Humboldt had scientific disagreements with Goethe (especially because Humboldt converted to the "vulcanist" theory of geology, whereas Goethe remained a "neptunist"), he always held Goethe in great esteem.[36] To understand the whole of nature in its living unity was the central goal Humboldt and Goethe held in common. Yet their methods differed. Humboldt, who in his youth was somewhat closer to Naturphilosophie, soon rejected speculation in favor of empirical observation and distanced himself from Goethe's approach to science, which remained closer to Naturphilosophie (Linden 1940). He would not countenance a unity created deductively from a few fundamental principles given by reason; instead, he preferred a more empirical approach.

Humboldt also associated with Schiller, but that poet was less than appreciative of Humboldt's scientific bent. In a 1797 letter to his friend Körner, Schiller denigrated Humboldt for his "bare, trenchant reason, which would have nature (which is always impenetrable and in all its aspects venerable and unfathomable) shamelessly subjected to measurement, and which, with an impudence I do not comprehend, fashions its own formulas, which often are merely empty words and always are merely narrow concepts, into nature's yardstick" (Bruhns 1872, 1: 212–13).

Humboldt's original project was a comprehensive physical geography of the earth, but the inclusion of astronomic material led him to choose the title *Kosmos* for his work (Humboldt 1978, xvii). Although he did, on one occasion, use Weltanschauung in its literal sense (see above), his understanding of Weltanschauung was, on the whole, more modern—for instance, when he wrote of "the history of the physical Weltanschauung as the history of the knowledge of a totality of nature, that is, as the history of the notion of unity behind the phenomena and of the concert of forces in the universe" (Humboldt 1847, 138).[37] Here, Weltanschauung is the knowledge of the totality of nature—what we call Weltbild. It is the grand thought of the basic unity of the diverse elements of the world. The most important mission of research was, according to Humboldt, "to recognize unity in diversity, to encompass all the individual facts that the discoveries of the current ages present to us, to separate and examine the details, but not to succumb to their mass, and, in consideration of humankind's noble destiny, to capture the essence of nature, which lies hidden under the cover of appearances" (1845, 6).

Over time, Humboldt thought, this Weltanschauung would advance and approximate the truth: "The history of the Weltanschauung, as I see it, describes . . . the main stages in a gradual approximation to the truth, to the correct concept of the terrestrial forces and of the planetary system" (1847, 139). The three major driving forces of scientific progress are, according to Humboldt, better theory, better instruments, and special events that broaden the horizon of observation. Whereas the first two are quite obvious, the last may sound somewhat unusual. But if we take into consideration Alexander von Humboldt's background as an explorer and physical geographer, it becomes apparent that this third driving force points to the discovery of unknown lands and to the exploration of unknown phenomena in them. He specifically mentioned migrations, navigation, and war expeditions among those events that broaden the horizon of observation.

For Alexander von Humboldt, the complete scientific Weltanschauung (or Weltbild, in our terminology) was an ideal goal that he considered out of reach for the science of his day—and perhaps a goal that would never be fully achieved. Nonetheless, he regarded it as the purpose of natural science to keep progressing toward it and to do so through the careful study of individual natural phenomena.[38] "My main motivation was the desire to perceive the phenomena of the material world in their general interconnectedness, to perceive nature as a whole that is alive and moved by inner forces.

Through my association with highly talented men, I early on gained the insight that, without the serious quest for knowledge of the details, every big and general Weltanschauung can only consist of thin air" (Humboldt 1845, vi). In sum, Humboldt stuck to the goal of Weltanschauung—or, in our terminology, Weltbild—and he strove for a synthetic view of the totality of nature, but he rejected the speculative systems of the German *Naturphilosophien* of his time. For him, empirical examination must form the foundation of a Weltbild: Science starts with devotion to the diversity of empirical details and progresses from that basis to grasping the unifying principles that underlie that diversity.[39]

In the later part of the nineteenth century, a major effort to demarcate the scientific Weltbild from any more comprehensive Weltanschauung, and even to designate specific areas that will forever be closed to scientific inquiry, was undertaken by Emil du Bois-Reymond (1873 [1872]).[40] He was one of the exponents of what has been called the School of 1847 (Cranefield 1957, 1966), whose members also included Rudolf Virchow, Ernst Brücke, Carl Ludwig, and Hermann Helmholtz (Anderton 1993).[41] The core program uniting these physiologists (and the pathologist Virchow) was "organic physics"—the reduction of physiological processes to physical ones. The main targets of their mechanistic Weltbild were speculative Naturphilosophie and vitalism (the idea of an immaterial life force in living organisms). When du Bois-Reymond met Helmholtz in 1845, he enthusiastically wrote to his friend Eduard Hallmann, "In the meantime, I have made the acquaintance of Helmholtz, and it has been a great pleasure indeed. He is (sauf la modestie) after Brücke and my own insignificance the third organic physicist in our circle. A guy who has gorged himself on chemistry, physics, and mathematics, who entirely shares our viewpoint of Weltanschauung, and who is prolific in thoughts and novel imagination" (Swoboda 1978, xxxiv).[42] By mid-century, thus, du Bois-Reymond and the circle of young organic physicists had also appropriated the term *Weltanschauung* and filled it with contents radically opposed to Romanticism and Naturphilosophie.

After Naturphilosophie declined, the organic physicists came to face new opponents—scientists who now strove for pure empiricism and rejected any theorizing that went beyond observational data. In a 1856 letter to du Bois-Reymond, Ernst Brücke complained that "the closed phalanx of pure morphologists and demure observationalists is still doing its utmost to vilify us" (Swoboda 1978, xxxvii). The organic physicists decidedly did not want to give up the grand synthesis of a Weltbild in favor of an exclusive focus on empirical observation. Helmholtz emphasized in a 1862 speech, "But our knowledge should not remain merely in the form of catalogs. . . . It is not enough to know the facts; science comes into existence only when the law and causes of the facts are revealed" (1884, 1: 129). A little later in this speech, he precisely paraphrased the ideal of the scientific Weltbild when he described the sole mission of science: "Complete knowledge and complete understanding of the action of the natural and intellectual forces is the only goal that Wissenschaft can seek" (ibid., 142).[43]

Doubtlessly such a Weltbild based on mechanistic reductionism was anathema not only to any remaining sympathizers of Naturphilosophie among the scientists but also to large segments of the typically idealistic Kulturträger. Yet, at a more general level, the very quest for a unified Weltbild formed the common bond that tied the seeming antagonists together. As Anne Harrington aptly put it, "At the same time—and in an irony that largely eluded the consciousness of the actors themselves—there is also a sense in which the mechanistic push for 'unification' was driven by some of the broader cultural concerns that historians like Reddick have seen in the original Romantic-era preoccupation with Wholeness" (1996, 10).

Du Bois-Reymond's famous speech at the 1872 meeting of the association of German scientists and physicians, titled "Über die Grenzen des Naturerkennens" (On the Limits of the Knowledge of Nature), posited two such limits.[44] Du Bois-Reymond was not interested in pragmatic restrictions that at some point in the future might be overcome by scientific progress. Rather, he focused on limits in principle—limits that would even apply to someone who possessed Laplace's world formula.[45]

The first limit concerned the fine structure of matter and force. Du Bois-Reymond argued that both the idea that indivisible atoms exist, and the antithetical idea that matter can be divided into smaller parts ad infinitum are problematic:

> For if the inertial and in itself inactive substrate that cannot be divided further is to have real existence, it must fill a certain, if very small, space. Then one cannot understand why it should not be divisible further. . . . Conversely, if one, with the dynamists, regards only the centerpoint of the central forces as substrate, then the substrate no longer fills any space, for the point is the spacially conceptualized negation of space. Then there is nothing from which the central forces originate, and which could be inertial like matter. . . . Nobody who has thought about it a little more deeply misunderstands the transcendental nature of the obstacle that resists us here. No matter how one might try to bypass it, one always encounters it in one or another form. (du Bois-Reymond 1873, 13, 15)

The second limit concerned consciousness: "With the first impulse of pleasure or pain that the simplest organism at the origin of animal life on Earth experienced, this insurmountable cleft came into being; and the world has now become doubly incomprehensible" (ibid., 21). He argued, "As to mental processes themselves, we find that, even under astronomical knowledge [i.e., under perfect Laplacean knowledge] of the organ of the soul, they would be just as incomprehensible to us as they are now. In the possession of this knowledge, we would stand before them just like today—as before something entirely separate. The astronomical knowledge of the brain, the best we can get, reveals nothing more in it than matter in motion. No conceivable pattern or motion of material particles allows us to construct a bridge into the realm of consciousness" (ibid., 27–28). By contrast, du Bois-Reymond considered the limits to understanding organic life to be merely temporary and rejected the vitalists' concept of a special life force.

He finished with a flourish—with a highly evocative phrase for which he should become famous among his scientific contempo-

raries as well as among German scientists of the following generations: "Concerning the riddles of the material world, the scientist
has long been used to proclaiming his '*Ignoramus*' with manly self-
restraint. . . . Concerning the riddle of what matter and force are,
and how they are able to think, he must choose a motto that is
much harder to utter: '*Ignorabimus!*'" (ibid., 38). (The Latin *ignoramus* means "we do not know," whereas *ignorabimus* means "we
shall not know," or, in loose translation, "we shall never know.")

However, juxtaposed to this relatively weak variant of a mechanistic Weltbild that kept all phenomena of human consciousness
beyond its limits, a stronger variant of a mechanistic Weltbild and
even a mechanistic Weltanschauung could also be found among the
same scientists—a Weltanschauung that extended deep into the social
and political realms. Virchow, in his youth, was its most radical proponent. He wrote, for instance, "If medicine is to fulfill her great task,
then she must enter the political and social life" (quoted in Anderton
1993, 114). And "the last task of [medicine] is the consitution of
society upon a physiological basis" (quoted ibid.). For him, political
liberalism flowed naturally from the mechanistic Weltanschauung.
The idea of freedom from dogmatic and irrational restraints—central
to scientific inquiry—applied analogously to society and politics.[46]
After the failure of the German revolution of 1848, however, the
members of the School of 1847 on the whole seemed to have become
somewhat more moderate and sedate; and they drifted in the direction of du Bois-Reymond's "weak" scientific Weltbild.[47]

Du Bois-Reymond's *Ignorabimus* became a major irritant and target
for scientists and philosophers who did not accept absolute limits for
science and insisted on a complete scientific Weltbild, if not Weltanschauung. Forty years after du Bois-Reymond gave his speech, for
instance, an English-language report in the "Notes and News" section
of the *Journal of Philosophy, Psychology and Scientific Methods* (9 [1912]:
420) on the previously mentioned "Aufruf" for the formation of the
Gesellschaft für positivistische Philosophie explicitly chided du Bois-
Reymond's position, "The present day desires the solution of general
problems, which research itself throws up, and is not to be put off with
an *Ignorabimus* for which there is no evidence." Almost another forty

years later, Philipp Frank still chafed under du Bois-Reymond's slogan: "And this word 'ignorabimus' became the motto for a whole period, the motto of defeatism in science, the motto that was the delight of all antiscience tendencies of that era" (1949a, 81–82).

By the turn of the century, the coupling of Weltbild and science was well entrenched. Early in the 1900s (between 1904 and 1912), John Theodore Merz (1965) published his monumental *History of European Thought in the Nineteenth Century*. The first two volumes of this four-volume work focused on scientific thought. There Merz distinguished several "views of nature"—which are, in effect, scientific Weltbilder. He thought that the development of these comprehensive views of nature indicated that the nineteenth century had achieved a new stage of intellectual progress: "Two processes have helped to determine the intellectual progress of mankind. These two processes have often been apparently opposed to each other in their operations; but in reality neither of them can proceed very far without calling the other into existence. They are extension and condensation of knowledge"(Merz 1965 [1904], 1: 27). According to Merz, nineteenth-century thought in England, France, and Germany accomplished an improved method of extending knowledge and "a peculiar conception of its possible unity. . . . [O]ur age was elaborating a deeper and more significant conception of this unity of all human interest, of the inner mental life of man and mankind" (ibid., 29, 33). Merz's list of views of nature comprised the astronomical, atomic, kinetic or mechanical, physical, morphological, genetic, vitalistic, psychophysical, and statistical views. In his preface, Merz acknowledged an intellectual debt to Ernst Curtius, a noted German archaeologist and historian (ibid., vii). Perhaps this indicates one of the channels through which the German predilection with Weltanschauung and Weltbild passed into the English-language history of ideas.

The Weltbild issue was particularly dear to Max Planck.[48] In a famous speech of 1908, titled "Die Einheit des physikalischen Weltbildes" (The Unity of the Physical Weltbild), Planck enthusiastically subscribed to the goal of scientific unification: "As long as science has existed, it has had this ultimate, highest goal in mind—to subsume the variegated diversity of the physical appearances into a unified system, if possible into a single formula" (1970d, 28). In "The Scientist's Picture of the Physical Universe," Planck specified the task of the physicist: "Taking it, then, that the external world of reality is governed by a system of laws, the physicist now constructs a synthesis of concepts and theorems; and this synthesis is called the scientific picture of the physical universe. It is a representation of the real world itself in so far as it corresponds as closely as possible to the information which the research measurements have supplied" (1932, 85).

Planck also addressed the issue of Weltanschauung, and it is evident that he distinguished the terms Weltanschauung and Weltbild in a way very similar to ours. In "Die Physik im Kampf um die Weltanschauung" (Physics in its Struggle for Weltanschauung) of 1935, Planck acknowledged that the Weltanschauung of the researchers influenced their scientific work and vice versa (1970a, 285). However, Planck emphasized, science does not provide a complete Weltanschauung: It cannot give an answer to the ethical question, how should I act? (ibid., 297). It is thus mainly limited to what we call Weltbild, but it does include some ethical principles, such as truthfulness, honesty, and justice. Planck also spoke of the national roots of science: "Hence every science, just like every art and every religion, has grown on national soil. The fact that this could be forgotten for a while has come back to haunt our people bitterly enough" (ibid., 298). He added, however, "A science that is not able or willing to have an impact beyond the own nation does not deserve its name" (ibid.).

The Physical Weltbild at the Turn of the Century

When Max Planck was a beginning student at Munich University in 1875, physics professor Philipp von Jolly tried to dissuade him from pursuing that discipline, because, Jolly said, nothing essentially new remained to be discovered in it (Badash 1972, 55; Gillispie 1970–80). The professor did not stand alone with his opinion in the physics community of the late nineteenth century. Quite a few physicists of that era shared the view of physics as a closed science that had already attained its mature state in which all its principal laws were known (Badash 1972).[49] For instance, Albert Michelson, the great American experimental physicist, stated in 1894, "it seems probable that most of the grand underlying principles have been firmly established and that further advances are to be sought chiefly in the rigorous application of these principles to all the phenomena which come under our notice.... An eminent physicist has remarked that the future truths of Physical Science are to be looked for in the sixth place of decimals" (quoted ibid., 52).[50] Rather ironically, this pronouncement adumbrated a decade (between 1895 and 1905) in which a breathtaking bumper crop of spectacular advances in physics occurred: x-rays, the electron, radioactivity, the quantum, and finally, relativity.

To be sure, the belief in a completed theoretical edifice of physics, expressed by Jolly, Michelson, and others, was not the predominant, let alone consensus, view in the late nineteenth-century physics community (Kragh 1999). Even before the string of new discoveries made obsolete any talk of physics as a complete science, a number of physicists began to discover and scrutinize anomalies and open questions within the dominant theory of Newtonian mechanics.[51] Among the German-speaking physicists, the discussion about these open issues was increasingly conceptualized and framed as a debate about the Weltbild of physics.

We now briefly survey the state of physics around the turn of the century—the state of physics to which Einstein was exposed during his training, and which he was going to alter irretrievably through his own contributions. We address two aspects: the conflict between rival physical Weltbilder (primarily between the mechanical and the electromagnetic Weltbilder), which, as we shall see, crucially influenced Einstein's physical research program, and the debates about some fundamental issues implied by the physical Weltbilder (e.g., the nature of causality, determinism vs. probabilism, and the status of scientific theories), which defined the intellectual landscape in which Einstein's views about those topics evolved.

Mechanical Weltbild vs. electromagnetic Weltbild. The mechanical Weltbild, which ultimately rested on Newton's grand principles, was the orthodoxy that was either received as the final word on physics or assailed by alternate approaches.[52] As its principal challenger rose the electromagnetic Weltbild. Before we turn to that most serious rival, we mention, at least in passing, some more obscure Weltbilder. There was, for instance, the hydrodynamic Weltbild. Championed by the Norwegian physicist Carl Bjerknes and the German Arthur Korn, it tried to explain electromagnetism as well as gravity in a hydrodynamic way, based on spheres moving in an infinite, incompressible, and frictionless fluid. Bjerknes posited that two harmoniously pulsating balls moving through that frictionless fluid would attract or repel each other as if they were electrically charged. He never succeeded in his goal of developing a comprehensive theory that would ground Maxwell's general theory of electrodynamic phenomena in his hydrodynamic model.

Related to the hydrodynamic Weltbild were the vortex atomic theory and the ether squirt theory. The vortex atomic theory, which was developed by William Thomson (Lord Kelvin) and had a following among British physicists, based the whole Weltbild on the ether. Atoms were vortical modes of motion of the ether, which was conceptualized as a perfect fluid. In the ether squirt theory devised by Karl Pearson, the atom was regarded as a point in the ether from which new ether continuously flowed in all directions. Interestingly, Pearson's theory also contained the ideas of a hyperspace and of antimatter (ether sinks). The ether in these conceptions still had mechanical properties, and Bjerknes's, Thomson's, and Pearson's theories could be subsumed under the mechanical Weltbild, broadly understood. Hydrodynamics was seen as the application of mechanical principles to fluids and bodies in fluids, not as something qualitatively different.

A more ambitious challenger was the thermodynamic Weltbild founded on heat and energy as its primary concepts. A major issue of contention between the mechanical Weltbild and the thermodynamic Weltbild was the second law of thermodynamics. The mechanical laws of motion can be thought of as reversible in time, but the second law posits an irreversible trend in the direction of entropy. Whereas Ludwig Boltzmann devised a statistical-mechanical theory to explain the second law within the mechanical Weltbild, the exponents of the thermodynamic Weltbild, notably Georg Helm and Wilhelm Ostwald, rejected Boltzmann's work and took the opposite tack.[53] Their program was to reduce mechanical laws to energy laws and thus to make thermodynamics the base of a comprehensive physical Weltbild they called *energetics*. At the 1895 meeting of the association of German scientists and physicians in Lübeck, Ostwald lectured on "Die Überwindung des wissenschaftlichen Materialismus" (Overcoming wissenschaftlichen Materialism), claiming that energy had displaced matter as the basic concept, and Helm also gave a report on the state of energetics. Many distinguished scientists immediately challenged the views of the two energeticists in the ensuing controversial debate (Jungnickel and McCormmach 1986, 2: 220–22). Soon, the thermodynamic Weltbild was eclipsed by the electromagnetic Weltbild as the most serious threat to the mechanical orthodoxy.[54]

In the electromagnetic Weltbild, mechanical phenomena were subordinate to the electromagnetic field, or to the electromagnetic ether, which played a central role in this Weltbild as the medium in which the electromagnetic waves would propagate. The electromagnetic Weltbild aimed at the complete reduction of mechanics to electromagnetism: All laws of nature would be reducible to properties of the ether as defined by electromagnetic field equations. The electron, for example, was considered a property of the ether, and thus the mass of the electron was viewed to be of electrodynamic origin. In extreme versions of the electromagnetic Weltbild, all matter was an epiphenomenon of electromagnetism.

An early proponent of the electromagnetic Weltbild was Emil Wiechert, who in 1894 argued for the primacy of the electromagnetic ether from which electric particles—and all matter—were built up (McCormmach 1970a, 460). Wilhelm Wien (1901 [1900]) explicitly stated the program of providing an electromagnetic foundation to mechanics: the mechanical laws were to be deduced from electromagnetic laws. Max Abraham, Adolf Bucherer, leading experimentalist Wilhelm Kaufmann, Paul Lagevin, and Henrik Lorentz were among the major proponents of this view, of which they offered different versions.[55] For instance, whereas Abraham thought of electrons as rigid spheres, Lorentz believed that electrons were deformable: they would contract in the direction of their motion. By 1904, Lorentz's theory was widely recognized as the leading representation of the electromagnetic Weltbild, although Lorentz himself was very guarded about making far-reaching claims (McCormmach 1970a, 483–84).[56]

The clash of the up-and-coming electromagnetic Weltbild with the traditional mechanical Weltbild greatly agitated the physics community Einstein entered. The turn-of-the-century physicists' preoccupation with achieving a comprehensive physical Weltbild was, of course, not an entirely singular event. At all times and in all cultures, some thinkers have been fascinated by the intellectual challenge of finding a comprehensive "theory of everything"—a theory that in a few elegant strokes would explain the universe. Representatives of all countries in the international physics community participated in the debates about physical theory around 1900. Yet the German physicists collectively exhibited a particularly strong urge to generalize theories into a grand synthesis, a complete Weltbild. For instance, as Helge Kragh pointed out, Rutherford's belief that all matter was electrical in nature "does not mean, however, that Rutherford and his colleagues were adherents of the electromagnetic world view or electron theory in the Continental sense. It is no accident that most of the contributors to electron theory were Germans and very few, if any, were British. There were subtle differences between the British and the German attitudes. It was one thing to believe in the electric theory of matter, as many British physicists did, but another to eliminate all mechanical concepts and laws in favor of electromagnetic ones such as those Abraham and his allies aimed at" (1999, 116).

It may illustrate the particularly German affinity to the notion of the Weltbild that the American Carl Snyder's (1903) *New Conceptions in Science* appeared, in its German translation, as *Das Weltbild der modernen Naturwissenschaft* (1905). In its substance, Snyder's work was about the scientific Weltbild—indeed a scientific Weltanschauung in its more comprehensive scope. He asserted that science would fully explain the natural universe and that it was going to regulate all facets of human life: "Physical science will not stop short of a reduction of the universe and all it contains to the basis of mechanics; in more concrete terms, to the working of a machine. . . . [The scientific methods] will make possible the scientific organization of industry, of politics, of morals—in brief, of the whole scheme of our daily lives" (Snyder 1903, 32–33). In the battle of rival physical Weltbilder, the American came out in favor of the orthodox Weltbild based on mechanics, which, in his opinion, was able incorporate electromagnetism. For Snyder, "mechanics is the foundation upon which the whole superstructure of science rests" (ibid., 31). Thus, he believed, physics could be reduced to mechanics. "And, finally, the whole domain of physics—electricity, heat, light, magnetism—is coming under the bondage of one or two mechanical conceptions, a mechanics of the ether, or a mechanics of ultimate particles, the corpuscles, which are both matter and electricity at once. Perhaps eventually it will come to a compromise between the two. Either way, mechanics reigns" (ibid., 32).[57]

Einstein's own desire of creating a comprehensive Weltbild thus overlapped with the ambitions of many of the leading physicists—he joined a group of like-minded seekers. It seems likely that his ideas were given much freer play in the physics community around the turn of the century than they would have been given in the more compartmentalized settings of later periods. In an autobiographical account, Einstein (1979a [1949]) explained how his own research was motivated by his growing dissatisfaction with both the orthodox mechanical Weltbild and its electromagnetic rival.[58] He wrote that, around 1900, he realized that both these Weltbilder were flawed and that he became convinced that only the discovery of a "allgemeinen formalen Prinzips [universal formal principle]" would lead to valid results (ibid., 20). In a 1907 paper in the *Annalen der Physik*, titled "On the Inertia of Energy Required by the Relativity Principle," he noted that "our current electromechanical [should probably read: electromagnetic] Weltbild is unable to explain the entropy characteristics of radiation, the laws of the emission and absorption of radiation, and those of specific heat" (*CP*, 2: 415).[59] In the same paper, he also also said that "a complete Weltbild that corresponds to the relativity principle," though it did not yet exist, would allow for a general solution of the issues under consideration (ibid., 2: 414–15).[60] A subsequent chapter will explore in detail how the juxtaposition of rival physical Weltbilder inspired Einstein's research program, which was devoted, from early on, to a comprehensive synthesis.

The Planck-Mach controversy. Closely connected to the struggle between rival physical Weltbilder were debates about fundamental epistemological and methodological questions (the status of scientific theory, causality, determinism, etc.).[61] A brief but poignant controversy at the beginning of the twentieth century pitted two of the foremost scientists of the era against each other, Max Planck and Ernst Mach.[62] It will be instructive for us to retrace that argument because it foreshadowed the issues that Einstein tackled and that led him from being a Machian to a different viewpoint, which was more in line with Planck's.

Planck, as we have seen, wholeheartedly subscribed to the goal of a complete synthesis, of a comprehensive scientific Weltbild. He thought this goal could be achieved only by deviating from the methods Mach (and later the Vienna Circle) advocated. In the aforementioned 1908 speech "The Unity of the Physical Weltbild," he noted, "the characteristic feature of the whole development of theoretical physics up to now has been a unification of its system, which has been achieved by a certain emancipation from the anthrophomorphic elements, especially the specific sense perceptions" (Planck 1970d, 31).

Criticizing Mach's sensualism and his emphasis on the economic functions of Weltbilder, Planck proposed the following goal for the scientific Weltbild: "[T]his goal is—not the complete adaptation of our thoughts to our sense perceptions, but—the complete separation of the physical Weltbild from the individuality of the creative mind" (ibid., 47, 49).[63] In taking this stance, Plack reasserted unity as the supreme aspiration: "Which is the particular element that, despite these obvious disadvantages [i.e., loss of immediate grounding in sense experiences], has given the future Weltbild such a decisive advantage that it can prevail over all former Weltbilder?—

It is nothing else but the unity of the Weltbild. The unity in respect to all individual features of the Weltbild, the unity in respect to all places and times, the unity in respect to all researchers, all nations, all cultures" (ibid., 45).

In his rejoinder, Mach again characterized the activity of the researcher as an economic one: "The farther and the more thoroughly one analyzes the scientific methods, the systematic, classifying, simplifying, logical-mathematical structure, the more one recognizes scientific activity as economic" (1910, 600). Mach made the Darwinian principle of the "survival of the fittest" the foundation of his evolutionary concept of knowledge. For Mach, the Weltbild reflects both nature and the subjectivity of the researcher, but the influence of subjectivity is being reduced in the course of scientific progress: "Of course, the human, socially maintained Weltbild becomes increasingly independent of idiosyncrasies in the succession of researchers, and it thus becomes an increasingly pure expression of the facts. In general, however, both the environment and the observer express themselves in every observation and in every view" (ibid., 602). Mach further identified the "belief in the reality of atoms" as a fundamental difference between Planck and himself (ibid., 603).

Earlier, Einstein's colleague Friedrich W. Adler (who in 1916 would assassinate the Austrian prime minister) had come to Mach's defense. In his detailed analysis of the "anthropomorphic elements" from which, according to Planck, physics needed to emancipate itself, Adler (1909) distinguished five subcategories of anthropomorphic elements and argued that, on four of these subcategories, Mach actually agreed with Planck's goal of distancing physics from anthropomorphic aspects. It was the remaining subcategory—sensory perceptions—where Planck and Mach differed fundamentally: "The 'physical Weltbild' does not consist, as Planck portrays it to us, in an acoustics without sounds, in a thermodynamics without heat, in an optics without colors, but it is, according to Mach's view, much wider and more comprehensive. It unites the facts of all sense areas; it does not restrict itself to the isolated picture of spatial sensation, but it presents the comprehensive picture of total experience" (Adler 1909, 821–22). "Planck's 'Weltbild' is not the picture of the world;

neither is it the picture of the physicist's discipline, but only a segment of it. If we regard it as the picture of the world, then this discipline turns into a nursery of metaphysics" (ibid., 822).

In his reply to Mach, Planck (1910) asserted that Mach's natural philosophy fell short of fulfilling its own promise of eliminating all metaphysical elements from scientific knowledge. Planck pointed out that Mach, on the one hand, had posited that scientific knowledge had biological-economic roots—it helps humans survive—but that, on the other hand, he had also stated that the economy of thinking ("*Denkökonomie*") was not tied to practical economic needs. This surreptitious shift in the concept of economy, Planck concluded, meant that, contrary to its own claim, Mach's theory retained a metaphysical element. According to Planck, economy cannot be the only guiding principle of science.

> The principle of economy, even in its most general form, is useless for giving direction to physical research—for the simple, known reason that one can never know in advance from which standpoint the economy is preserved in the best and most permanent way. Therefore, if the physicist wants to advance his science, he must be a realist, not an economist, that is, he must search, among changing appearances, for the permanent and everlasting, for that which is independent of human senses, and emphasize it. In this, the economy of thinking serves as a means, not as an end. This has always been so, and it will remain the case, despite E. Mach and his supposed antimetaphysics. (Planck 1910, 1190)

Einstein's pivotal speech on "Motiv des Forschens," which he delivered in 1918 on occasion of Planck's sixtieth birthday, showed how far he had moved from an earlier Machian stance toward a Planckian viewpoint. This speech will be discussed in detail later. Yet we might add here that, in a letter to Paul Ehrenfest of the same year (September 4, 1918), Einstein demonstrated that he had not entirely lost all Machian instincts: "Planck does not let go of his metaphysical concept of probability. When I look at minds of his kind, an irrational rest remains that I cannot assimilate (then I always have to think of Fichte, Hegel, etc.)" (*CP*, 8: 865).

Later German Physicists

We now briefly touch upon how leading German physicists kept the discussion of the scientific Weltbild, and particularly the physical Weltbild, alive in the twentieth century. Einstein's theory of relativity had achieved only a partial unification, and a new major rift appeared between the physical Weltbild based on his theory and the quantum-mechanical Weltbild.[64] The following will demonstrate that the Weltbild issue remained a major preoccupation for some of the most distinguished physicists. Even while those scientists disagreed over certain elements of the physical Weltbild, they all agreed that such a Weltbild was the noblest and most worthwhile goal of physics, and this shared understanding provided the very basis for their discourse. Again, we will find the two familiar themes in their ponderings about the Weltbild: the delimitation of the scientific Weltbild vis-à-vis areas that must remain beyond its grasp and the reconciliation of rival Weltbilder. The first theme focuses increasingly on the limitations of scientific objectivity, the second, and related, one on relativity versus quantum mechanics. We start with one of Einstein's contemporaries, distinguished physicist Max von Laue, whom Einstein held in high esteem.

In a 1921 speech called "Das physikalische Weltbild," Laue discussed the state of physics after the monumental discoveries by Einstein and others by contrasting classical physics and quantum mechanics. In a somber analogy, he likened the political turmoil of the day (in the wake of military defeat and revolution) to turmoil in the scientific world, where research in the field of elementary particles had threatened the unity of the physical Weltbild. He described the obvious contradictions between classical and quantum-mechanic approaches as extraordinarily irritating: "Here we are in a highly embarrassing position" (Laue 1961, 46). For Laue, the contemporary Weltbild of physics was rather precarious: "How about the truth of today's Weltbild? Obviously, it is incomplete" (ibid.). However, the coexistence of apparently contradictory empirical findings may foreshadow a great breakthrough, Laue surmised. Just as Einstein's theory of relativity had synthesized apparent contradictions, a new synthesis may come about that would reconcile classical physics and quantum mechanics. Thus, Laue ended on a hopeful note: "Because it must eventually be possible to solve the contradiction, we can assert: A truly fundamental breakthrough in physics is imminent" (ibid., 47).

The limitations of the physical Weltbild were at the forefront of Erwin Schrödinger's and Werner Heisenberg's musings about the topic. In 1930, Schrödinger (1984b) gave a talk titled "The Change of the Physical Weltbegriff," in which he discussed the two basic concepts of continuity (waves) and discontinuity (quanta). He argued that extrapolation is inevitable in any measurement. Thus, continuity is an underlying assumption; it cannot be measured directly. For him, the deterministic "Naturbild" had failed (Schrödinger 1984b, 606). "The wave functions do not describe nature in itself, but the *knowledge* we have of it on the basis of the actually made observations" (ibid., 607). Following Heisenberg's argument, Schrödinger thought that an absolute and purely objective description of nature was impossible, even in principle, because observations themselves would interfere. Instead, he said, our scientific statements represent only a subject-object relation. According to Schrödinger, that situation is not as bad as it sounds in terms of relativism. It suffices that the subject-object relation finds a "firm, clear, completely unequivocal expression" (ibid., 608). Schrödinger's point seems like a variant of Kant's transcendental argument in which the focus has shifted from the categories of perception to the intersubjective uniformity of perceptual outcomes: If only our observations of the world are stable enough, we need not care for absolute knowledge of the world in itself.

In his 1948 article "The Distinctiveness of the Weltbild of Science," Schrödinger (1984a) traced the scientific worldview back to the ancient Greeks. Here his perspective was different: He focused on the fundamental differences between scientific and nonscientific Weltbilder rather than on the latest epistemological conundrums in cutting-edge physics. The two central features of the scientific Weltbild were, according to Schrödinger, the "Verständlichkeitsannahme" (the assumption that the world is comprehensible), and

"Objektivierung" (objectivization, i.e., removing subjective elements from the perception).[65] "Whereas all material for the Weltbild is provided by the senses as organs of the mind, whereas the Weltbild itself is and remains for each person a creation of his own mind and beyond this has no verifiable existence at all, the mind itself remains a stranger within this Bild, it has no place in it, one can find it nowhere in it" (Schrödinger 1984a, 442). Schrödinger also stated that the natural Weltbild contains no values, thus implicitly agreeing with the distinction of the limited scientific Weltbild and the more comprehensive Weltanschauung: "There is not the slightest chance that we can detect the meaning of it all from purely scientific inquiry" (ibid., 452).

In *Science and Humanism*, Schrödinger considered scientific specialization a necessary evil. The scientist, in his opinion, has a responsibility to communicate with others and to seek an integrated totality of knowledge. However, absolute truth of scientific models, in terms of totally reflecting reality, is impossible. "Probably we cannot ask for more than just adequate pictures capable of synthesizing in a comprehensible way all observed facts and giving a reasonable expectation on new ones we are out for" (1951, 24). This appears to be a somewhat more pessimistic version of the idea that the scientific Weltbild is a work in progress.

Werner Heisenberg's "The Representation of Nature in Contemporary Physics" restated his famous principle: "[W]e can no longer talk of the behavior of the particle apart from the process of observation" (1958, 99). He shared with Schrödinger the view that our "laws of nature" refer to our knowledge of elementary particles, not to the particles themselves.

> The atomic physicist has had to come to terms with the fact that his science is only a link in the endless chain of discussions of man with nature, but that it cannot simply talk of nature 'as such.' . . . When we speak of a picture of nature provided by contemporary exact science, we do not actually mean any longer a picture of nature, but rather a picture of our relation to nature. The old compartmentalization of the world into an objective process in space and time, on the one hand, and the soul in which this process is

mirrored, on the other—that is, the Cartesian differentiation of *res cogitans* and *res extensa*—is no longer suitable as the starting point for the understanding of modern science. (ibid., 100, 107)

Thus, Heisenberg emphasized the role of the observer in the creation of the scientific Weltbild.

This section concludes with Carl Friedrich von Weizsäcker, who reviewed the development from classical physics to "neuere Physik." "Hence the developmental path of modern physics is the history of a more and more radical abandonment of certain central hypotheses, while consciously conserving the respectively unscathed rest. In this development, we can distinguish three phases: the theory of relativity, the quantum theory, and the today only emerging theory of elementary particles" (1945, 63–64). For Weizsäcker, quantum mechanics did not make the traditional view of causality obsolete but limited it: "Quantum mechanics recognized the relational character of the categories. Substance, causality, etc., do not signify realities as such, but realities that are perceived by humans. . . . Ontologically, this means that the concept of the object can no longer be used without reference to the subject of knowledge" (ibid., 79, 92). Equipped with the results of quantum mechanics, Weizsäcker revisited Kant's transcendental philosophy and found, "Thus only modern physics has filled out the empty scheme of Kant's theory of nature, although of course with the changes and modifications that were inevitable in such a case" (ibid., 122).[66]

The central disagreement that separated Einstein from many of his colleagues involved the notions of causality and determinism. Einstein had an unwavering allegiance to determinism, even when quantum mechanics yielded results that appeared to undermine it.[67] In light of these findings, Niels Bohr and Werner Heisenberg developed the probabilistic Copenhagen interpretation of quantum mechanics, which was recognized by many other distinguished physicists, among them Max Born and Wolfgang Pauli. Einstein, however, was among the most vociferous opponents. Having grown up in the spirit of scientific Weltanschauung that the books of

Büchner, Bernstein, and others imparted to him, he never gave up the idea of a deterministic universe, and he simply could not accept that God would play dice, as he put it.

5.2. DIFFERENT GROUPS OF SCIENTISTS— DIFFERENT STRATEGIES

The German scientists at the close of the nineteenth century were certainly no homogeneous social group. As the scientific sector expanded and became professionalized during the course of the century (Ben-David 1984, 1991; Cahan 1985; Mendelsohn 1964a), the larger group of scientists also became internally more differentiated. At the apex, the science professors at the universities belonged, through their positions, to the very elite of the Bildungsbürgertum. At the margins, one encountered a smaller group of potentially subversive *freischwebende* intellectuals. And there was a larger, and growing, social class of people whose livelihoods in some way or another depended on science and technology—engineers, teachers, and technicians of various sorts. These people tended to come from an emerging parallel system of educational institutions (*Realschulen, polytechnische* institutes, etc.) that provided a scientific and technical education at various levels.

The views these groups held on science and Kultur, and the strategies they pursued, also varied. We first distinguish five sets of leading nineteenth-century German scientists and allied scholars who articulated these views and strategies.[68] (It should go almost without saying that this is a gross simplification, yet it is one that seems suitable for our purposes.) Then we look at the larger social strata that identified with the ideas put forward by those intellectual leaders.

(1) The first group, in early and mid-nineteenth century, consisted of philosophical materialists, starting with Feuerbach, and then, of course, Marx and Engels. These were primarily philosophers steeped in Hegelian theories, rather than empirical researchers, but they were committed to a wissenschaftliche Weltanschauung and advocated empirical science as the basis of that Weltanschauung.

(2) The second group, in the early nineteenth century, comprised empirical scientists, such as Liebig, Ohm, and Alexander von Humboldt. They suffered somewhat under the rule of speculative "Naturphilosophie," and, in counterreaction, emphasized an empirical approach that shied away from grand philosophical systems. However, as Jungnickel and McCormmach (1986, 1: 23–33) showed in their careful study, early nineteenth-century textbooks in physics did address philosophical and epistemological issues, presenting an alternative to Naturphilosophie. In this group, Alexander von Humboldt was perhaps the scholar with the strongest emphasis on the scientific Weltbild.

(3) Starting in the mid-1800s, a group of scientific materialists, including Ludwig Büchner, Karl Vogt, and Jacob Moleschott, popularized the materialistic Weltanschauung (Gregory 1977). Typically having some background in medicine, they were better versed in actual science than the philosophical materialists of the first group were. The radical views of this third group jeopardized their membership in the Bildungsbürgertum. Büchner, for instance, was forced to quit his chair at the University of Tübingen after his publication of *Kraft und Stoff*.

(4) In the second part of the nineteenth century, a fourth group formed, which we might call the scientific Kulturträger establishment. This group included prominent scientists, such as du Bois-Reymond and Brücke, who had succeeded in making inroads into the top echelons of academe. They became the leaders and role models of a substantial part of the scientific community. Although they were methodological reductionists, they stayed away from militant philosophical materialism and radical political claims and favored a circumscribed scientific Weltbild. Yet they issued occasional visible attacks on the mandarin orthodoxy.

(5) Around the same time, also in the late nineteenth century, a fifth group emerged, including Haeckel, Ostwald, and their Monistenbund, as well as Mach (from whom the Vienna Circle descended).[69] Although many of them held high posts in academe, they were more radical than the fourth group and

more forward in confronting the larger group of Kulturträger
with far-reaching claims of a scientific Weltanschauung.

These groups differed in their strategies of advancing the cause
of science in society. On the one hand, some scientists, especially
those of the fourth group, tried hard to establish their credentials as
Kulturträger and thus to legitimize their membership in the
Bildungsbürgertum. Russell McCormmach wrote of the "self-
appointment of Wilhelmian academic scientists to the class of
'culture-bearers'"—following the footsteps of such people as du
Bois-Reymond and his allies (1976, 158). Their strategy could be
characterized as making more or less gentle reform attempts from
the "inside." In his famous "Ignorabimus" speech, du Bois-
Reymond (1873 [1872]) acknowledged principal limits to scientific
inquiry and thus implicitly delineated a coexistence between science
and the humanities. Such a compromising stance toward the other
Wissenschaften may have accelerated the acceptance of scientists in
the ranks of the Kulturträger and the Bildungsbürgertum. Rather
than trying to boost the prestige of the "Real-" and technical institutions, this part of the scientific elite wanted to fortify the position of
science *within* the university—a testament to both the overwhelming
social prestige of the university and these scientists' acceptance of
the *Bildungsbürger* ideal. (We shall discuss the relationship of the scientific Kulturträger to the wider Kulturträger group shortly.)

On the other hand, there was, of course, a much more radical
opposition from the "outside"—the mass popularization of science
as a scientific Weltanschauung that opposed much of what the
establishment stood for and that advocated social and political
reforms on the basis of science (Bayertz 1985). This materialist/sci-
entific countercurrent to the prevailing idealism in nineteenth-century
Bildung was led by the first, third, and, to some extent, fifth groups
of intellectuals and scientists described. These scientists wanted
nothing less than to make science the basis of a new Kultur and a
new society. In contrast with the scientific Kulturträger's establishment in the Bildungsbürgertum, some of the leaders of this movement never quite shed the flavor of *freischwebende Intelligenz*

(although other leaders in that group had arrived at academic top positions). The ideas developed by these scientists and intellectuals, along with popular educational works by all types of scientists, reached wider segments of the population, especially the new scientific/technological strata. One needs to remember that some of the popular scientific books were extraordinary best-sellers in Germany. At times, Alexander von Humboldt's *Kosmos* ranked second in sales only to the Bible. Bernstein's *Naturwissenschaftliche Volksbücher* and Büchner's more radical exposition of a materialistic Weltanschauung in *Kraft und Stoff* also found scores of avid readers. One of them was, of course, young Albert Einstein, who was introduced to this genre of literature by Max Talmey, a medical student and weekly dinner guest in the Einstein home.

Science became a battle cry with which several social groups assailed the sociopolitical status quo and the supremacy of the entrenched Bildungsbürgertum. In addition to the growing group of engineers, technicians, and other people whose jobs were based on technology, the entrepreneurial wing of the Bürgertum also had a natural affinity to science in its connection with technology. Albert Einstein grew up in this segment of society. The electrical business of his father and uncle was at the leading edge of technology in that era (Pyenson 1985); today, we would call them high-tech entrepreneurs. Father Hermann was a "freethinker" and probably held certain sympathies for the Weltanschauung pouring from the pages of the *Kosmos*, of the *Naturwissenschaftliche Volksbücher*, and even of *Kraft und Stoff* into his son's mind, or at least may have considered it a welcome antidote to young Albert's early religious inclinations. Hermann was an alumnus of the Stuttgart *Realschule*, and, as he did, many businessmen, technicians, and members of similar social strata graduated from one of the alternate educational institutions that typically carried the words *technisch* or *real* in their names.[70] Consequently, they tended to be sympathetic to the protracted struggle that these alternate institutions waged for parity with the traditional Gymnasium-university educational path.

Finally, under the influence of Marxism, the working-class movement made science the explicit basis of its Weltanschauung and politics. Far from being retrograde Luddites, most workers, and certainly their organizational leadership, embraced scientific and technological progess, but they wanted to bring that progress under their own control.

Kulturträger Ecology

In the last third of the nineteenth century, one might well capture the general situation of the academic scientists (mainly belonging to groups four and five) vis-à-vis the wider Kulturträger group in the metaphor that, at a Kulturträger family reunion, the scientists would certainly be invited but asked to sit at the children's table. Fritz Ringer, who explicitly excluded scientists from his famous study of German academics, noted, "Strictly speaking, German professors of physics and chemistry were as much mandarin intellectuals as their colleagues in the social studies and humanities. . . . It is my impression that in their attitudes toward cultural and political problems, many German scientists followed the leads of their humanistic colleagues" (1969, 6). Yet closer scrutiny will yield a more differentiated picture. In the following, we will sketch in greater detail where the members of the scientific Kulturträger establishment agreed with the other Kulturträger and, to begin with, where they disagreed.

Issues of contention. At times, leading German scientists (even those of the more moderate fourth group) did not shy away from attacking major icons of Kultur. For instance, Helmholtz publicly attacked Hegel, and du Bois-Reymond took on even Goethe in his *Rektoratsrede* titled "Goethe und kein Ende" on October 15, 1882 (du Bois-Reymond 1912, 2: 157–83). The latter speech was perceived as highly sacrilegious and created quite a stir. Its organization may have contributed to the scandal. At the beginning, du Bois-Reymond berated Goethe as a weak-willed procrastinator who compensated for these tendencies by overglorifying the life of action. Du Bois-Reymond, of course, defended theoretical, nonutilitarian knowledge. Next, he launched into a literary critique of *Faust*, which was probably the most inflammatory part, as it appeared to attack Goethe, the lionized poet. Only then did du Bois-Reymond move

onto Goethe, the scientist. He assailed Goethe's dislike of the experimental approach, his dismissal of mechanical causality, and his
theory of color. Although du Bois-Reymond recognized Goethe's scientific contributions in morphology, he concluded that "even
without Goethe, science in general would have come as far as it has,
and German science would perhaps have come farther" (ibid., 2:
174). He felt that the cardinal defect of Goethe's science was not
idiosyncratic; rather, it was a "flaw in the basic German character
that must be recognized so that it can be opposed. This flaw, which
ultimately of course is related to traits of greatness, is the tendency
to favor deduction over induction, and speculation, whose heavily
inflated balloon easily explodes as it rises, over empirical science,
which remains on safe ground" (ibid., 2: 170). The final part paid
Goethe, the poet, high compliments—but by that point, the audience was probably too scandalized to take note.[71] In a later comment on his speech and the ensuing controversy, du Bois-Reymond
reiterated his reverence for Goethe's poetry and recalled that he had
learned the first part of *Faust* by heart (ibid., 2: 181).[72]

There were indeed some areas of disagreement between the scientific and nonscientific Kulturträger (Paul 1984): Many scientists
rejected the pervasive idealism among the Kulturträger. They
opposed the speculative method and criticized the emphasis on
neohumanism in the Gymnasium. It bears repeating that typically
they did not advocate the primacy of the emerging system of scien-
tific/technical educational institutions over the traditional Gymnasium, but they merely wished to open up the Gymnasium for more
and gentle, as du Bois-Reymond's example will now demonstrate.[73]
Nonetheless, they evoked stiff resistance from many classicists and
philologists.

Under the motto "Conic sections! No more Greek script!" du
Bois-Reymond in 1877 proposed a shift toward mathematical and
scientific education at the Gymnasium (1912, 1: 620). At the same
time, he made clear that he wished to preserve the substance of the
Gymnasium education: "But I do not at all intend to turn the Gymnasium into a scientific school. . . . In the higher grades, I would like
to see . . . not even physics and chemistry with experiments, but

mechanics, and introductions into astronomy and into mathematical and physical geography, for which one hour more than currently could be scheduled" (ibid., 1: 617). To compensate for the increase in science education, du Bois-Reymond suggested reducing school hours in religion and in Greek. "The Gymnasium best prepares itself for the struggle against the intrusions of *Realismus* [the pro-science educational movement], when it itself, within certain limits, provides space for Realismus" (ibid., 1: 620). His goal was to adapt Bildung to the changing times, not to subject it entirely to Realismus, and not at all to replace it with something completely different, such as the American "neobarbarism" he strongly abhorred.

Most scientific Kulturträger also opposed the growing strands of Neo-Romanticism, irrationalism, and mystic-racist nationalism, that is, the ascendance of the backward-facing Janus face of Kultur. It may be significant that, based on large-scale empirical studies, Virchow's *Beiträge zur physischen Anthropologie der Deutschen* demonstrated the realities of race mixture in Germany that clearly contradicted any myths of the "blond beast" and the pure Nordic race. He concluded that "this [Germanic] general type is not a uniform one to the extent previously assumed" (1877, 370). Furthermore, Virchow commented, "And still there remained the need to reconstruct such types [of European Culturvölker]. Strangely, this need itself has been stimulated time and again by considerations that are entirely extraneous to Wissenschaft, especially by politics. . . . Luckily, Wissenschaft has not succumbed to these efforts. On the contrary, it has increasingly deconstructed also the existing political and linguistic units" (ibid., 2). A calm and confident feeling of the empirical scientist's superiority over the politicians' follies pervades these lines, yet we now know that these follies gained the upper hand a few decades later. Einstein himself became targeted as one of the primary foes of that particularly virulent strain of racist and irrationalist Weltanschauung that undergirded the National Socialist regime.

Communalities. Beyond the previously noted differences in outlook, however, scientists and nonscientific Kulturträger shared a common allegiance to the university as their social milieu and also to Wissenschaft—an overarching and unifying concept, even though initially modeled after the classics. As portrayed in the case of du Bois-Reymond, the disagreements about the neohumanistic Gymnasium curriculum, if they existed at all, were mostly rather mild. Many scientists were genuinely proud of their Gymnasium education and held it in high esteem. Heisenberg's discussion "On the Relationship Between humanistische Bildung, Science, and the Occident" illustrates the enduring respect many German scientists had for the Gymnasium even in the twentieth century. According to Heisenberg, one of the most valuable benefits for the scientist of a humanistische Bildung is the acquaintance with principled thinking. "[A]nyone who wants to get to the bottom of things in whatever discipline, be it technology or medicine, will sooner or later encounter these sources in antiquity, and he will draw many advantages for his own work from having learned principled thinking and principled inquiry from the Greeks. I believe that one can clearly see in the work of Max Planck, for instance, that his thinking was influenced and made productive by the humanist school" (1955, 43). Heisenberg thought he himself reaped similar benefits: "In this situation, my training in principled thinking, which we had received at school, has helped me enormously, and has in any case led me to not being satisfied with partial pseudo-solutions. A certain knowledge of Greek natural philosophy, which I had acquired then, was also of great use for me" (ibid., 44).

In an 1877 speech titled "Kulturgeschichte und Naturwissenschaft," du Bois-Reymond addressed the threats to modern Kultur (1912, 1: 567–629). He acknowledged that the supremacy of Naturwissenschaft came at a cost: "[B]ut where [science] dominates exclusively, it cannot be overlooked, and it can easily happen, that the mind has fewer ideas, the imagination has fewer images, and the soul has fewer feelings. The result is a narrow, dry, and hard mindset that has been abandoned by the Muses and the Graces" (ibid., 1: 604). He lamented a decline in literary knowledge among the young: "There was a time in Germany when people no longer quoted from the first part of *Faust* because the quotation was beaten to death. Are we really entering a time when people can no longer quote from it because the allusion would not be understood?"

(ibid., 1: 611). As mentioned, du Bois-Reymond himself had memorized the first part of *Faust*.

A related attitude shared widely among the Kulturträger was a disdain for narrow specialization within one's professional field (*Fachidiotentum*). The idea of Bildung and the role of Kulturträger encouraged a broader outlook. With respect to physics, Cathryn Carson pointed out, "Until the 1970s, the tradition of quasi-philosophical meditations on physics had been cultivated primarily by older, established scholars and had taken root in the somewhat staid setting of Central European academia" (1995, 649). Sometimes, of course, this tradition of quasi-philosophical meditations held an even wider reign over a scholar's life course. Einstein's case provides a prime example of intense philosophical interest displayed by a *young* physicist who faced a precarious job situation and lived a distinctively Bohemian lifestyle. Mary Jo Nye also emphasized that many scientists considered themselves part of the wider group of intellectuals and therefore pursued widespread interests: "Until the 1930s many chemists and physicists (or chemical philosophers and natural philosophers) thought of themselves as first of all *intellectuals* and only secondarily as *scientists*. For the most part, their education and interests were broadly gauged and wide-ranging" (1996, 226).

Finally, there was the high regard for pure, nonutilitarian Wissenschaft that had become engrained in the German universities since Wilhelm von Humboldt's days (Gregory 1989; Turner 1971). In the previously quoted speech of 1877, du Bois-Reymond also warned against science degenerating into applied and technological activities (1912, 1: 604). In his view, America was the country where this utilitarian trend—"Neobarbarei"—was most pronounced (ibid., 1: 605). He diagnosed "Amerikanisierung" of German Kultur and emphasized classical Bildung as an antidote: "May Hellenism keep Americanism away from our intellectual borders" (ibid., 1: 608). As McCormmach noted, most scientists of that era adopted this general outlook and did not like to be drawn too closely into technology or applied research (1976, 162). In his 1862 speech "On the Relation of the Sciences to the Whole of Wissenschaft," Hermann von Helmholtz expressed this widespread sentiment succinctly: "Those who strive for immediate practical benefits in their pursuit of the Wissenschaften can rest quite assured that they will strive in vain. Complete knowledge and complete understanding of the working of the natural and mental forces is the only thing that Wissenschaft can seek" (1884, 1: 142).

Part of the emphasis on basic science may have been the desire to find acceptance in the eyes of the other Wissenschaften by aspiring to the exacting standards of pure Wissenschaft that emerged from the then world-renowned seminars in philology and similar disciplines. In the early nineteenth century, it was the humanities that first gained international admiration for German Wissenschaft, and science was compelled to follow their model, in institutional organziation as well as in the ethos of curiosity-driven research.[74] As the religious scholar Joachim Wach put it, "A Wissenschaft that exists and is valued only because it serves our most immediate needs of the day, does not deserve to be" (1930, 203). If science was to be too closely allied with technological applications, its place among the Wissenschaften might be jeopardized, it might be relegated from the universities to the distinctly less prestigious *Technischen Hochschulen,* and the scientists might, in the eyes of German society, sink to the lower status of mere technicians rather than be accepted as part of the Kulturträger elite. But, on the whole, scientists also saw the benefits of applied work. "By and large, Wilhelmian scientists steered a middle course. They upheld the values of nonutilitarian research and rational and empirical modes of inquiry; at the same time, they promoted industrial technology in limited contexts, as in the creation of applied scientific disciplines, and they recognized nonscientific components in the collective culture and in the world-view of the individual" (McCormmach 1976, 157).

Within the larger process of the decline of the mandarins that Ringer (1969) described, we should take note of some shifts in the ecology of this class. In mid-nineteenth century, the emerging natural sciences occupied the children's table of the Wissenschaften, as previously noted, whereas the humanities held forth from the seats of honor. The professors of these disciplines were the prototypes and role models of what it meant to be a German university professor. In the Weimar era, after the loss of World War I, the internal situation of the Kulturträger group became somewhat more complicated. The first aspect we turn to is the general international framework; here the science professors gained more respect as the humanists lost it. Significantly, in the aftermath of World War I, the physicist Albert Einstein became a centerpiece of German wissenschaftliche Kulturpolitik. The usual German Kulturträger had become thoroughly discredited abroad, and science (with its stronger presumption of objectivity, i.e., supranational validity) was now presented as the flower of German Kultur. A second aspect to consider is the German academics' position within the international communities of their respective disciplines. In the natural sciences, there can be little doubt that interwar Germany was among the leading nations, if not the leading nation, in the world. In physics, people like Einstein and the other physical luminaries were recognized among the foremost thinkers of their time (and by forcing them and numerous other first-rate scientists into emigration, National Socialism destroyed an unquestionably flourishing enterprise). The general intellectual level of the Weimar-era humanities, by contrast, was comparatively perhaps less superior. This situation contributed to making the science focus of German wissenschaftliche Kulturpolitik abroad an attractive and credible strategy. Within Germany, however—and this is the third aspect—the scientists had to contend with powerful antirationalistic and antiscientific currents in Kultur that prevented them from occupying a generally recognized leadership position within the wider Kulturträger group, even in spite of their extraordinary achievements.[75]

This section has shown how most of the academic scientists in nineteenth-century Germany positioned themselves within the larger Kulturträger group. The very root of the communalities was a shared emphasis on synthesis—Weltanschauung and Weltbild. This was the vehicle by which the scientists could most seriously and meaningfully claim their place as Kulturträger, and we will elaborate this point in the concluding section of this chapter.

Legitimization of Science in Kultur

The core element of agreement between orthodox Kulturträger and scientists, even those scientists who were part of the freischwebende Intelligenz, was unity and synthesis. As Robert Paul put it, "German natural scientists remained deeply imbued with holistic values emphasizing the importance of a comprehensive world-view" (1984, 4). Among those scientists, Russell McCormmach detected the "German mind with its penchant for unity, idealism, and abstraction. . . . [The German scientists] were committed to the unity of science and beyond that to the unity of intellectual culture" (1976, 159, 163). Dietrich von Engelhardt noted, "Generally of a more encyclopedic outlook than most of the Geisteswissenschaftler of their time, the scientists of the nineteenth century envision the connection of humanistische and realistische Bildung" (1990, 112).

The nineteenth-century German scientists' quest for a scientific Weltbild or even scientific Weltanschauung was in accordance with a prime imperative of Kultur, and thus science was meant to be anchored in Kultur.[76] In this project of establishing a place for science in Kultur, Goethe played a major role. For the German Kulturträger, Goethe was a fascinating and inexhaustible part of their imaginative lives, and the immense cultural prestige he carried in those circles could be tapped by the scientists. We have already seen that Goethe was an active scientist and proud of it, although this portion of his work was greeted with ambivalence, if not contempt, by most members of the subsequent generations of scientists. What was much more important than Goethe's concrete scientific contributions was his giving expression and legitimacy to the deep ambitions of German science, and thus drawing it into the realm of Kultur.

Whole books have been written about the interest among scientists in various aspects of Goethe's work. There is now even a large bibliography called *Goethe in the History of Science*, edited by Frederick Amrine (1996). (See Amrine, Zucker, and Wheeler 1987.) Among the scientists connected in some way to Goethe are, for instance, Johann Bernhard Stallo, Wilhelm Ostwald, the physiologist Arnold Adolphe Berthold, the neurophysiologist Rudolf Magnus, the botanist Gottlieb

Haberlandt, the physical chemist Gustav Tammann, the bacteriologist Robert Koch, the psychologist Georg Elias Müller, Emil von Behring, Carl Jung, and the English scientist William Henry Fox Talbot. A particularly intriguing case is that of Nicola Tesla, who, though not German by descent, was caught up so much in the German style of Bildung that he claimed, and sometimes demonstrated, that he knew the whole of Goethe's *Faust* by heart. By Tesla's own account, as reported in the *Dictionary of Scientific Biography*, Goethe's poetry even propelled his scientific discoveries in a curiously immediate way. While Tesla was reciting a passage of *Faust* to a friend in a Budapest Park, the concept of the polyphase induction motor, based on the principle of the rotating magnetic field, occurred to him instantaneously and completely—as if in a lightning flash.

We already mentioned the controversial Goethe speech by du Bois-Reymond, but Goethe was also talked and written about by physicists such as Hermann von Helmholtz, Erwin Schrödinger, Wilhelm Wien, Arnold Sommerfeld, and Max Born. Hermann von Helmholtz—not only physicist but also biologist and philosopher, and thus the very prototype of the Kulturträger at his best—provides a fitting example. Throughout his brilliant career, he gave popular lectures to educated audiences on general, though mostly scientific, subjects. Collected in books with such suggestive titles as *Science and Culture* (1995), each lecture was thoroughly laced with quotations from the literary canon chosen so as to reflect the search for the Holy Grail of unity in science. Helmholtz had excellent credentials for such a celebration. After all, he was the co-discoverer of the law of energy conservation, which reaches out and gives common ground to all physical and biological sciences and engineering, and he was in reverent accord with Darwin's theory that interconnected all living being. In the mentioned volume of public lectures, the first (1853) and the last (1892) were on Goethe.

Fragments of Goethe's poetry could be encountered routinely not only in the exhortations by politicians or in the popular lectures of other Kulturträger but even in the textbooks of advanced science itself. Thus Arnold Sommerfeld, in the third volume (*Electrodynamics*) of his *Lectures on Theoretical Physics*, sent his readers of the general relativity theory off with a quotation from *Faust*, part 2 (1952, 311).[77] Another example occured on two pages of a textbook by one of Einstein's own scientific predecessors, whom in 1900 he had called "quite magnificent"—namely, Ludwig Boltzmann.[78] His *Vorlesungen über Maxwells Theorie der Elektricität und des Lichtes* had two parts (part 1 in 1891 and part 2 in 1893), each preceded by a short epigraph. Boltzmann could count on every German reader to know the lines he quoted and that they referred to the early pages of Goethe's *Faust*. The first epigraph is a very loose rendition of lines 380–81. But the most intriguing aspect of this epigraph is that it stops where it does—right before two of the most celebrated and programmatic lines of the play. The four lines (380–83) read, in rough translation:

> That I may no longer, with sour labor,
> have to teach others that which I do not know myself,
> and that I may perceive what holds the world
> together in its innermost.

Boltzmann's own version of the first two lines, however, is more like, "Well then, I shall with sour labor teach you that which I do not know myself." While these lines humbly dwell on the intellectual struggles and uncertainties inherent in the topic, Boltzmann almost seemed to rely on the readers, intimately familiar with *Faust*, to add the much more upbeat next two lines that had become emblematic of the quest for the ultimate insight into the fabric of the universe.

The second epigraph (italicized below) is from the following passage, only a few lines further down, which is quoted here in Stewart Atkins's (1984) prose translation: Faust opens the book of Nostradamus, seeking a guide to the force that holds the world together, and there he sees the "sign of the Macrocosm." He exclaims:

> Ha! as I gaze what rapture suddenly
> begins to flow through all my senses!
> I feel youth's sacred-vital happiness
> course with new fire through every vein and fiber.
> Did some god inscribe these signs

that quell my inner turmoil,
fill my poor heart with joy,
and with mysterious force unveil
the natural powers all about me?
Am I a god? I see so clearly now!
In these lines' perfection I behold
creative nature spread out before my soul. . . .
How all things interweave as one
and work and live each in the other. (lines 430–41, 447–48)

By the divine signs, Boltzmann meant, of course, Maxwell's equations, which were the key to Maxwell's synthesis of electricity, magnetism, and optics. It is quite significant that Boltzmann's version of Goethe's lines in the second epigraph is, in fact, also just a bit wrong. He too was no doubt quoting from memory, going back to his school days. Used constantly, such verses tend to be taken for granted and get fuzzy around the edges. Boltzmann's errors indicate that the lines have entered the stock of Kultur. To be sure, many of the Bildungsbürger used quotes from the classics merely as status symbols, and often got them from Büchmann's (1926) *Geflügelte Worte.* (See Frühwald 1990.) The most outstanding members of the Kulturträger, however, were truly inspired by the classic works they quoted.[79]

Like Spinoza, Goethe saw God and Nature as two aspects of the same basic reality. One of his papers, titled "Study after Spinoza," starts with the sentence, "The concept of being and of completeness is one and the same" and goes on to ponder the meaning of the infinite (1962, 13: 7–10). But significantly, the main point of that work is to argue for the primacy in scientific thinking of *unity* and for the wholeness "in every living being." In the same vein, the quantification and subdivision of natural phenomena, he thought, missed the whole point of the organic unity of man and nature in the explanation of phenomena. Albrecht Schöne (1987) documented in his book, appropriately titled *Goethes Farbentheologie,* that Goethe passionately abhorred the Newtonian idea of splitting light because it offended his ideal of the unity of Nature that sprang from a deepfelt religious sentiment.

Goethe's cosmotheistic belief coincided with that of many German scientists—and of Einstein.[80] The main point for us is the strong resonance between the Goethean, or Faustian, drive toward a unified fundamental understanding of nature and the preoccupation of many German scientists, among them Einstein, with the search for one single, completely coherent world picture, a Weltbild encompassing all phenomena. Physical science too would progress by the discovery of ever fewer, ever more all-encompassing fundamental concepts and laws. On this quest, the physicists, from Boltzmann to Einstein, could inscribe their banners with Goethe's words, "If a few grand formulas are accomplished, all must become One, all must originate from One and return to One" (quoted in Schöne 1987, 96).

We have now surveyed the sociocultural and intellectual landscape in which nineteenth-century German science existed. Before we examine, in the following chapters, how one particular individual, Albert Einstein, fit into this landscape of large-scale, collective features, let us restate our main hypothesis. Despite some undeniable rebellious tendencies, Einstein was far from being a radical countercultural iconoclast. Rather, he nourished himself on the deepest ambitions of German science and Kultur. To an extraordinary degree, he was able to bundle and express these cultural energies, and to fulfill the quest prescribed by his Kultur.

NOTES

1. Not all philosophers of science would, of course, use the specific term *Weltbild* in those deliberations.

2. E.g., Nietzsche 1987 [1872] and Spengler 1918–22.

3. On this topic, see Cotgrove 1978.

4. The word *Romantic* goes back to the old French *romanz* which originally referred to writings in the vernacular (*lingua Romana*), then specifically to novels (i.e., *romans*). Because the topics of the novels were commonly adventurous, miraculous, and incredible, the adjective *romantic* meant "unreal" when it was first evidenced in English in 1650. With a similar meaning, the word appeared in French (1694) and German (1698) (*Der Große Brockhaus* 1981).

In German, *romantisch* first meant "novel-like" but then described a more general poetic and enchanted or—in a negative connotation—weird and overwrought quality. *Romantisch* was also used for describing wild and scenic landscapes. At the end of the eighteenth century, *romantisch* became the label for the Romantic school of poetry, art, and Weltanschauung in general (Grimm and Grimm 1955). "Romanticism was probably at its most self-conscious in the Berlin-Jena coterie, led by Schlegel among others, which propagated the term and eventually drew fire from Goethe" (Wiener 1968–74). To Friedrich Schlegel, Romantic meant "anticlassical." Neither Coleridge nor Wordsworth adopted the label (ibid.).

5. In so doing, we follow the lead of scholars such as Fred Gregory, who wrote, "One must not forget, therefore, that the formulation of scientific and dialectical materialism, both rooted in the impact Feuerbach's work had on German society in the 1840s, grew originally from deep within German idealism. Long before our study of Vogt, Moleschott, Büchner and Czolbe is over, it will be clear that they were never able to rid their materialism of its idealistic heritage" (1977, 28).

6. On Romantic science, see Poggi and Bossi 1994.

7. As to the Naturphilosophen, Barry Gower rightly pronounced, "There is probably no single individual who could usefully and accurately be described as a typical *Naturphilosoph*" (1973, 304).

We should also emphasize that Romanticism, although predominantly associated with German thinkers, was not a purely German movement. For instance, the poets William Blake, Lord Byron, John Keats, and Percy Bysshe Shelley, as well as the scientist Humphry Davy, were among the major representatives of Romanticism in Britain.

A major interest of Humphry Davy's was to understand the nature of light and heat. Davy also demonstrated photosynthesis in marine vegetation and the presence of carbon dioxide in venous blood. He made seminal discoveries in electrochemistry, as well as practical inventions, such as a safety mining lamp that would not ignite the combustible gases often found in mines. Davy shared the Romantic belief that everything in nature was interconnected. Typical of the Romantic distaste for compartmentalizing knowledge, Davy read philosophy and wrote poetry in addition to his scientific pursuits. Finally, he resembled the Romantics in his approach to work. He was not given to steady and systematic efforts, but rather he worked in outbursts of high intellectual energy, the objects of which were often whimsical or accidental.

8. Similarly, Lord Byron wrote, "That Knowledge is not happiness, and science/But an exchange of ignorance for that/Which is another kind of ignorance" (*Manfred*, 4th act, 2nd scene, in Byron 1975 [1817], 489).

9. In the twentieth century, the National Socialists took up these anti-mechanistic, antipositivistic, and anti-Newtonian sentiments (e.g., Krieck 1942).

10. We must add that Newton himself may well have objected to how the Romantics (mis-)represented his physics.

11. Hegel 1975, 1: 519; cf. 1985, 1: 500.

12. See Churton 1997 and Jonas 1958, 1964.

13. In the early seventeenth century, Robert Fludd (1574–1637), for instance, published an important work on the microcosm/macrocosm idea titled *Utriusque cosmi maioris scilicet et minoris metaphysica, physica atque technica historia* (1617–21).

14. The idea of the universe as an organism has a rich tradition. One of its more recent incarnations is Alfred North Whitehead's (1925, 1929) philosophy of organism, or doctrine of organic mechanicism.

15. Indeed, his scientific production was not negligible. The collection of his scientific work runs to some 540 pages in one of the editions (Goethe 1962, 13).

16. On the Romantic affinity to chemistry, see Kapitza 1968.

17. On this topic, see Mendelsohn 1964b, 1965.

18. The group of Romantic scientists also included, among others, Henrik Steffens (geology), Karl Adolf Eschenmayer (physics and chemistry), Karl Ernst von Baer, Johann Hermann von Autenrieth, Ignaz Döllinger, Ignaz Paul Vital Troxler (biology), and Gotthilf Heinrich v. Schubert (natural history and anthropology) (Hennemann 1967–68, 116–18).

19. 1841 letter to Varnhagen von Ense, quoted in Snelders 1970, 193.

20. See Culotta 1974; Knight 1966–67, 1975, 1990; Lenoir 1980, Snelders 1970.

21. The choice of "naturalistische Weltanschauung" in the fourth edition appeared to be fleeting. In the ninth edition of *Kraft und Stoff*, Büchner quoted the passage in question, which, according to him, was included in the first four editions (1867, 263–64). However, this quotation, given in the version of the first edition, in fact deviated from the "naturalistische Weltanschauung" of the fourth edition (1856, 263) and instead read "mechanischen und materiellen Naturanschauung" (1867, 264). (See also the twelfth edition [1872, 282–83].)

22. Heinrich von Treitschke similarly, albeit somewhat disapprovingly, used the word *Weltbild* in connection with materialism, when he

wrote of the "mechanische Weltbild" of French Enlightenment philosophers (1879, 1: 93).

23. On the scientific Weltanschauung, see Dennert 1907, 1909; Hallier 1875; Kuhlenbeck 1899; Noiré 1874; Portig 1904; Steiner 1894; and Troels-Lund 1900.

24. Against Ostwald and the Monistenbund, Oscar Schmitz polemicized in his *Weltanschauung der Halbgebildeten* (1914).

25. On the Vienna Circle, see Johnston 1972, 181–95.

26. In this shift to methodology, a certain book on the physical Weltbild may have considerably influenced some members of the Vienna Circle. It was *La théorie de la physique chez les physiciens contemporains*, published in 1907 by Abel Rey. (See Holton 1993, 18–19.) In this book, Rey described the crisis in physics that was caused by the failure of the mechanical Weltbild: After the ontological basis of science was undermined, faith in science was greatly diminished.

27. Neurath 1973, 301; see Verein Ernst Mach 1929, 9.

28. Likewise, Ludwig Boltzmann called metaphysics a "migraine of the human mind" (quoted in Flamm 1983, 261). Although Boltzmann and Mach did not at all get along with each other at a personal level, they shared certain elements in their basic scientific outlook.

29. Neurath 1973, 305–306; see Verein Ernst Mach 1929, 15–16. Richard von Mises (1930) described the naturwissenchaftliche Weltbild from the viewpoint of this group. In exile, he completed the *Kleines Lehrbuch des Positivismus* (1939), which was translated into English and released in 1951 as *Positivism: A Study in Human Understanding*. Also see Frank 1928.

30. The same position was also propounded in the 1929 pamphlet *Wissenschaftliche Weltauffassung: Der Wiener Kreis*: "The scientific world-conception knows *no unsolvable riddle*. Clarification of the traditional philosophical problems leads us partly to unmask them as pseudo-problems, and partly to transform them into empirical problems and thereby subject them to the judgment of experimental science" (Neurath 1973, 306).

31. Regardless of whether scientists explicitly address the fundamental issues implied in the Weltbild, many philosophers of science would be quick to point out that scientists cannot but work on the basis of a Weltbild, even if it remains implicit and unquestioned.

32. See, for instance, Lecher 1912, *Physikalische Weltbilder*; Dingler 1951, "Das physikalische Weltbild"; Neuberg 1951, *Das Weltbild der Physik. In seinen Grundzügen und Hauptergebnissen dargestellt*; Reichenbach 1930, *Atom und Kosmos: Das physikalische Weltbild der Gegenwart*; and—replacing

Weltbild with Naturbild—Haas 1920, *Das Naturbild der modernen Physik*. (Tellingly, the title of the 1933 English edition of Reichenbach's book avoided the term *Weltbild* or its English equivalent: *Atom and Cosmos: The World of Modern Physics*.) The Weltbild has, of course, been a topic also outside of physics, e.g., v. Bertalanffy 1949, *Das biologische Weltbild*; Linser 1948, *Chemismus des Lebens: Das biologische Weltbild der Gegenwart*. The title of Heim's 1951 *Die Wandlung im naturwissenschaftlichen Weltbild* refers to the "scientific Weltbild" in general.

33. In addition to scientists, various nonscientists (such as theologians, philosophers, historians, or social scientists) have examined the scientific Weltbild. They have commonly contrasted "the" scientific Weltbild with other Weltbilder. Many of these scholars have portrayed the scientific Weltbild as a newcomer that threatens the more long-standing ones. Those with a normative bent see it as their responsibility to adopt this new scientific Weltbild, fight it, or put it in its proper place vis-à-vis the traditional ones.

34. In his own terminology—which, unfortunately, differs from ours—Humboldt called this scientific Weltbild "Weltanschauung."

35. On "Humboldtian science," see Cannon 1978 and Nicolson 1987.

36. For the relationship between Humboldt and Goethe, see Bruhns 1872, 1: 185–235.

37. Humboldt's emphasis on the natural science aspect is evident in the term of "*physische* Weltanschauung."

38. "We are still far from the time when it might be possible to synthesize all our sense perceptions into a unified concept of nature. One may doubt whether this time will ever come. The complexity of the problem and the immensity of the cosmos almost thwart our hopes. Yet even if complete success is unattainable, a partial solution of the problem, the quest for the comprehension of natural phenomomena, remains the highest and eternal purpose of natural science" (Humboldt 1845, 67–68).

39. "It is a sure indicator of the amount and value of discoveries that are to be expected in a science, if there exist facts that are still unconnected and almost without interrelation, or if several facts, which have been observed with equal meticulousness, seem to contradict each other" (Humboldt 1978, 21).

40. His chosen terminology is different from ours. The "mechanistische Weltanschauung" corresponds to our scientific Weltbild (du Bois-Reymond 1973, 31).

41. Another of Johannes Müller's students, Theodor Schwann, could also be counted as an associate of this group (Mayr 1965).

42. On Helmholtz and du Bois-Reymond, see Hörz and Wollgast 1986. Helmholtz's letters to his parents were edited by Cahan 1993.

43. Helmholtz also extolled the esthetic beauty of the scientific Weltbild (1884, 1: 142–43). We might note here that Henri Poincaré grounded the scientific Weltbild in esthetics. Against Tolstoy, who found science for its own sake absurd, he argued, "The scientist does not study nature because it is useful; he studies it because he delights in it, and he delights in it because it is beautiful. If nature were not beautiful, it would not be worth knowing, and if nature were not worth knowing, life would not be worth living" (1958 [1904], 8). Yet beauty and usefulness coicide. "We see too that the longing for the beautiful leads us to the same choice as the longing for the useful" (ibid.). Poincare tentatively located the cause for this fortuitous coincidence in the workings of evolution.

44. Three years before du Bois-Reymond, Carl Hebler had publicized similar ideas, though his work received much less attention. Echoing Kant's third antinomy of pure reason (Kant 1925 [1781], 286–88), Hebler asserted that the modern Weltanschauung contained two main components. "Two major elements belong to our modern Weltanschauung: First, the complete subjugation of our thinking under general epistemological norms, which are equally valid for all phenomena and domains; and, as a result, the belief in an unassailable lawfulness of all events. This lawfulness, inseparable from the nature of things, demands for every event a full causal explanation through its precedents, and always links same preconditions to same outcomes. The second major element is the equally strong conviction that humanity, despite this lawfulness and within and through it, is able and destined to develop its spirituality freely and comprehensively, only with those limits that lie in our nature or in the conditions for the existence or development of that spirituality" (Hebler 1869, 22). For a more recent treatment of the limits of science, see Wigner 1950.

Thus, whereas the first element corresponded closely to the scientific Weltbild, Hebler augmented it with a second element—also called *ethischer Idealismus* (1869, 27). His was an early attempt of recognizing the scientific Weltbild while at the same time delimiting its sphere. Du Bois-Reymond pursued a similar purpose with his lecture nicknamed *Ignorabimus* three years later.

45. According to this notion, someone who knew all physical laws, and perfectly knew the state of all matter in the universe at a given time, would then be able to calculate the state of the universe at any time in the past or the future.

46. In this, he resembled the exponents of a radical materialist Weltanschauung, such as Büchner and his allies Jacob Moleschott and Karl Vogt.

47. In his 1877 speech at the assembly of the German scientists and physicians, Virchow argued that the sciences should exercise self-restraint in using their freedoms because otherwise religion and the state might crack down on them.

48. On Planck, see Heilbron 1986.

49. At the end of the nineteenth century, one could also hear voices that were highly critical of science and pronounced the "bankruptcy of science." (See MacLeod 1982.)

50. According to Robert Millikan, Michelson here probably referred to Lord Kelvin (Badash 1972, 52).

51. Among the early departures, made by Wilhelm Eduard Weber, Rudolf Clausius, Bernhard Riemann, and Carl Neumann, from the orthodox mechanical Weltbild in the direction of electrodynamics, McCormmach listed: "the replacement of Newtonian instantaneous action by the finite progagation of electric force, the violation of Newton's law of action and reaction by electrodynamic forces, the stipulation of a nature-imposed upper limit on the possible relative velocity of particles, the suggestion of the need for a completely symmetric use of space and time in descriptions of electrodynamic phenomena, the proposal of a new concept of energy conservation suited to electrodynamic rather than mechanical needs, and the recognition, at least in a mathematical sense, of a velocity-dependent apparent mass for electric particles" (1970a, 472). At first, these departures were not seen as a threat to the mechanical Weltbild, but Maxwellian electrodynamics put them in a larger context, and, toward the end of the century, they became the seeds for the electromagnetic Weltbild.

52. A terminological note: We speak of the mechanical (*mechanische*) Weltbild when we have a body of specific physical theories in mind that are based on Newton's laws. By contrast, the mechanistic (*mechanistische*) Weltbild or even Weltanschauung is understood as a much more broadly defined philosophical concept that likens the world to a big maschine and denies the existence of any nonmechanical forces or phenomena. Our brief survey of those various physical Weltbilder is indebted to Jungnickel and McCormmach 1986, 2; Kragh 1999; McCormmach 1970a; and the relevant entries in the *Dictionary of Scientific Biography*.

53. Among the sympathizers, one could count Pierre Curie and Ernst Mach.

54. Einstein shared Planck's negative opinion about energetics as a Weltbild. In the article "Max Planck as Researcher," which he wrote in 1913 on occasion of Planck's appointment as Rektor of Berlin University, he endorsed Planck's harsh polemics against energetics. (The article appeared in 1913 in *Die Naturwissenschaften* 1, 1077–79 and is reprinted in *CP*, 4: 561–63.) On the other hand, Einstein had great respect for thermodynamics, whose law about the impossibility of a *perpetuum mobile* he considered a model for the universal principle he was going to look for in his own theoretical efforts (1979a, 20).

55. For a survey of the various strains of the electromagnetic Weltbild, see McCormmach 1970a.

56. When Einstein first published his theory of relativity, it was frequently confused with Lorentz's theory—some scientists spoke of the Lorentz-Einstein theory. On the relationship between Einstein's and Lorentz's physics, see McCormmach 1970b. For a recent argument emphasizing similarities between Einstein's theory and that of Lorentz (and Poincaré), see Darrigol 1996.

57. Around the same time, in 1902, the philosopher Eduard von Hartmann published *Das Weltbild der modernen Physik*, which focused more narrowly on overcoming the divergence of a number of different physical Weltbilder and on establishing the one Weltbild he deemed superior. Encouraged by kindred tendencies toward a natural philosophy (Ostwald), his goal was a "Naturphilosophie des Unorganischen" that went beyond the day-to-day business of physical science and investigated the basic issues underlying science (Hartmann 1902, iv). He spoke of three rival "fundamental systems: 1. the energy-space-time system, 2. the mass-space-time system, 3. the force application-space-time-system. The first leads to the energetical, the second to the hylokinetic, the third to the dynamist Weltanschauung" (ibid., 187). The third Weltanschauung was the one Hartmann himself espoused. In it, scientific laws and dynamic action together form the world: "The totality of the laws of the world exhausts the 'world as idea,' the world thought; the totality of dynamic action or force application elevates the mere world thought to the 'world as action' ['Welt als That'] and only thus makes the world a 'fact' ['Thatsache']. Hence law-like determination and force application are the two sides of the existence of the world and the events in it; only the former poses ever new tasks to knowledge, whereas the latter remains the same and receives all content including atomic structure from the former" (ibid., 209). Hartmann noted that physics is only dealing with the first aspect, tacitly assuming the dynamic function. He also strongly

supported the "ether hypothesis as one of the best-proven scientific hypotheses" (ibid., 137). Although the philosopher's contributions had relatively little impact on the physics community, they were more widely discussed among the intellectuals at large. Lenin, for instance, criticized Hartmann as "another German idealist of a more reactionary tinge than [Hermann] Cohen" (1927, 243). This outlook, according to Lenin, was the reason why Hartmann preferred dynamism while being correctly aware of the implications of such a decision: "Either materialism, or a universal substitution of the psychical for the physical" (1927, 244).

58. Also see Einstein's 1920 talk "Ether and Relativity Theory," *CP*, 7: 305–23.

59. The "electromechanical (elektromechanisches) Weltbild" makes no sense here; it is apparently a glitch. What Einstein probably meant to write was *electromagnetic*. Another mix-up was writing *Thermodynamik* instead of *Elektrodynamik* in the afore-mentioned central passage of his autobiographical notes where Einstein recounted his dissatisfaction with both the mechanical and the electrodynamic (not: thermodynamic!) Weltbilder (1979a [1949], 19–20). This error was spotted and corrected in later editions, with the approval of Einstein himself (Abiko 2003, 203–205). Nonetheless, the error had far-reaching consequences, because, according to Seiya Abiko, it misled some Einstein scholars to identifying him more closely with the electrodynamic Weltbild than is warranted.

60. Max Planck quickly realized that Einstein had made significant strides toward overcoming the division between the mechanical and electromagnetic Weltbilder. In 1908, he said, "In general, the original opposition of ether and matter seems to me to be somewhat receding. Electrodynamics and mechanics do not oppose each other as uncompromisingly as it is usually assumed in larger circles, where people even speak of a struggle between the mechanical and the electrodynamic Weltanschauungen" (1970d, 32). In Planck's opinion, physics was already well on its way toward a final Weltbild: "[I]n physics, we are justified in asserting that already our current Weltbild, though it is still iridescent in very different colors according to the researcher's individuality, contains certain features that no revolution, neither in nature nor in the human intellect, will ever be able to wipe out" (ibid., 49).

In "Die Stellung der Physik zur mechanischen Naturanschauung" of 1910, Planck discussed the "Revolution" and deep crisis in the wake of the demise of the traditional mechanische Naturanschauung—which is clearly a synonym for the mechanische Weltbild. He acknowledged the contribution of the theory of relativity to advancing the scientific Weltbild. "But physical questions are not decided according to esthetic considerations, but by experiments. . . . Exactly herein lies the high physical significance of the principle of relativity—that it provides a very precise answer, and one that can be checked by experiments, to a number of physical questions that previously were entirely unfathomable" (Planck 1970c, 67).

61. In his work on the Weltbild of physics, Eduard von Hartmann also discussed the epistemological underpinning of physics. He rejected any strict foundation of physics in sense experiences. In his view, physics constructs nature *deductively* from general concepts and axioms; induction from experience is merely an aid to create these concepts (1902, 210). He furthermore objected to the reduction of physics to description (Kirchhoff, Mach) and asserted that abstractions are necessary: Natural sciences, in Hartmann's view, are based on hypotheses—even though scientists often tend to ignore this fact (ibid., 211, 218). Hartmann also urged physicists to give probability estimates of their results. The transition to a statistical science would be a monumental advance for physics. "The task of a future physics will be to apply the principle of quantitative precision also to the calculation of the probability coefficients of the inductions of laws and causes, and, for each hypothesis, immediately to determine the size of the probable error of the probability coefficient that was calculated for it. Such a physics would differ from today's physics just as much as today's physics differs from a physics in which no mathematics at all was yet applied" (ibid., 222). This may sound like a curious anticipation of the uncertainty argument, but the thrust of the argument was the reverse: Rather than to focus on the fact that certain and total knowledge is impossible, the task of physics is to estimate probabilities as exactly as possible, and thus to *improve* precision. Hartmann further emphasized that the irreversibility of time was the foundation of causality in the world (ibid., 224–25).

62. On this controversy, see Toulmin 1970.

63. Elsewhere, too, Planck emphasized the role of speculative hypotheses against positivism (1932, 92, 86–87).

64. On issues related to the quantum-mechanical Weltbild, see d'Espagnat 1989.

65. This assumption had been particularly emphasized by Helmholtz (Anderton 1993, 245).

66. From a late-Marxist point of view, Hager and Röseberg (1977) also distinguished a sequence of Weltbilder of the physical sciences: Their starting point was the mechanical Weltbild. The integration of mechanics, thermody-

namics, and electrodynamics then brought about the transition to the Welt-bild of classical physics. The authors thought that the present physical Welt-bild was in flux and that it remained to be seen how the diverse debates in modern physics would turn out and whether there would be a new synthesis.

67. Planck was also concerned about the measurement problem and the ensuing uncertainty in the quantum area. While he favored a strict dynamic causality, he took a pragmatic view: "In his attempt to build up his hypothetical picture of the external universe the physicist may or may not, just as he likes, base his synthesis on the principle of a strict dynamic causality or he may adopt only a statistical causality. The important question is how far he gets with the one or the other" (Planck 1932, 99).

68. See a similar grouping by Engelhardt 1990, 114.

69. Virchow should probably be put somewhere between the fourth and fifth groups. Whereas at times he seemed more "radical" than his fellow organic physicists of the fourth group, he moderated his views considerably. Furthermore, a controversy flared up between Haeckel and Virchow who attacked Haeckel's *Deszendenzreligion* (Weingart, Kroll, and Bayertz 1988, 117–18).

70. *Realistische* Bildung, which focused on science, was contrasted with the *humanistische* Bildung, which focused on the classics.

71. Du Bois-Reymond acknowledged that Goethe was one of the greatest poets of all times: "Goethe's poetry is an inexhaustible source of universally human, that is, German Bildung, and a wing that is always ready for the flight out of the confines of the daily work of one's scientific discipline into the realm of the eternally true and beautiful. Hence, it should be present on every German student's bookshelf" (1912, 2: 179).

72. He also possessed the manuscript of an unpublished poem by Goethe, which he had published in 1878 for the Goethe centenary (du Bois-Reymond 1982, 158–59).

73. For a harsher critique, see Ostwald 1909, 182–83; 1910, 342–54.

74. Several factors promoted the astounding rise of research to the highest professional priority among the German professoriate in the early nineteenth century. As Steven Turner (1971) documented in the case of Prussia, these factors included a *Wissenschaftsideologie* that prized research, competition among and within the universities, and the various governments' hiring policies for professorial positions. (The universities were state run.) Also see Paulsen 1902, 68–69.

75. Forman (1971) asserted that those antirationalistic currents in the Weimar Republic were so strong that they made a number of physicists

espouse a-causal interpretations of quantum mechanics. For critical views on Forman's thesis, see Brush 1980; Hendry 1980; Kraft and Kroes 1984.

76. The quest for comprehensive and logically consistent systems pervades German academe even nowadays and even beyond the sciences. In his essay "What Bildung do Future Students Need? Reflections of a Jurist," German law professor Reinhard Zimmermann emphasized "that the legal matter does not form an insoluble and arbitrary jumble of individual rules and individual cases, but it can be traced back to a rational system. [Legal scholarship] strives to present the law as a logically consistent whole. And it attempts to demonstrate how individual rules and the solutions of individual cases can be deduced from general propositions, and how they thus can be understood and related to each other" (1996, 16). Therefore, he argued, the training of law students must be oriented more toward inculcating the awareness "that our law is more than merely the sum of individual legal rules, but forms a system that is rationally comprehensible and amenable to wissenschaftliche penetration" (ibid., 19). This kind of quest for a comprehensive and logical system contrasts with the focus, in English and American law, on individual precedents and cases. The analogies between this ethos in German law and the German scientists' pursuit of a Weltbild are evident. But, of course, legal systems are created by humans and can easily be changed, whereas scientists seek to discover immutable principles—"laws"—of nature.

77. "Göttinnen thronen hehr in Einsamkeit,/Um sie kein Raum, noch wen'ger eine Zeit./Von ihnen sprechen ist Verlegenheit./Nichts wirst du sehn in ewig leerer Ferne,/Den Schritt nicht hören, den du tust,/Nichts festes finden, wo du ruhst."

[Goddesses throne in solitude exalted,/Around them no space, much less time./To speak of them is awkwardness./You will see nothing in an eternally void distance,/You won't hear the step you take,/You won't find anything fixed where to rest.]

78. Letter to Marić, September 13(?), 1900, in *CP*, 1: 260.

79. The following passage, which Freud wrote in the very first of his collected *Letters*, illustrates that such allegiance to the classics in some cases even significantly influenced one's career. In that letter (June 16, 1873), Sigmund Freud, just turned eighteen, bragged to his friend Emil Fluss that he had passed the commonly dreaded battery of final exams, the Matura. He had done well on the whole, but he had particularly excelled in one exam marked "praiseworthy." It was the one in Ancient Greek, where he had to deal with a lengthy passage from King Oedipus. Freud stated proudly, "I had read that passage also on my own, and I made no secret of it" (1960, 6).

80. As Walter Moore put it in his biography of Erwin Schrödinger, "all German-speaking scientists [were] imbued with the spirit of Goethe. . . . They have absorbed in their youth Goethe's feeling for the unity of Nature" (1989, 49).

6.

ALBERT EINSTEIN'S BACKGROUND

Listening to his father reading from the German classics, especially Schiller and Heine, at the light of a lamp must have been one of young Albert Einstein's profoundly formative influences (Reiser 1930, 26). For father Hermann, the ideals of Kultur and Bildung clearly superseded those of traditional Judaism. Hermann Einstein was far from unique in this outlook; rather, he shared it with many German Jews of his generation. This chapter will locate Albert Einstein's nineteenth-century ancestors within the profound social changes occurring during that time. How and why did they come to embrace Kultur and Bildung? Before we turn to Einstein's family, however, we need to address the issue of Jews and Kultur more generally.

6.1. JEWS AND KULTUR

Before the age of Enlightenment, the Jews were a segregated minority in Germany (as in other countries) who came under special rules and were often treated harshly. Only when the Enlightenment gave rise to the idea that all persons, regardless of who or what they were, possessed certain fundamental rights, the "Jewish question" arose. The major products of Kultur spearheaded a universal humanism and thus legitimized the quest for emancipation. Just as the Bildungsbürgertum used Bildung and Kultur to denounce the privileges of the aristocracy, so too could the Jews base their demands for their own equality on these concepts. Hence, leading Jewish thinkers eagerly embraced Kultur and Bildung. For Moses Mendelssohn, Bildung was the vehicle of Jewish emancipation (Botstein 1991, 23); similarly, Berthold Auerbach later called for religion to become Bildung (Mosse 1985a, 4)—thus urging Judaism to emulate the development that was taking place especially among the Protestant Bildungsbürgertum.

Christian Wilhelm Dohm's (1781) influential book *Über die bürgerliche Verbesserung der Juden* (On the Civil Improvement of the Jews) triggered an intense debate about the Jewish question and marked the beginning of the process of Jewish emancipation. This process was protracted and differed from one German territory to the other. Yet, on the whole, the major advances came in two main phases, from 1780 to 1815 and from 1840 to 1870.[1]

Moreover, belonging to the ranks of Kulturträger and Bildungsbürger provided the opportunity for social assimilation—to the chagrin of the more orthodox Jews who lamented the rapid loss of Jewish traditions. Kultur was able to replace traditional Judaism, whose prescripts tended to reinforce self-segregation, and Kultur provided a shared framework in which Jews communed with Gentiles who likewise had abandoned traditional Christianity. Bildung and Kultur thus advanced "also the personal integration through friendships between gebildeten Jews and gebildeten Germans."[2] The Kulturreligion of the Protestants had its parallel in the Kulturreligion of the Jews. (See Graupe 1977, 95.)

As soon as they were given the opportunity, most Jews eagerly absorbed German Kultur.[3] "Yet there are ample indications that Jewish families even of modest means read German literature once they had been introduced to it in modern schools. . . . Preference was given to the German classics—Goethe, Christoph Martin Wieland, Friedrich von Schiller, Heinrich Heine" (Katz 1985, 87–88). Goethe was certainly lionized in the Bildungsbürgertum in general, but he was, if that was possible, even more fervently celebrated among its Jewish segment.[4] Most of Goethe's biographers were Jews (Barner 1986), and in the mid-1920s, Jews almost became the majority among the members of the Berlin Goethe-Gesellschaft (Mosse 1990, 171). German Jews also revered the other members of the classic pantheon of Dichter und Denker. They celebrated Schiller as the protagonist of human rights and liberty and, of course, Lessing, whose *Nathan der Weise* was the manifesto of religious tolerance (Toury 1977, 181–82). Of special significance was the converted Jew Heinrich Heine, who evoked strong reactions, both positive and negative, among the German Jews (Kahn 1985; Kahn and Hook 1992). George Mosse noted that some Jews placed great emphasis on the myth that Goethe and Heine liked each other, so as symbolically to cement the integration of Jews and Gentiles at the very Olympus of Kultur (1990, 171).[5]

Bildung became a major avenue of the social rise into the Bürgertum; it became what Jacob Toury called the "expression of the Jewish will to join the Bürgertum" (1977, 171). No longer attending segregated Jewish schools, Jewish pupils entered higher secondary education in enormous numbers. In 1864, the proportion of Jews among students in higher secondary education in Württemberg, Baden, and Bavaria was about twice the proportion of Jews among the general populations in these states (ibid., 173). The overrepresentation was even more drastic in Prussia, where the proportion of Jews among the students was more than six times as large as the proportion of Jews in the population. Furthermore, in this state, one in four Jewish elementary school pupils went on to higher schools in the 1860s—but only one in twenty Protestants, and one in thirty-eight Catholics (ibid.).

Once the old barriers had fallen, numerous Jews became successful in business, moving particularly into the new and expanding branches of the economy (Graupe 1977, 233–35). But even many of those who had secured a place in the Besitzbürgertum by virtue of their wealth aspired to Bildung and Kultur. As if with Pauline Koch, Albert Einstein's mother, in mind, Toury wrote, "Many, a very high percentage indeed, of the members of the Jewish middle class desired to be counted not only among the Besitzbürgertum, but also, and above all, among the Bildungsbürgertum" (1977, 181).

Thus, in general, the Jews enthusiastically aspired to and joined the ranks of Kulturträger and Bildungsbürger. Many worked in the legal and medical professions, both of which had roots in old Jewish traditions. Others entered journalism, literature, the arts, and other areas of Kultur. Although there was lingering discrimination (especially for government service, which included university positions), the names of Jewish Kulturträger in all Kultur fields amounted to impressive lists.[6]

The Jewish Kulturträger congregated in the larger cities whose social atmosphere they found most congenial (ibid., 180). Most of them were Zufallsjuden—"accidental Jews," for whom their Jewishness was nothing more than an accident of birth (Graupe 1977, 256). As George Mosse (1990) pointed out, Jews adapted the klassische concept of Bildung because it was open and progressive and because its universal humanism transcended the categories of religion, ancestry, and custom. This idea of Bildung was strongly imbued with the spirit of the Enlightenment, and it corresponded to the forward-looking elements of Kultur. As noted above, wide parts of the German Bildungsbürgertum in the late nineteenth and early twentieth centuries abandoned these notions and moved toward Neo-Romanticism, nationalistic mysticism, and the glorification of the German race. For obvious reasons, most of the Jewish Bildungsbürger could not participate in this shift, and they retained their classically inspired vision of Bildung and Kultur (which, for them, was the foundation of their own emancipation). According to Mosse, "the [classical] concept of Bildung, which to a large extent allowed the German Jew entry into the Bürgertum, became a part of their own Jewish identity, and . . . they held on to it even when the German concept of Bildung had drastically changed" (1990, 170, also see 1985b). In 1934, Arnold Zweig observed, "The Jew is the vanguard of this Western (that is, world) morality in a country that at this time denies everything that unites all humans" (quoted in Sombart 1989, 37).

The German Jews were, of course, no monolith. There were pockets of a more traditional life, especially in the smaller towns and villages, and there were the new arrivals from Eastern Europe, who often had distinct outlooks and lifestyles. Not all Jews subscribed to the universalist, or what they called liberal, concept of Bildung and Kultur, and the Jewish reactions to the change in German Bildung and Kultur and to the rise of anti-Semitism varied.[7] (See Meyer 1967.) Donald Niewyk described three typical responses of Jews in Weimar Germany to the increasing anti-Semitism: liberalism, Zionism, and German chauvinism—of which liberalism was by far the most common. He counted about three-quarters of the German Jews in this category (1980, 96).

6.2. EINSTEIN'S FAMILY IN THE NINETEENTH CENTURY

A fragmented patchwork of essentially rural territories at the beginning of the nineteenth century, Germany had transformed into a unified industrial superpower by its end. Large segments of the German people were caught up in the sweeping socioeconomic, legal, and political changes, but modernity came at a particularly rapid clip to the Jewish population. In 1832, 93 percent of the Württemberg Jews, for example, still lived in rural communities (Pyenson 1985, 66). Then a huge wave of migration depleted the Jewish communities in the countryside of Württemberg as well as of other German regions. From rural backwaters, Jews in great numbers moved to the urban centers, where they eagerly embraced the opportunities of modern life. Jebenhausen, for instance, the home of Albert Einstein's maternal ancestors, lost almost nine-tenths of its Jewish population within forty-one years; it had 550 Jewish inhabitants in 1845, but only sixty-one in 1886 (Tänzer 1927, 97). With but fifteen Jews left in 1897, the *israelitische Kirchengemeinde* could no longer function.[8] It was dissolved in 1899, its remaining members and its property being absorbed by the Jewish community in the larger town of Göppingen nearby. By contrast, Christian numbers in Jebenhausen did not experience such a precipitous decline; in fact, they increased from 832 in 1846 to 1,461 in 1869 (ibid.). Geographic mobility went hand in hand with socioeconomic mobility. Among the German territories, Württemberg was considered the most successful in aiding the Jews to abandon their traditional livelihood of peddling. Whereas in 1812, 85.5 percent of the Württemberg Jews were peddlers, the number fell to 38.9 percent in 1837 and to 17.7 percent in 1852 (Rürup 1975, 27).

To understand why most German Jews lived in the countryside before the nineteenth century, we take a brief glance back in history. (For more historical information, see Graupe 1977, 13–26.) There used to be a strong presence of Jews, many of whom were active in large-scale finance and trade, in the larger cities during the Middle Ages. Most large cities, however, expelled their Jews by the end of that period, whereas many small German territories welcomed *Schutzjuden* (protected Jews), offering them a safe haven in return for the *Schutzgeld* (protection money) and other taxes.[9] These small, often tiny, territorial entities regarded the settling of Jews as a way of increasing revenue and fostering economic development, given the scarcity of land under their control. The Jews were forced to specialize in commerce, such as peddling and the trading of cattle and horses and of agricultural products; they were not allowed to own land and could not enter the artisan guilds. The rural Jews, thus, were neither peasants nor artisans but traders, occupying a special niche in the agricultural economy. Once the legal restrictions that bound the Jews to this role were lifted, they left the countryside at a much faster pace than did the Christian peasant population with its stronger ties to the land. Pyenson described the culture of the rural Jews, which differed markedly from that of the Christian rural population (1985, 65): As the three traditional sensibilities of southern German Jews, he named "adaptability with regard to earning a

living, respect for learning, and desire for progress." These characteristics made them uniquely prepared for leaving their surroundings
once they could.

The history of Albert Einstein's ancestors exemplifies this large-
scale and multifaceted process of mobility: from the rural community to the city, from small-time trading to cutting-edge business and
industry, and from traditional Judaism to secular, universal Bildung
and Kultur.

Jebenhausen: Einstein's Maternal Ancestors

Jebenhausen, near Göppingen, was a *reichsunmittelbares Rittergut*
(a knight's estate that was subject only to the empire), which
belonged to Imperial knights, the Freiherrn von Liebenstein. In
1777, Freiherr Philipp issued a letter of protection for Jews to
settle in Jebenhausen (Tänzer 1927, 8). A few decades later, in
1805, Jebenhausen came to Württemberg.

In the early nineteenth century, a protracted process of turning
Schutzjuden into emancipated citizens began. The landmark law of
1828 (*Erziehungsgesetz*) improved the legal status of Jews in Württemberg, without, however, giving them full equality. This was
achieved only in 1869 (ibid., 58). Among other things, the 1828 law
promoted professional desegregation. Jews were allowed and
encouraged to quit peddling. A few Jebenhausen Jews became artisans; others chose to step up the scale of their trading operations.
And some became entrepreneurial pioneers in the textile industry—
a rapidly expanding branch of the economy in the early stages of the
Industrial Revolution. Typically, most of these textile businesses
soon migrated to other places, especially the nearby town of Göppingen, whose location on a railway mainline favored its burgeoning industry.

At the beginning of the Jewish settlement in Jebenhausen, the
education of the children was entirely up to the Jewish community.
Originally, there was an emphasis on Hebrew and religious instruction, and the rabbi was heavily involved in schooling. Increasingly,
private teachers were hired. When, in 1818, a Protestant minister

wrote a report about the Jewish school, a certain degree of secularization was already observed (ibid., 165–67). The pupils were taught reading in Hebrew and German, writing in Hebrew and German, and arithmetics. "Furthermore, there is no memorization. And neither is there any regular instruction in religion and ethics" (ibid., 167).

In the 1820s, the Württemberg government started to intervene in the Jewish schools. As the general idea of the equality of all citizens was taking hold, the government concluded that it should regulate the Jewish schools in the same manner as it did the Christian schools. In 1823, the provincial government (*Regierung des Donaukreises*) asked the Jews of Jebenhausen to consider establishing a public Jewish school. The Jewish community leaders strongly supported such a plan. According to a report from the Göppingen *Oberamt*, the Jewish leaders

> have no objections against the higher-level [school] administration designing the curriculum, prescribing the school regulations, and selecting the text books, as long as their religion is not interfered with. . . . From all this follows that this Jewish community, despite the slowing commerce and the increasing poverty of several families, recognizes the concern that the higher-level administration shows for the Bildung and the so necessary moral improvement of their children. The local rabbi, an enemy of all reforms, who, from a financial point of view, prefers that everything would remain the same, will try hard to persuade a few members of the Jewish community to support the preservation of the current educational institution, but the largest part of the Jews feels it necessary that they too must progress in Cultur. . . . (quoted ibid., 169–70)

Soon thereafter, in March 1824, the Jewish school in Jebenhausen became a public school. It is striking how rapidly the new concepts of Bildung and Kultur had diffused even to the provincial backwater.

Rabbi Tänzer, the historian of the local Jewish community, commented, "In the Jebenhausen community, the educational tendency of the law of 1828 had astonishingly swift and manifold effects in that a Bildungs-friendly attitude, which was substantially nourished by the excellent Jewish school, appeared very quickly in all areas of public and private life. Of course, this attitude also had harmful, but inevitable consequences for the future of the community. The modern Bildung that had entered this quiet, out-of-the-way village soon made the place look too limited in the eyes of many, especially younger, people" (ibid., 88). *Wandergeist* became rampant among the younger members of the Jewish community, and a sizable group of Jews left Jebenhausen for the United States and other far and near places (ibid., 36).[10]

How thoroughly and rapidly, in the course of a few decades, the German Jews assimilated Kultur can also be gauged from the outlook of the typical nineteenth-century German-Jewish migrant to America (Cohen 1984; Glanz 1970). In the new country, the immigrants tended to maintain a dual identity of being both Jews and Germans. Tenaciously many clung to a German Kultur, which to them meant the culture of universal emancipation, in the face of what they considered an inferior American culture. Emphasizing the role of the German language for the collective identity of the German Jews in America, the pithily named publication *Zeitgeist* proclaimed in 1880, "By the medium of the German language, that kind of noble Jewish spirit can be preserved that was first conceived by our immortal Moses Mendelssohn" (Cohen 1984, 61).

The history of the Koch family strikingly illustrates the migration process that emptied out most of the rural Jewish communities. In 1842, still in Jebenhausen, the brothers Sigmund and Hermann Koch founded an optical business, later to be joined by their younger brother Albert, the youngest of eight siblings.[11] We can fix the year when the Koch family resolutely entered modernity. In 1852, when their father Zadok, a peddler, died, five Koch sons left Jebenhausen in the pursuit of economic opportunities.[12] Sigmund and Hermann took their business to Stuttgart. Albert went to Paris to found an optical firm of his own. Two other Koch brothers, Heinrich, a grain and cattle trader, and Julius, a baker, left for Cannstatt near Stuttgart, where they ran a large-scale grain-trading business. Julius Koch (Albert Einstein's grandfather) closely matched the stereotype of the entrepreneurial pioneer—dynamic, shrewd, and strong-willed. Maja Einstein described him as "of a particularly practical intelligence and great energy. Any theorizing was alien to him" (Winteler-Einstein 1987, xlix). He also had a "choleric temperament," and Maja surmised that Albert's bad temper as a child was inherited from this grandfather (ibid.). The two Koch brothers combined their families in an unusual communal household (ibid.). The wives took turns cooking for everybody. One week, Jette Bernheimer, Julius's wife, was in charge of the kitchen; the next, it was Fromet Rothschild, Heinrich's wife.

Julius and his brother prospered and became purveyors to the royal court of Württemberg—an important mark of social distinction in those days. Although Julius was a rich and successful entrepreneur, he was also someone many Bildungsbürger would despise because to them he represented money without Bildung, having attended only the Jewish school in Jebenhausen. Julius probably recognized his lack of Bildung as a social drawback, and his dabbling in art collecting could be seen as an attempt to make up this deficit. Granddaugther Maja remembered, "With his wealth, he developed the inclination to become a patron of the arts; but he did so in a petty style and according to the principle of his profession, that is, to spend as little as possible for it" (ibid.).

In any case, his daughter Pauline was well versed in the elements of Bildung that were expected of a woman of the better families. Being a girl, she did not attend the formal institutions of Bildung, Gymnasium and university, but she was taught to play the piano and was acquainted with classical German literature. She was groomed for, and probably came to expect, life among a wealthy entrepreneurial Bürgertum that had acquired the essence of Kultur and Bildung. Pauline Koch possessed "a breath of genuine culture, a love of music which was to be inextricably entwined with her son's work and, in the pursuit of her ambition for him, a touch of the ruthlessness with which he followed his star. She appears to have had a wider grasp than her husband of German literature, and while for him Schiller and Heine were an end in themselves, for her they were only a beginning" (Clark 1971, 6).

Buchau: Einstein's Paternal Ancestors

Lewis Pyenson described Buchau in the middle of the nineteenth century as a "small bucolic town. Storks nested on the town roofs and fished in the nearby marshes of the Federsee. . . . Most of Buchau's residents engaged in agriculture" (1985, 64). The settlement was founded on an island in the lake called Federsee, which afforded good natural protection from attacks and raids. Over time, the shoreline has shifted, particularly because of land reclamation projects, and Buchau now lies on the lakeshore (Schöttle 1977 [1884], 27).

Constitutionally, the small town of Buchau was a slightly bizarre oddity. It was an imperial city (*Reichsstadt*) and, in this respect, similar to such other imperial cities as Frankfurt and Nürnberg (Müller 1912). It could not begin to compare itself, however, with these cities in importance or size. In 1760, it counted 1,170 inhabitants (Ladenburger 1987, 90). Next to the imperial city was the *Stift* Buchau, a small convent for aristocratic women, which also was *reichsunmittelbar* (not subject to any sovereign, except the empire). With the dissolution of the empire in the Napoleonic wars, both small entities lost their independence. From 1803 to 1806, they belonged to the Fürsten von Thurn und Taxis. Thereafter, Buchau came to Württemberg, and the Fürsten kept the Stift as private property (Mohn 1970, 24).

The Jewish community in this tiny city in the lake was of long standing. Pfarrer Schöttle, an early local historian, noted 1577 as the year when the archival record first mentioned Jews in Buchau (1977 [1884], 179–80; also Tänzer 1931). Joseph Mohn dated the Jewish presence in Buchau even back to 1382, thanks to a record from that year that was discovered in 1937 (1970, 10; also Ladenburger 1987, 88, 94). Albert Einstein's ancestral family had continually lived in Buchau since 1665, as proclaimed by a commemorative plaque on the house in Hofgartenstraße 14, where his parents lived until 1878.[13]

The story of the Buchau Jews was similar to that of the Jews of Jebenhausen and of other communities in the Kingdom of Württemberg. Traditionally, they had been peddlers traveling around the vicinity. When the barriers that confined them to that occupation fell, members of the Buchau Jewish community swiftly availed themselves of the new opportunities. From 1822 on, they were allowed to move into houses outside of the Jewish quarter (Mohn 1970, 21). "Especially after 1828, the Jews dealt in real estate, stocks and bonds, cattle, horses, leather, textiles, and household items. A few stores sold crucifixes, rosaries, and holy water fonts" (ibid., 49). In 1834–35, two textile factories were founded by Jews. In 1838, Hermann Moos opened a store that developed into a clothing factory (ibid., 34). "Since the creation of the German free trade zone (*Zollverein*), Jewish merchants from Buchau could be found in all parts of the member states. At present, they have been involved primarily in land and real estate trade. They rose so high that it is almost exclusively they who have money in abundance" (Schöttle 1977, 183).

In 1884, Pfarrer Schöttle commented on the rapid changes in the Jewish community: "Since the law of 1828 permitted the Jews to acquire citizenship, the local Jews, too, became citizens, and now participate in all public offices and positions of honor. In their present businesses, they keep a very large number of Christian workers and servants. Hence their influence is substantial. And they thus have attained a social position of which one could hardly have dreamt fifty years ago" (ibid., 184). An outward indication of the increasing stature of the Jewish community was that, in 1839, a new synagogue was officially opened in the presence of the king (ibid., 189). The event was widely reported in the press. The article in the *Frankfurter Ober-Postamts-Zeitung* on October 29, 1839, contained a heavy measure of Enlightenment rhetoric featuring religious tolerance and univeral brotherhood. It indicates how easily the tenets of classical Kultur merged with the project of Jewish emancipation: "Oh! Would that our fathers rose from their graves; that they saw it with their own eyes. How would they delight in this view; they who passed away in the horrible time of benighted intolerance; they who always heard the names 'Jews' and 'Christians' and never the word 'human being' or 'God's image'; they whose fate was always fear and never peace. Would that they rose from their graves, our fathers and the fathers of our Christian brethren, and be witnesses to the love of their children that they exhibit today, in the presence of the heavenly father of all human beings" (quoted in Mohn 1970, 26). With the best of intentions, though apparently somewhat underinformed about Jewish custom, King Wilhelm I of Württemberg donated a bell for the occasion. His noble gesture made the Buchau synagogue famous as the only synagogue in Germany with a belltower (Einstein 1991a, 168).

In the old days, the rabbi educated the Jewish children of Buchau. From 1770 to 1803, they went to the *Stiftsschule*, thereafter to the *katholische Stadtschule*, while they received private instruction in Hebrew. In 1825, a separate *israelitische Elementarschule* was instituted, upon the request of the Jewish parents (Mohn 1970, 22). A Martin Einstein graduated as a doctor of medicine in 1842 and practiced medicine in Buchau from 1862 until at least 1884 (Schöttle 1977 [1884], 194, 214). Pyenson counted him among Albert Einstein's "forebears" (1985, 65). Although Martin was definitely not a direct ancestor of Albert, he very probably was related.

Buchau was a small community. In 1821, it had 1,635 inhabitants; at the end of the century, in 1895, the number rose to 2,266. The Jewish group among the residents, however, dwindled almost by half between these two years, from 504 to 298 (Ladenburger 1987, 97). The process of Jewish out-migration also took place in Buchau, albeit at a less dramatic scale than in Jebenhausen. There were efforts to make Buchau itself a viable industrial location. Under pressure from Jewish entrepreneurs, the Württemberg railways provided the town railway access by opening a narrow-gauge line from Buchau to Schussenried in 1896. The town of Buchau supported this project with a sizable subsidy. Some twenty years later, another line was constructed to Riedlingen. When the National Socialists took power in 1933, there were still about two hundred Jews in Buchau.

Albert Einstein's paternal grandfather, Abraham, was a merchant (born 1808) who spent most of his life in Buchau. Maja noted that he "was said to have enjoyed great respect near and far as an intelligent and law-respecting man" (Winteler-Einstein 1987, xlix). Although his own education was apparently limited to what he had learned in the Buchau schools, he provided higher education for his sons Hermann and Jakob. Hermann attended the *Realschule* in Stuttgart. A bright and intelligent student, with a special talent in mathematics, he obtained the *Mittlere Reife* diploma. This entitled him to do his compulsory military service in a special program that lasted only one year and functioned as an officer-training course. With two additional military exercises, one could become a *Leutnant der Reserve*, which carried a heavy dose of social prestige. Hermann, however, never undertook those exercises (Fölsing 1993, 17–18). Jakob, about three years Hermann's junior, had more distinguished educational and military careers. He graduated as an engineer from the Stuttgart *Polytechnische Schule* and participated in the war against France as an *Ingenieuroffizier* (Hermann 1994, 73).

Einstein's Parents

After Hermann completed an apprenticeship as a merchant, he joined two cousins as a partner in an Ulm bed-feather business. Part of the Kingdom of Württemberg since 1810, Ulm had earlier been an imperial city and was famed for its tall-steepled cathedral.[14] It was located on the Bavarian border and had about thirty-two thousand inhabitants in 1879—not a big city, but much larger than Buchau with its two thousand residents (Hermann 1994, 72). In 1876, Hermann married Pauline Koch, eleven years his junior. For about a year, the couple lived in Buchau. Then they moved back to Ulm, where, with capital from his in-laws, Hermann set up a small electrical and engineering business that soon folded (Clark 1971, 6). Yet Hermann's interest in this expanding "high-tech" industry of his era continued unabated. He joined his brother Jakob's firm in Munich, which started as a business for water and gas installations and soon turned into a small factory for electrical equipment. (For a detailed description, see Pyenson 1985.) Jakob was in charge of the technical side; Hermann handled the business side.

Hermann had an easygoing and kind personality. Albert remembered him toward the end of his own life as "exceedingly friendly, mild, and wise" (quoted in Clark 1971, 5–6). Philipp Frank wrote,

"By nature, Hermann Einstein was an optimist and loved life. . . . His lifestyle and his Weltanschauung were in no way different from those of the average citizen of an average-size town in southern Germany" (1949a, 9). Pauline was tougher and more strong-willed than Hermann, and, despite the considerable age difference, she was clearly the dominant personality of the couple.

Hermann Einstein never lived up to the level of business success and wealth that prevailed in his wife's family. The business in Munich failed, and the other ventures he then undertook in northern Italy were similarly precarious. The electrotechnical firm the brothers started in Pavia failed in 1896 (Winteler-Einstein 1987, liv; also Pyenson 1985, 50). Without Jakob, Hermann set up another electrotechnical firm in Milan, which was liquidated in 1898 (Winteler-Einstein 1987, lv n. 28). Then he obtained concessions to install and run electric lighting systems in two Italian towns (ibid., lv n. 29). These businesses did not founder, but now Hermann's chief creditor saw his chance to recover some of his money. This creditor was the husband of Fanny, his sister-in-law—a man named Rudolf Einstein. Rudolf, a textile entrepreneur in Hechingen, did not share his last name with Hermann by coincidence. He was his first cousin (and also the father of Elsa Einstein, who was to become Albert's second wife). In a letter to Mileva Marić, Albert complained about how much "my dear uncle Rudolf ('The Rich')" nagged his parents to get his money back (Renn and Schulmann 1992a, 38). The almost permanent atmosphere of crisis that pervaded Hermann's business career appeared to have taken its toll. Hermann died in 1902 from a heart ailment, only fifty-five years old.

Hermann Einstein so stubbornly persevered in entrepreneurial activities largely because he wanted to spare his wife the social *declassement* that his becoming an employee would have implied. As Maja put it, "In particular, he did not want to hurt his wife who would have had a very hard time accepting any potential demotion on the social scale" (Winteler-Einstein 1987, liv). Pauline was a Koch, used to belonging to the wealthy entrepreneurial Bürgertum, and she always appeared to have been able to round up capital from her family for her struggling husband's business ventures—a person whom, one supposes, some Kochs might have suspected of "black sheep" tendencies.

Hermann Einstein still knew what life in Buchau was like, but he also received a modern education in the city of Stuttgart. He preferred a modern lifestyle by far. For Pauline, the rural setting was one more generation removed; she grew up in an entrepreneurial, cosmopolitan family. She shared Hermann's general outlook, and, if anything, she may have been more eager to acquire the symbols of Bildungsbürgertum than was Hermann, who gravitated toward the gemütliche concept of the good life that suffused the lower classes of southern Germany.

Ludwig Feuchtwanger's description of the Munich Jews in the late nineteenth century appears to fit well for the Einstein family: "They pushed the religious and cultural values of their parents to the periphery of their lives, embracing instead a modernist worldview. Theirs was an optimistic vision of the future, one strongly coloured by humanitarian impulse. They sought material success in their trades and careers, and looked forward to attaining respect from the community at large. . . . While not especially devout, they valued donating time and money to social causes, especially Jewish charities" (quoted in Pyenson 1985, 67).

Anton Reiser—a pseudonym for Albert's son-in-law Rudolf Kayser—characterized the Einsteins' home life as a "well-to-do philistine atmosphere" (1930, 27). This sounds almost as if it were a direct quote from Albert because *philistine* was indeed one of his favorite words for disparaging his family's lifestyle. According to Reiser, Hermann "loved literature and in the lamplight, evenings, read aloud Schiller and Heine. He was not concerned about politics, yet the rising power of the new Empire and Bismarck's mighty form impressed him" (ibid., 26). The family members had self-consciously distanced themselves from traditional Judaism. "Albert longed for a religious life and for religious instruction. But he heard only ironic and unfriendly talk about dogmatic ritual. His father prided himself on the fact that he was a free-thinker. His belief harmonized with the thought of his time, which was controlled by the philosophy of materialism. Albert's father was proud that Jewish rites were not practiced in his house" (Reiser 1930, 28).

For understanding the family's social position and aspirations, it is very significant that Albert was sent to the Gymnasium. This was apparently an unusual choice for a manufacturer's son. A study of Rhineland entrepreneurs showed that most of them sent their sons to *höhere Bürgerschulen* or Realschulen, which were more technically and scientifically oriented (Pyenson 1985, 49; Kocka 1975, 62). To a practical-thinking entrepreneur, the benefits of a classical education in a Gymnasium were not entirely obvious. Indeed, the philosophy of the Gymnasium emphasized nonutilitarian knowledge. Its students were supposed primarily to acquire Bildung, rather than job skills, so that they could become the Kulturträger of the next generation.

By the end of the nineteenth century, many Kochs were already firmly entrenched in the entrepreneurial wing of the Bürgertum and sought to add the trappings of Bildung and Kultur to their wealth. With pride, they supported Albert's education. Aunt Julie (the wife of Pauline's brother Jakob Koch), who lived in Genoa, Italy, sent her "little professor" a monthly allowance of a hundred Swiss Francs to Zürich (Highfield and Carter 1993, 46). This benefactor was, in Albert's eyes, a prime example of the dreaded philistine. Recall that, in a letter to Mileva, Albert described Julie as "a veritable monster of arrogance and insensitive formality" (Renn and Schulmann 1992a, 11).

Albert

How did young Albert react to his background? As indicated previously, he definitely disliked the elements of philistine lifestyle he saw around himself. He did not regard wealth as a supreme value, perhaps in part because he experienced the dark side of capitalism, the (literal) heartbreak of a struggling entrepreneur. He also was much more interested in basic science than in applied technology, which Hermann and Jakob Einstein, as well as a branch of the Koch family, had chosen as their career. Despite his father's hopes that Albert would become an electrical engineer and would join the family business, Albert stayed clear of engineering and focused on theoretical physics instead. Albert Einstein also was less than enthusiastic when his son Hans Albert reverted to the family tradition by going into engineering (Einstein

1991b, 41). Hans Albert became a distinguished professor of hydraulic engineering at the University of California at Berkeley.

In his penchant for basic science, Albert Einstein was precisely the product of Bildung that his family had sought for him as the capstone of a remarkable intergenerational process of social mobility. By attending the Gymnasium, Albert was poised to become the first certified Kulturträger in his immediate family. But the nonutilitarian, idealistic, and rebellious elements of Bildung (described before) were the basis from which he could criticize as trite and vulgar, that is, "philistine," the way his family and relatives lived.

When Albert Einstein arrived in Berlin, he had come a long way from the small town of Ulm, where he was born. Berlin in the early twentieth century held a strong claim to being the most avant-garde and cosmopolitan city in the world (Hall 2001). As a famed Kulturträger, Einstein could move in the premier intellectual, artistic, and political circles of this metropolis.[15] Yet Albert and Elsa never fully adopted the ways of the sophisticated and cynical urban elite and retained a simpler demeanor. Count Harry Kessler, a rich diplomat and patron of the arts, commented on a dinner party at Einstein's place with a group of society figures, "A kind of emanation of goodness and simplicity removed even this typical Berlin party from the ordinary, and transfigured it through something almost patriarchal and fairytale-like" (1982, 289).

While a number of observers noted this trait of simplicity, not everybody shared Count Kessler's enchantment with it. Thomas Mann's wife, Katia, had this to say about the physicist, whom she saw frequently in Princeton, where the Einsteins and Manns lived in the same neighborhood: "He was very nice and not particularly stimulating. Einstein had indeed something child-like in his personality, such big baby eyes; he had something naive about him, a sweet person; and he definitely was an extremely one-sided genius, wasn't he? Really an enormously specialized talent, so in everyday life he was not a very impressive person" (Mann 1974, 122).

Einstein remained a stranger to the sophisticates and probably had little desire to impress them or even to fit in. He was the antithesis of the social climber. For him, issues of social status and prestige paled against his quest for the physical Weltbild.

NOTES

1. For the history of Jewish emancipation, see Rürup 1975. This book also includes a short history of the term *Emanzipation* (Rürup 1975, 126–32).

2. Mosse 1990, 169; see Mosse 1985a, 10–11.

3. See Katz 1985 and Mandelkow 1990, 192–94. On the Jewish reaction to Kultur, see Reinharz and Schatzberg 1985.

4. There are, of course, exceptions. One of Goethe's most notable detractors, Ludwig Börne, was also a Jew.

5. For example, see Fritz Friedlaender's *Heine und Goethe* (1932) and also Walter Robert-tornow's *Goethe in Heine's Werken* (1883). Robert-tornow was related to Rahel Varnhagen, a converted Jew and ardent admirer of Goethe, who was a major driving force behind the Goethe cult (Frühwald 1990, 207).

6. See especially the monumental compilation of Kaznelson (1962), *Juden im Deutschen Kulturbereich*, Toury (1977, 178–210), and a recent German volume of conference papers, titled *Juden als Träger bürgerlicher Kultur in Deutschland* (Schoeps 1989).

7. Note that this use of the term *liberal* differs from contemporary American usage.

8. The traditional minyan prescript required the presence of a minimum of ten adult Jewish males at services.

9. In the fourteenth century, the originally royal prerogative of protecting Jews passed to the electors (*Kurfürsten*), and later to all components of the German empire (*Reichsstände*).

10. Utz Jeggle observed that, in Württemberg, the poorer Jews emigrated to the United States, whereas the wealthier ones relocated their businesses to nearby towns (1969, 196–99). For details on the Jewish emigration to America, see Cohen 1984 and Glanz 1970.

11. Albert Einstein may have been named after this maternal great-uncle, rather than after his grandfather Abraham, as is generally assumed (*CP*, 1: 1n2; Hoffmann and Dukas 1972, 8; Fölsing 1993, 15).

12. Originally Zadok Löb; in 1821 the name changed to Dörzbacher, in 1842 to Koch. See Tänzer 1927, 48, 52.

13. A guidebook about *Bad Buchau und der Federsee* had a Jost Einstein among its contributors (German, Filzer, and Einstein 1987).

14. In the *Reichsdeputationshauptschluss* of 1803 (in which several German princes received church-ruled territories, abbeys, and imperial cities

in compensation for their territorial losses to France west of the Rhine), Ulm lost its independence and became part of Bavaria. A treaty between Bavaria and Württemberg in 1810 transferred Ulm to Württemberg.

15. On Einstein's interactions with the cultural milieu of Berlin, see Stern 1999 and Levenson 2003.

7.

REBELLION AND TRADITION

Roots of Einstein's Physical Weltbild

Having become acquainted with the central intellectual elements of German culture, we are now equipped to revisit the paradox posed in the first chapter. There we had drawn up seemingly contradictory images of Einstein the rebel and Einstein the traditionalist. How exactly are these opposing elements intertwined? Might we have to conclude that in Einstein we are dealing with a split personality? Or that the opposite kinds of Einstein, hopelessly contradictory, cancel out into impotence? No, these are two different components of one coherent mental structure that uses the apparently conflicting parts to support each other. It is the very synthesis of these elements that allowed Einstein's mind to be so extraordinarily creative.

7.1. SYNTHESIS

Considering the assorted rebellious and traditionalist aspects of Einstein, we would miss the point if we merely concluded that in Einstein's persona there was something like a demarcation line that divided Einstein into part rebel, part traditionalist. In our discussion of Kultur, we found that it contained in itself a rebellious element that, precisely because of its uncompromising devotion to the central ideals of Kultur, riled against a cultural reality that was found wanting. Similarly, Einstein's peculiar brand of rebellion did not condemn Kultur from the outside, and it did not reject its core values. It could draw on a tradition of rebelliousness that was a subterranean current within Kultur itself.

As we could already witness in several of Einstein's letters, he frequently complained about the philistines, while extolling Bohemian life as a "gypsy." In this, he closely resembled the poets of "Sturm und Drang" and "Romantik," for whom, too, the philistines were the primary bogeymen. For many German poets, Goethe among them, their prized ideal of individuality had two aspects: On the one hand, the individual despised and defied convention; on the other, the individual revered other exceptional individuals and geniuses of all times. Individuals thus were not merely rebels; they also became loyal members of a supratemporal community of geniuses—that existed in a parallel universe to that of the philistine masses. This mixture of rebellion and reverence for great individuals also characterized Einstein. His ticket to this exclusive community of geniuses lay in his unrelenting quest for truth, which "frees us from the fetters of the ego and makes us companions of the best" (Einstein 1929, 29).

From Einstein's perspective, the community of geniuses contained Goethe himself, of course, along with the other great poets of similar persuasion. When Kultur was officially canonized, it was impossible to avoid certain elements of rebelliousness because they were integral parts of the works of most of the "great German poets." Thus, we find another fault line in Kultur, in addition to the internal cleavages we already encountered. Owing to its in-built aspect of rebellion, conventional Kultur contained the seeds of its own destabilization. Most philistines would, of course, never notice, but rebels could draw inspiration from Kultur itself. To give a twist to our Kulturträger metaphor: The pillars of Kultur were laced with explosives; every now and then a fuse was found. In many cases, therefore, rebelliousness did not completely negate Kultur but invoked the rebellious tradition of Kultur against its own calcification.

Einstein thus drew on an extended tradition of rebellion that was embedded in Kultur itself (and found its societal substrate among the "free-floating intellectuals"). He often felt compelled to speak out frankly on political issues, and he did so without regard to the potential repercussions. When his worried friend Max von Laue urged him to be more circumspect, Einstein replied: "Where would we be if Giordano Bruno, Spinoza, Voltaire, and Humboldt had thought and acted this way" (quoted in Stern 1980, 44). In a way, thus, Einstein saw himself as a member of the community of defiant geniuses and laid claim to a tradition of rebelliousness nourished by Kultur.

Einstein's rebellion was motivated by, and remained true to, the core ideals of Bildung and reason, as they were defined in the late eighteenth and early nineteenth centuries. It differed drastically from that of radical rebels, nihilists and existentialists of various stripes, who began to emerge by the late nineteenth century—those who, after realizing that the brilliant ultimate goal of a comprehensive and rational Weltanschauung was unsustainable, turned around in bitter disappointment and completely rejected all values for which Weltanschauung stood. (See Camus 1958.) This dark offshoot of Kultur played no role in Einstein's intellectual makeup, but it severely affected his life circumstances.

7.2. THE CULTURAL THEMA OF EINSTEIN'S PHYSICS

We now examine how this synthesis of rebellion and traditionalism was achieved in Einstein's physics. We start by showing how Einstein's physical research program resonated with Kultur. There was, first of all, the shared centrality, in Kultur and Einstein's work, of the quest for a unified world picture (for a Weltbild and a Weltanschauung).

As we have seen, German-speaking scientists, prepared by a common core of humanistische Bildung that linked them to the other Kulturträger, tended to identify with the quest for the grand synthesis of a Weltbild or Weltanschauung that Goethe's *Faust* typified. This was also true, to a particularly high degree, for the environment in which Einstein moved. His colleague and friend Aurel Stodola (1931), for instance, a professor of engineering at ETH, published *Gedanken zu einer Weltanschauung vom Standpunkte des Ingenieurs* (Thoughts on a Weltanschauung from the Engineer's Point of View.)[1] And to Stodola's *Festschrift*, Marcel Grossmann, Einstein's lifelong friend, contributed the essay "Fachbildung, Geisteskultur und Phantasie," replete with typical Kulturträger rhetoric. Grossmann advocated "Zivilisation [as] the substrate of Kultur," and he lamented the high level of specialization in the training of engineers in particular and of academics in general (1929, 187–88). Einstein's friends were also familiar with the imagery of science that, as we showed earlier, permeated the German scientific community in the nineteenth century. A letter from Besso (February 23, 1939), written in a somewhat overwrought style, for instance, was replete with the religious metaphors of science that hark back to the nineteenth century, for example, "cult of the Goddess Reason" and "You have served and you still serve: in the chain of riddles, in the building of the temple, in the ascent to the mountain" (Einstein and Besso 1972, 342, 343).

With the wider scientific community as well as with his friends, Albert Einstein shared the goal of a grand, all-encompassing Weltbild. He professed his fundamental, even existential, allegiance to it eloquently in the address he gave in 1918 in honor of Max Planck's sixtieth birthday, which was originally titled "Motiv des Forschens," and in English was published under the title "Principles of Research."[2] That Einstein gave a speech on this occasion was particularly fitting because Planck was perhaps the one among Einstein's colleagues who resembled him the most in his ardent quest for the Weltbild. One might presume that, largely because of this shared goal, Max Planck, editor of the *Annalen der Physik*, had immediately understood the significance of the work submitted by an unknown Swiss patent officer, had taken an interest in the brilliant young physicist, and later was instrumental in persuading Einstein to move to Berlin, the center of the physics world at the time. Without Planck's support, Einstein's career in physics might well have had a less spectacular trajectory.

A reference to the religious metaphor of the "Tempel der Wissenschaft" started the talk, and then Einstein solemnly stated that, into the shaping of a world picture, every artist, philosopher, or scientist, each in their own way, "places the center of gravity of their emotional life" (1955a, 107, 108). Einstein called that search for a world picture *"the supreme task"* of the physicist—the task "to arrive at those universal elementary laws from which the cosmos can be built up by pure deduction" (Einstein 1954, 221). This is what, in his view, distinguishes the physicist's Weltbild from the other possible *Weltbilder*: It is in principle, though not in actuality, comprehensive because "the general laws on which the structure of theoretical physics is based claim to be valid for any natural phenomenon whatsoever. With them, it ought to be possible to arrive at the description, that is to say, the theory, of every natural process, including life, by means of pure deduction, if that process of deduction were not far beyond the capacity of the human intellect. . . . There is no logical path to these laws; only intuition, resting on sympathetic understanding of experience, can reach them" (ibid.). Thus, he leaned toward Planck's side in the Mach-Planck controversy discussed earlier. Rather than letting theory emerge strictly from sense perceptions of the phenomena, he spoke of a "pre-established harmony" à la Leibniz between the theory and the phenomena it explains (ibid., 222).

For Einstein, this talk about the Weltbild was more than merely nice words to be ritualistically uttered at celebratory occasions. As we will show now, his life's work as a physicist, starting with his very first publication on capillarity in 1901, was consciously devoted to that Faustian search for the Weltbild.[3] In hindsight, in his autobiographical notes, Einstein explained how his research program rose from his dissatisfaction with the mechanical Weltbild that with "dogmatic rigidity" dominated the physics of his student days (1979a [1949], 7). He found it seriously flawed by several shortcomings, one of which he labeled "besonders häßlich" (particularly ugly) (ibid., 10). The choice of such a strong word—in German, its literal root is *Haß* (hatred)—indicates that he took it almost as a personal affront that the physical Weltbild at that time fell short of its own ideal. Furthermore, when Einstein realized, shortly after 1900, "that neither mechanics nor electrodynamics (except for marginal cases) can claim exact validity," he fell into despair (ibid., 19). "More and more I despaired of the possibility to find the true laws by constructive efforts that are based on known facts. The longer and the more desperately I labored, the more I became convinced that only the discovery of a general, formal principle could lead us to certain results" (ibid., 19–20). Again, this personal anguish signifies Einstein's deep commitment to the quest for the Weltbild.

This commitment shaped Einstein's scientific career from the beginning. In his first paper, Einstein tried to remove a duality between Newtonian gravitation, which directs the motion of macroscopic objects downward, and capillary action, which drives the molecules of the submicroscopic world of the liquid upward. In its way, this is indeed a search for the connection between the macrocosm of observable gravitation and the microcosm of molecular motions. Here was a case where apparently opposite phenomena could be brought into a common vision. Right there, Einstein's lifelong interest in the program of a unification of the forces of nature made its tentative first appearance. In a letter to his friend Marcel Grossmann (April 14, 1901), he exclaimed about his work with Keplerian enthusiasm: "It is a magnificent feeling to recognize the unity (*Einheitlichkeit*) of a complex of phenomena, which to direct observation appear to be quite separate things" (*CP*, 1: 290–91). Even though Einstein later discussed the context of that first paper as not worthwhile, he never turned his back on the substantive goal.

In 1902, Einstein sent from Bern a paper on extending Boltzmann's ideas in thermodynamics and statistical mechanics and for the first time used the word *Weltbild* in a publication, even as he was beginning to try to build his own.[4] And Einstein kept using the term *Weltbild* also in his correspondence, for instance, in a letter to Lorentz (January 23, 1915).[5]

Of course, Einstein did not always agree with the way others tried to achieve the Holy Grail, but he respected their attempts. A revealing example is a review he published in 1917, right in the middle of his second miraculous burst of creativity (Einstein 1917). For even while quite sick—successively dealing with general fatigue, liver ailments, stomach ulcer, and jaundice (Pais 1982, 299–300)—during those war years, he succeeded in creating what he called a *generalized* theory of relativity, with its cosmological conclusions. In addition, he was then publishing a stream of other results, ranging from the experiment with de Haas on the gyromagnetic effect to the quantum theory of induced emission (which was the key idea later making possible the laser). Yet he found time and felt the necessity to review a new book that must have looked attractive to him. It was a volume that reprinted two lectures by Helmholtz on Goethe, and the review gave Einstein the opportunity to express his solidarity with his fellow seekers of a coherent world picture, even if he could not accept all of Goethe's science. "The second lecture will be read with delight by everybody who is able to enjoy the wissenschaftliche view of the world. At the end of his life spent struggling for wissenschaftliche insight, Old Helmholtz shows what Goethe's Weltbild is all about" (Einstein 1917, 675).

The search for a comprehensive physical Weltbild was Einstein's lifelong intellectual quest. A few days before his death, Einstein probably meant this quest, whose ultimate goal he realized had eluded him, when he told Janos Plesch, "My whole life I have tried to think one single thought to its end. Not once have I succeeded" (quoted in Hermann 1994, 546). Incidentally, now we can understand Einstein's wrath in a letter of 1934 to his publisher that to a mere collection of his diverse articles had been given, without his permission, the exalted title *Mein Weltbild*.

In his program of physical research, Einstein thus wholeheartedly adopted the *goal* of a unified Weltbild that he found in his cultural environment. But, almost by definition, a widespread cultural goal that was shared by most of the science community cannot sufficiently account for the singular quality of Einstein's work. Einstein's rebellious streak was a crucial additional factor, just as Gardner's (1993) theory maintains. The marginality of the rebel allowed Einstein to operate at what one might call an optimal distance from the accepted theories. Einstein retained his childlike wonderment about the most basic questions and simply would not stop asking them and thinking about them. He never lost his youthful insistence on getting the full truth, rebelling against half-truths and convenient conventions, against the realism and pragmatism often considered the hallmark of a mature adult. His approach was, to borrow Jerome Kagan's term, "optimally discrepant," neither too rebellious nor too obsequious to received wisdom, and from harnessing together his seemingly centrifugal tendencies sprang his creative advances (2002, 159).[6] In order to build the comprehensive Weltbild, he first had to clean the slate of existing unsatisfactory attempts, go back to the basics, and then begin a fresh, better attempt.

In this endeavor, he could draw on a somewhat marginal school within the science of his day, a school that, like Einstein, believed that much of the physics of the late nineteenth century was standing on clay feet and that a thoroughgoing critique was needed.[7] Ernst Mach thus became an important companion at the outset of Einstein's heroic journey—"a man of rare independence of judgment," as Einstein (*CP* 6: 278) praised him in his Nachruf (1916).[8] Mach, for instance, was critical of the conventional notions of an absolute space and an absolute time, and this may have inspired Einstein's own bold revisions of the concepts of space and time, which were at the heart of the theory of relativity. In a letter to Moritz Schlick (December 14, 1915), Einstein wrote: "You also realized correctly that this school of thought [positivism] greatly influenced my efforts, that is, E. Mach and even more so Hume whose *Enquiry Concerning Human Understanding* I studied with diligence and admiration shortly before discovering the theory of relativity. It is very well possible that I would not have found the solution without these philosophical studies" (Einstein Archive #21610). In his "Autobiographisches," Einstein again gave credit to Mach and Hume: "The critical thinking that was necessary for discovering this central point [of special relativity] was decisively advanced especially by my reading of David Hume's and Ernst Mach's philosophical writings" (1979a [1949], 20; see p. 5).

Eventually, however, there was a parting of ways. Mach's stinging criticism of the conventional physical thinking of the day served Einstein well in that first rebellious, destructive phase. Yet for Einstein, this was not an end in itself, just the necessary preparation of clearing the ground. Einstein's goal remained the achievement of a unifying world picture, and therefore he had to leave behind what he considered the mainly negative forces of Mach. A splendid illustration of this can be found in the Besso-Einstein correspondence. Michele Besso (letter dated May 5, 1917), still apparently a Machist, chided Einstein for his critical stance toward Mach: "As to Mach's pony, let us not scold it; for has it not seen you through the hellish journey across the relativities? And who knows, it might even carry its horseman, Dom Quixote de la Einsta, through at the evil quanta!" (Einstein and Besso 1972, 110). Einstein replied (May 13, 1917), "I do not scold Mach's pony; you know, what I think about it. Yet it cannot give birth to anything living, it can merely exterminate harmful vermin" (ibid., 114). This exchange indicates quite clearly Einstein's dissatisfaction with mere critique and his desire to create something positive—the physical Weltbild.[9] On this quest, Einstein journeyed away from Mach and toward rationalism and realism (Lanczos 1965, 14). As Don Howard (1984, 1990, 1994) showed, Einstein moved, probably under the influence of Duhem, toward a more conventionalist and holistic view of epistemology. Thus he diverged from the positivism of people like Schlick and Reichenbach, with whose positions he earlier had enthusiastically agreed.

Having seen that Einstein's physical research program was deeply committed to the culturally pervasive *goal* of creating a comprehensive physical Weltbild, and that his rebellious independence helped him in radically clearing obstacles, we now turn to a third influence on Einstein's physics. Here we look at the major *tools* that enabled Einstein to advance toward his goal—at the predispositions and strategies that shaped his actual scientific work. To describe these factors that influence the way scientists do their work, Gerald Holton (e.g., 1973) introduced the very useful concept of *themata*. Again, we will see that some of Einstein's most fundamental guiding themata came from the surrounding Kultur.

To a considerable, and perhaps surprising, extent, Einstein's research was *not* driven simply by empirical facts. For example, it used to be commonly assumed, as seems offhand so reasonable, that the famous Michelson-Morley experiment was the crucial influence leading Einstein to the relativity theory. Robert A. Millikan, for instance, after describing the Michelson-Morley experiment, concluded with the sentence: "Thus was born the special theory of relativity" (1949, 343–44). Einstein disagreed. In 1952, R. S. Shankland, then the chairman of the physics department at the Case Institute of Technology in Cleveland, contacted Einstein about the talk he (Shankland) was shortly to give in honor of the hundredth anniversary of Michelson's birthday. He hoped, of course, to be able to embellish his commemorative speech with a reference to Einstein's acknowledgment of Michelson's critical contribution to the theory of relativity. But Einstein replied frankly that the influence of the Michelson-Morley experiment on his own thinking was rather indirect, and he recounted the real influences on his theory of special relativity. He said that he had relied on well-established, earlier findings—experiments by Faraday, Bradley, and Fizeau. "They were enough" (in Shankland 1963).

Similarly, Einstein replied to F. G. Davenport's inquiry about the influence of Michelson's work on the creation of the relativity theory (February 9, 1954), "In my own development Michelson's result has not had a considerable influence. I even do not remember if I knew of it at all when I wrote my first paper on the subject (1905). The explanation is that I was, for general reasons, firmly convinced how this could be reconciled with our knowledge of electro-dynamics. One can therefore understand why in my personal struggle Michelson's experiment played no role or at least no decisive role" (quoted in Holton 1973, 325–26).

Moreover, Walter Kaufmann's experimental "disproof" of the relativity theory did not faze Albert Einstein. There was a two-year lag (1905–1907) of responding to the distinguished experimentalist's findings. Then Einstein basically said that more data would be necessary to decide whether Kaufmann's study was erroneous or whether the theory of relativity was wrong. His intuition told him that Kaufmann's data must have been faulty because they supported rival theories that had a much narrower scope than the theory of relativity. "In my opinion, those theories have a rather small probability, because their fundamental assumptions concerning the mass of moving electrons are not explainable in terms of theoretical systems which embrace a greater complex of phenomena" (*CP*, 2: 461).

What is perhaps most astonishing about Einstein's work as a physicist is his courage to place his confidence, often against much of the available evidence, on a few fundamental guiding ideas, which he called categories in a non-Kantian sense, that is, freely choosable (Einstein 1979b, 500; see 1949, 674). Gerald Holton (1973) identified the same courageous tendency among many major scientists who put their chips early on a few nontestable but highly motivating presuppositions, which Holton named *themata*. In Einstein's case, the set of themata included *simplicity*, harking back to Newton's first rule of philosophy, that "Nature is pleased with simplicity, and affects not the pomp of superfluous causes" (1934 [1687], 398). Einstein wrote veritable hymns to the concept of simplicity as a guide in science, and exemplified it in his own lifestyle (Holton 1986, 15). In a letter to Cornelius Lanczos (January 24, 1938), Einstein acknowledged that a deeply held belief in the mathematical simplicity of nature had superseded the Machian skepticism of his younger years: "Coming from the skeptical empiricism of the Machian type, I was transformed, by the problem of gravity, into a faithful rationalist, that is, into someone who searches the only reliable source of truth in mathematical simplicity" (quoted in Holton 1973, 241).

A second of Einstein's thematic presuppositions was *symmetry*— a concept that he introduced into physics in 1905, when most of his readers surely wrote it off as an aesthetic, optional choice, but which has become one of the fundamental ideas in modern physics.[10] Yet another thema was his belief in strict *Newtonian causality and completeness in description*, which explains why Einstein could not accept as final Niels Bohr's essentially probabilistic, dice-playing universe.

Einstein's strong belief in the *continuum* was an additional thema, as exemplified by the field concepts that enchanted him from boyhood on when he saw his first magnetic compass.

There were two or three more themata to which he also clung obstinately, but the most important thema was without doubt that of *unity, synthesis, wholeness* which in itself is closely linked to the goal of a Weltbild.[11] As he put it in a 1916 letter to the astronomer W. deSitter, he was always driven by "my need to generalize" (*CP*, vol. 8A: 359). That need continued uninterrupted, from his first paper on capillarity to his last ones on finding a general unified field theory that he hoped would join gravity and electromagnetism and even provide a new interpretation of quantum phenomena (Pais 1982, 9). In between, that preoccupation had led him from the special theory to what he first called typically the *verallgemeinerte*, the generalized, theory of relativity. He even gladly lent his name to join thirty-two other scholars in the publication as early as 1912 of a public manifesto on behalf of a new society aiming to develop across all sciences their "unifying ideas" and "unitary conceptions." (See Holton 1998, 26.)

To be sure, that unquenchable desire to find unifying theories led him astray in some instances, as had Galileo's analogous obsession with the primacy of circular motion. It is also clear that some splendid science is done by researchers who seem to have no need of thematic presuppositions. But in the case of Einstein, his thematic acceptance of unity or wholeness was definitely one of the demons that had got hold of the central fiber of his soul, to paraphrase an evocative phrase Max Weber had used in a different context (1967, 37). The thema of unity lies at the heart of an explanation of Einstein's first miraculous burst of creativity around 1905. A first hint can be found in the letter he wrote in the spring of 1905 to his friend Carl Habicht (Seelig 1954, 88–89). In a single paragraph, Einstein poured out a list of major works he was then completing—above all, the three most important ones. The first in this list is what is now known as the discovery of the quantum nature of light, as evidenced in the photoelectric effect. The second was his prediction and detailed description of Brownian movement, a random, zigzag movement of small bodies in suspension that are still large enough to be seen through a microscope. In this paper, he traced the cause in exact detail to the bombardment of these bodies by the invisible submicroscopic chaos of molecules. And the third of the papers was, of course, what became the original thirty-page presentation of Einstein's theory of special relativity. In it, he discarded the ether, which had been preoccupying the lives of a large number of prominent physicists for more than a century, with the nonchalant remark that it is superfluous; then dismissed the ideas of the absolutes of space, time, and simultaneity; showed that some basic differences between the two great warring camps, the mechanic and electromagnetic world pictures, were easily dissolved into a new, relativistic one; and finally wrote, as an afterthought, $E=mc^2$.

Each of these papers in 1905 is a dazzling achievement, and they clearly seem to be in three completely different fields. Was there a common thread tying together those three papers that had been published in rapid succession? That exuberant letter to Habicht did not give the full answer. The missing piece of the puzzle lay in a letter Einstein wrote to Max von Laue in January 1952, which indicated the hidden connection (quoted in Holton 1986, 65). Einstein's study of Maxwell's theory, which led him to the special theory of relativity, had also convinced him that radiation had an atomistic, that is to say, quantum, structure, exhibiting fluctuation phenomena in the radiation pressure, and that these fluctuations should show up in the Brownian movement of a tiny suspended mirror. Thus the three separate fireworks displays—relativity, the quantum, and Brownian movement—had originated in a common cartridge.

Moreover, Einstein's approach to the problem in each of these diverse papers had essentially the same style and components. He started by stating his dissatisfaction with what seemed to him asymmetries or other incongruities (which others could dismiss as being merely aesthetic in nature). He then proposed a principle of great generality, analogous to the axioms Euclid placed at the head of his "holy" geometry book. Then Einstein showed how to remove, as one of the deduced consequences, his original dissatisfaction, and at the end, briefly and in a seemingly offhand way, he proposed a small number of experiments that would bear out the predictions falling from his theory. Clearly this is the work of a physicist who is not guided primarily by experimental facts but by themata.[12]

Perhaps most significantly, the fundamental motivation behind each paper was the same that he had announced five years earlier in that letter to Marcel Grossmann we have seen, and which became his chief preoccupation in science for the rest of his life: "To recognize the unity of a complex of appearances which . . . seem to be separate things." It was the goal of unity that suffused Einstein's research and became his overriding thema. The paper on the quantum typically started with the sentence, "There is a deep formal difference between the theoretical understanding which physicists have about gases and other ponderable bodies, and Maxwell's theory of electromagnetic processes in the so-called vacuum" (*CP*, 2: 150). Here Einstein wondered why atomicity should not apply to both matter and light energy. On the Brownian movement, he hypothesized that if there is spontaneous fluctuation in the microcosm of classical thermodynamics, it must also show up in the macrocosm of visible bodies. And the relativity paper removed the old barriers between space and time, energy and mass, electromagnetic and mechanic *Weltbilder*.

But if it were his commitment to a few themata—first of all the thema of unity—that supported Einstein's forays into uncharted territory, often with the barest encouragement from the phenomena, one wonders what provided the courage to adopt these themata and to stick with them through thick and thin. Holton (1998) demonstrated the resonance between Einstein's thematic belief of unity in science and the topic of unity in certain literary works to which he had allegiance. Unity was indeed a central theme of Kultur, directly enshrined in its key goals of Weltanschauung and Weltbild.

Goethe's *Faust* was the quintessential representation of this quest for unity. The fact that Goethe straddled poetry, literature, and science added to his power as a transmitter of cultural themata into science, aiding the transformation of a central cultural goal into a scientific research program. Anne Harrington showed how Goethe's holistic impulse made a lasting impression on the German scientific community (1996, 10).[13] As she put it, "Goethe's resulting aesthetic-teleological vision of living nature would subsequently function as one of the later generations' recurrent answers to the question of what it 'meant' to be a holistic scientist in the grand German style" (ibid., 5). As already pointed out, the consensus between Goethe and Einstein did not extend to concrete theories—Goethe's scientific views were not taken seriously by Einstein or by most scientists of Einstein's generation. Yet the basic impulse was the same. Both held similar pantheistic or cosmotheistic views about nature. Faust's intense quest for "what holds the world together in its innermost," for a complete, unified Weltbild, epitomizes that of Goethe himself and that of Einstein. And for both, this central cultural goal of unity became the thema that underlay their scientific research programs.[14] David Cassidy perceptively summarized:

> Einstein's concern with a unifying worldview and adherence to a research program derived from it were not unique to him nor to physics. The 'unifying spirit,' as it was called, pervaded much of central European thought at the turn of the century. German idealism, neo-Romanticism, and historicism, stretching from Immanuel Kant and G. F. Hegel to Benedetto Croce and Wilhelm Dilthey, each pointed to some sort of transcendent higher unity, the existence of permanent ideas or forces that supersede or underlie the transient, ephemeral world of natural phenomena, practical applications, and the daily struggle of human existence. The scholar, the artist, the poet, the theoretical physicist all strove to grasp that higher reality, a reality that because of its permanence and transcendence must reveal ultimate 'truth' and, hence, serve as a unifying basis for comprehending, for reacting to, the broader world of existence in its many manifestations. (1995, 14)

While Gerald Holton (1998) focused on the cultural roots of Einstein's key thema of unity, Don Howard (1997) presented a parallel argument for another one of Einstein's themata. This was Einstein's firm belief in spatiotemporal separability, which, in part, underlay his obstinate opposition to the indeterministic interpretation of quantum theory.[15] Howard argued that this Einsteinian thema was rooted in

Schopenhauer's philosophy to which Einstein had a strong affinity. In addition, Einstein's thema of deterministic causality resonated with the work of Spinoza, Einstein's favorite philosopher (more about Einstein's relation to these two philosophers later).

Large segments of the German scientific community embraced themata that resonated with the surrounding Kultur, but this cultural resonance was particularly strong in the work of one of this community's most distinguished members, Albert Einstein. Einstein's physics grew from the soil of Kultur, the culture of his time and place. Although culture does, of course, not fully explain Einstein's scientific successes, it is impossible to imagine them independent of it.

NOTES

1. Also see Meisner's (1927) *Weltanschauung eines Technikers.*

2. Originally titled "Motiv des Forschens," the address was published in *Mein Weltbild* under the rather unfortunate title "Prinzipien der Forschung" (1955a [1934], 107–10). This led to "Principles of Research" in the English translation of the address in *Ideas and Opinions* (1954, 219–22).

3. Einstein's interest in a comprehensive *Weltbild* and his predisposition to fasten upon the big questions rather than the small pieces was already awakened at his earliest encounter with scientific literature, as Jürgen Renn and Robert Schulmann (1992b) point out. The scientific books Einstein read as a boy, such as the *Naturwissenschaftliche Volksbücher, Kraft und Stoff,* and *Kosmos,* typically did not dwell on details but provided an overview of science as a coherent corpus of understanding—they purveyed the scientific *Weltbild.*

4. This was the paper titled "Kinetic Theory of Thermal Equilibrium and of the Second Law of Thermodynamics," published in *Annalen der Physik,* 9 (1902): 417–33, and reprinted in the *Collected Papers,* 2: 57–73. Einstein devoted this paper to showing that thermodynamics could be grounded in mechanics, and he concluded that "the Second Law [of thermodynamics] thus appears as a necessary consequence of the mechanical Weltbild" (*CP,* 2: 72). He further suggested "that our results are more general than the mechanical representation used," thus displaying again his urge toward generalization (ibid., 2: 73).

5. "In describing the relative motion (of any kind) of two systems of coordinates, K1 and K2, it does not matter if I relate K2 to K1 or, reversely,

K1 to K2. If, in spite of this, K1 is distinguished by the fact that, relative to K1, the general laws of nature are supposed to be simpler than relative to K2, then this distinction is a fact without physical cause: One of two things that by definition are equal, K1 and K2, is distinguished without a physical cause (a cause that, in principle, is accessible to observation).—Against this, my trust in the consistency of natural events most vigorously objects. A Weltbild that does not need such arbitrariness should, in my view, be preferred."

6. Within Simonton's chance-configuration theory, the same phenomenon is described as the "essential tension between the iconoclasm necessary for uninhibited chance permutations and the traditionalism required to optimize the odds that any resulting configurations will earn social acceptance and thus prove influential" (1988, 233).

7. As Peter Galison (2003) pointed out, the position at the Swiss Patent Office was a congenial environment for the young, rebellious Einstein. His job as a patent examiner required him to subject patent claims to the most thorough and skeptical examination possible. His boss at the patent office, Friedrich Haller, instructed him, "When you pick up an application, think that anything the inventor says is wrong. . . . You have to remain critically vigilant" (quoted in Galison 2003, 243). Einstein's work thus depended on, and honed, the critical scientific thinking that he also employed, at a much more abstract level, in his critique of the physical Weltbild of the time.

8. On Mach's influence on Einstein, see Broda 1979 and Hiroshige 1976.

9. In another letter to Besso, written much later (January 6, 1948), Einstein reiterated his view: "Mach's weakness, as I see it, lies in the fact that he believed more or less strongly that science consists merely of putting experimental results in order; that is, he did not recognize the free constructive element in the creation of a concept" (Einstein and Besso 1972, 391).

10. The thema of symmetry also has deep cultural roots, going back to the Naturphilosophen and, in particular, to Oersted (Williams 1973).

11. The thema of unity and unification also played an important role in biology, as Vassiliki Smocovitis (1996) recently documented in a book titled *Unifying Biology*. It includes a reference to William Morton Wheeler's comment that it might take "a few super-Einsteins" to unify biology—a fitting use of Einstein as the icon of the theme of unification (quoted in Smocovitis 1996, 109).

12. In addition, Peter Galison (2003) argued, Einstein's work at the Swiss Patent Office may have brought him into contact with patents concerning the coordination of clocks—a major technological issue at the turn of the cen-

tury—and thus may have stimulated his thinking about what simultaneity means, a question that was central for his theory of special relativity.

13. The dominance of the thema of unity was, of course, not total, and its influence was dwindling at the close of the nineteenth and in the early twentieth centuries. Detailed studies by Jonathan Harwood (1993) of the German genetics community and by Pauline Mazumdar (1995) of the immunology community revealed major struggles between the adherents of the thema of unity and their opponents.

14. As was typical for students of Einstein's cohort, he was exposed to a heavy dose of Goethe in the course of his education. For instance, in his Matura essay in German, Einstein summarized Goethe's *Götz von Berlichingen* (*CP*, 1: 26–27), and in his second semester at ETH (summer 1897), he enrolled in Professor Saitschik's lecture course "Goethe (Werke und Weltanschauung [!])" (ibid., 364). Einstein apparently retained a life-long interest in Goethe. His library included a large number of Goethe volumes—among them two series of *Werke* and, in addition, three editions of *Faust* (two in German, one in English). Besides the shared commitment to the thema of unity, several other communalities of lesser significance were observed. Armin Hermann compared the two in terms of their relationships with women (1994, 112). In his view, they both were scared of deep emotional commitment and therefore chose inadequate partners. Both men admired Switzerland for its democratic and humane features. Of the various citizenships Einstein held (or was considered to hold) in the course of his life, he treasured the Swiss one the most. And Goethe once exclaimed, "I am glad that I know a country like Switzerland; come what may, now I have always a safe haven there" (quoted in Flückinger 1974, 50).

15. This is the idea that space and time are sufficient to separate individual physical systems.

8.

PHILOSOPHY AND COSMIC RELIGION

Einstein's Weltanschauung

Goethe's *Faust* is on almost everybody's short list of the most significant works of Kultur, as the celebrated embodiment of its deepest aspirations. In the play, during Faust's nocturnal monologue, an inserted line reads: "He opens the book and beholds the sign of the macrocosm" (between lines 429 and 430). There is no further description of the sign, and it seemed worthwhile to explore what kind of a sign this might be.

Sign of the Macrocosm

In *Faust*, the book containing the sign is described as "And this mysterious book,/By Nostradamus's own hand" (419–20).[1] Most commentators, however, have agreed that the reference to Nostradamus should not be taken literally (e.g., Beutler 1977, 754; Proskauer 1982, 37; Trunz 1949, 495; Trunz 1986, 517; Steiner 1981, 23).

Rather, they have said, it is meant to stand for a whole genre of mysterious books. These commentators note that the sign of the macrocosm stems from *Pansophie* (a program to gain comprehensive wisdom about everything) or related alchemistic, mystical, or Kabbalistic traditions, which all liberally used symbols and diagrams to describe the essence of the universe.[2] As we already mentioned, one of the central beliefs in this tradition was that macrocosm and microcosm were interconnected. The human organism, for instance, was a microcosm that mirrored the macrocosmic world—parallels would be drawn between the human organs and the planets, or the metals, and so on. One could even see Leibniz's monadology as a variation on this theme, as each monad is said to reflect the whole universe.

We know that Goethe was strongly interested in Pansophie and read works by Paracelsus and similar authors, such as Georg von Welling's *Opus Mago-Cabbalisticum et Theosophicum* (1784 [1719]) and Anton Johann Kirchweger's *Aurea Catena Homeri* (1723).[3] Rudolf Steiner surmised that *Faust's* sign of the macrocosm was inspired by a diagram in *Aurea Catena Homeri* (1981, 23; 1955). But perhaps we should take a clue from Goethe himself. In 1790, the seventh volume of his *Schriften*, which included the *Faust Fragment*, was published. For its frontispiece, Goethe had commissioned the artist Johann Heinrich Lips to reproduce a seventeenth-century etching by Rembrandt, in which a scholar beholds a magical sign. In particular, the sign itself was meticulously reproduced (one N was inverted).[4] Rembrandt made that etching, which in the eighteenth century became known as "Dr. Faustus," in the early 1650s when he was influenced by old Italian masters (Münz 1952).[5] Ludwig Münz thought that a woodcut by Domenico Campagnola inspired Rembrandt's etching. Whereas the composition of that woodcut has clear similarities, a human-shaped apparition takes the place of the sign. An alternative precursor of Rembrandt's etching is a 1529 woodcut by Lukas Cranach the Elder (Waal 1964). Still other possible archetypes might have been a woodcut by Hans Holbein, or the title page of the *Magia Naturalis* by Johann Baptista Porta (Bojanowski 1956; Behling 1964).

Rembrandt apparently did not invent the sign for his etching—Hans-Martin Rotermund (1957) found very similar signs on seventeenth-century dies for magical amulets. But what does the sign mean? There have been various attempts at decoding the anagram. Here is Martin Bojanowski's (1938, 1956) explanation, which, though not convincing, appears at least as plausible as those given by others (also see Büttner 1995): The inner circle obviously refers to Christ (INRI). The adjacent circle of letters was decoded as the Latin "ADAM TE ADGERAM" (Adam [human being], I will lead you closer [apparently, to INRI]). The outer circle, according to Bojanowski, transforms into "TANGAS LARGA LATET AMOR," which he translated as "Vieles magst du berühren, verborgen bleibt die Liebe"—"You may touch many things; [yet] [divine] love remains hidden [to you]."[6]

All this is still rather cryptic, but it might mean that the love of God is the root of true knowledge, beyond any superficial achievements.[7] We cannot be sure if that is indeed the meaning of the sign, but it is one to which Goethe might have gladly subscribed because it could be easily interpreted as supporting his pantheistic views of nature and his critique of a purely mechanistic science. In a characteristically obscure way, the Romanticist Novalis alluded to what seems to be a similar notion—that love is the focal point of the quest for knowledge: "Building worlds is not enough for the penetrating mind/But a loving heart satisfies the restive spirit" (1960d, 453).

And with Einstein, too, "TANGAS LARGA LATET AMOR," had he known it, might well have struck a chord. Albert Camus' observation that "rebellion cannot exist without a strange form of love" was perhaps true also for this rebel (1958, 304). Although Einstein did not believe in a personal, let alone Christian, God, he was close to Spinoza's pantheism in which God and nature coincide, and in which *amor Dei intellectualis* is the guiding principle. As Einstein put it,

> While it is true that scientific results are entirely independent from religious or moral considerations, those individuals to whom we owe the great creative achievements of science were all of them imbued with the truly religious conviction that this universe of ours is something perfect and susceptible to the rational striving for knowledge. If this conviction had not been a strongly emotional one and if those searching for knowledge had not been

inspired by Spinoza's *Amor Dei Intellectualis*, they would hardly have been capable of that untiring devotion which alone enables man to attain his greatest achievements." (1954, 61)

What Einstein said about creative scientists in general he obviously meant to apply also to himself. For Einstein's own religious conviction was similar, though with the added element of acknowledging that certain mysteries might elude rational comprehension by humans. His religiosity was inspired by his reverence for something whose existence becomes palpable to the scientist operating at the frontier of knowledge yet that transcends scientific knowledge: "If, with our limited means, one tries to penetrate nature, one finds that, beyond all the relationships we can perceive, there is something exceedingly fine, intangible, and inexplicable; the awe of this power beyond our grasp is my religion; in this sense I am indeed religious" (reported in Kessler 1982, 550).

In the following, we will examine more closely Einstein's ideas about religion and philosophy. It is obvious that, since his radical break with religious Judaism at the young age of twelve, Einstein's views on religion were not influenced by any organized religion. Rather, they were inspired by Einstein's readings of some favorite philosophers (Schopenhauer and Spinoza in particular). This chapter intends to demonstrate that Einstein's Weltanschauung, too—just like his Weltbild—was rooted in, and strongly resonated with, Kultur, and that Einstein, although he did not subscribe to the most optimistic and far-reaching pronouncements of a scientific Weltanschauung, was indeed one of the German "scientist-priests."

8.1. COSMIC RELIGION

Because Einstein rejected institutionalized religion, some of his detractors accused him of being an atheist or an irreligious person. Such claims rest on a rather narrow definition of religion. It appears more useful here to rely on the wider definition of religion given by the psychologist and philosopher William James: "Religion . . . shall mean for us *the feelings, acts, and experiences of individual men in their solitude, so far as they apprehend themselves to stand in relation to whatever they may consider the divine"* (2003 [1902], 29). According to James, what exactly is regarded as the divine can vary far beyond the bounds of organized religions. "The only thing that [religious experience] unequivocally testifies to is that we can experience union with *something* larger than ourselves and in that union find our greatest peace" (ibid., 440).

Once such a wider definition is employed, it becomes clear that Einstein was "a deeply religious unbeliever," as he put it, reveling in the paradox.[8] As in so many other areas of his life, he followed his own path, which he called Cosmic Religion. Einstein's peculiar brand of religion occasionally piqued the interest of scholars (e.g., Franquiz 1964; Morrison 1979), but we also have one major work on the topic, Max Jammer's (1995; see 1999) book *Einstein und die Religion*, to which the following is indebted. Einstein's definition of religion revolved around the big questions of life. "What is the meaning of human life, or for that matter, of the life of any creature? To know an answer to this question means to be religious" (quoted in Calaprice 1996, 151)—although one might allow that *searching* for an answer is already enough. From this more abstract perspective, religion in the conventional sense of organized religion appears as but one variant of religion in the wider sense—which, in the now familiar terminology, can be seen as a central element of Weltanschauung.

Einstein was aware of strong interrelationships between science and religion, and he thought that much of the antagonism between conventional religion and science stemmed from boundary violations on both sides. "A conflict arises when a religious community insists on the absolute truthfulness of all statements recorded in the Bible. This means an intervention on the part of religion into the sphere of science; this is where the struggle of the Church against the doctrines of Galileo and Darwin belongs. On the other hand, representatives of science have often made an attempt to arrive at fundamental judgments with respect to values and ends on the basis of scientific method, and in this way have set themselves in opposition to religion. These conflicts have all sprung from fatal errors" (Einstein 1954, 54).

For Einstein, science is about facts and religion is about goals and values—"to make clear these fundamental ends and valuations, and to set them fast in the emotional life of the individual, seems to me precisely the most important function which religion has to perform in the social life of man" (ibid., 51). He believed that, once this boundary was properly recognized, science and religion would be perceived not as rivals but as the complements they really were. "Science without religion is lame, religion without science is blind" (ibid., 55). By this, he meant that science needs a motivational drive from outside, whereas religion needs the factual knowledge that comes from science.

In addition to such boundary violations, Einstein identified two more (interconnected) obstacles that prevented a more harmonious relationship between religion and science. First, many conventional religions recognize some kind of anthropomorphic and personal God who is involved in the fate of individuals. Second, they also believe in free will, a belief that is used to legitimize divine reward or punishment. Einstein unequivocally rejected both ideas as delusions. "I cannot imagine a personal god who would directly influence the actions of individual creatures, or who would sit in judgment over his creatures" (Einstein 1929, 9). And, "I do not believe in free will" (quoted in Herneck 1976, 199).

In his essay "Religion and Science," Einstein expounded what could be called his developmental stage theory of religion.[9] In his scheme, the religion of fear occupies the earliest and lowest stage. "With primitive man it is above all fear that evokes religious notions" (Einstein 1954, 46). Another source of religion consists of social impulses. They dominate the second stage and lead to "moral religion" (ibid., 47). The third stage is a "cosmic religion," to which "the religious geniuses of all times" adhered (Einstein 1954, 47; 1955a, 16). Cosmic religion sheds all anthropomorphic elements and the belief in free will. "The individual feels the futility of human desires and aims and the sublimity and marvelous order which reveal themselves both in nature and in the world of thought. Individual existence impresses him as a sort of prison and he wants to experience the universe as a single significant whole. . . . Looked at in this light, men like Democritus, Francis of Assisi, and Spinoza are closely akin to one another" (Einstein 1954, 47–48). At this stage, the antagonism between science and religion vanishes.

For Einstein, science and religion—scientific Weltbild and quasi-religious Weltanschauung—were interdependent. He was convinced that the quest for the scientific Weltbild needed an extrascientific motivation, one that was furnished by his Weltanschauung centered on cosmic religion. He maintained "that cosmic religiosity is the strongest and noblest driving force of scientific research" (Einstein 1955a, 17). Conversely, advances toward the scientific Weltbild vindicated and reinforced the cosmic Weltanschauung by demonstrating that this Weltanschauung was at least partially viable. Moreover, through unveiling the beauty of nature, "science not only purifies the religious impulse of the dross of its anthropomorphism but also contributes to a religious spiritualization of our understanding of life" (Einstein 1954, 58). In this way, the scientific Weltbild serves as a motor for religious development—helping to overcome the blind spots of stage two religions and fostering evolution toward the third stage.

"A contemporary has said, not unjustly, that in this materialistic age of ours the serious scientific workers are the only profoundly religious people" (ibid., 49). According to this view, the scientists are truly religious in two respects. Societally, they function as the prophets and high priests of cosmic religion, and, as individuals, they are driven by an overriding religious impulse. "The interpretation of religion, as here advanced, implies a dependence of science on the religious attitude" (ibid., 61).

What Einstein saw as a common trait among scientists he also found in himself, defining his religion in the following way: "My religiousness consists of a humble admiration of the infinitely superior spirit that manifests itself in the few things that we, with our weak and feeble faculty of reason, can discern of reality" (Einstein 1929, 9–10). By "superior spirit," Einstein did not mean Platonic ideas beyond the empirical realm, or a personal God, but the constructive principle of the empirical universe. We will now elaborate Einstein's cosmic religion and thereby reveal the cultural roots of Einstein's Weltanschauung. They lay in philosophy, and there not primarily in the highly popular teachings of Kant, but in the work of two somewhat more marginal thinkers, Schopenhauer and Spinoza.

In this undertaking, William James will again be helpful. He asserted that the following two stages formed the common nucleus of all religions:

> [T]here is a certain uniform deliverance in which religions all appear to meet. It consists of two parts:—
>
> 1. An uneasiness; and
>
> 2. Its solution.
>
> 1. The uneasiness, reduced to its simplest terms, is a sense that there is *something wrong about us* as we naturally stand.
>
> 2. The solution is a sense that *we are saved from the wrongness* by making proper connection with the higher powers. (2003 [1902], 425)

We will identify two main stages of Einstein's development—associated, at the philosophical level, with Schopenhauer and Spinoza, respectively—that quite easily correspond to the two religious stages proposed by William James. Thus, James's abstract view of the religious experience will enable us to understand Einstein's intellectual life as a religious journey.

8.2. KANT

At the tender age of thirteen, we recall, Einstein made the acquaintance of Kant's philosophy during his frequent talks with dinner guest and family friend Max Talmey. After the reading of the *Critique of Pure Reason*, Talmey noted, "Kant became Albert's favorite philosopher" (1932, 164). We know of other encounters with Kant's philosophy. At the age of sixteen, when Albert attended the Aargau Kantonsschule, he again read Kant's *Critique of Pure Reason*, according to a school friend's recollections (Seelig 1954, 17). At the

Swiss Polytechnic Institute, he enrolled in Professor Stadler's lecture course "Die Philosophie I. Kants" (*CP*, 1: 364). During his stay in Prague, Einstein frequented Berta Fanta's salon, where music was played and philosophers were discussed, among them Kant (Illy 1979, 82; Seelig 1954, 146).

Yet on the extensive reading list of the Olympia-Akademie, Kant's works were conspicuously absent (Solovine 1987, 8–9). This is a first clue that Talmey's recollection of Einstein as a follower of Kant was either mistaken, or, more likely, that young Einstein's Kantian phase was rather short-lived. During his adult years, Einstein's relationship to Kant's philosophy was, as we shall see, deeply ambiguous. Einstein's lukewarm stance toward the philosopher was so much the more remarkable as, for Einstein's German-speaking contemporaries, Kant was easily the most highly esteemed and widely acclaimed among the philosophers.

Einstein appeared to become slightly more interested in Kant later in life. From a letter written to Max Born in 1918, we know that Einstein again read Kant around that time: "I am reading here Kant's Prolegomena, among other things, and I am beginning to grasp the enormous suggestive power that emanated, and is still emanating, from this guy. . . . At least, it is very nice to read, although not as beautiful as his predecessor Hume, who also had a much greater measure of sound instincts" (Einstein, Born, and Born 1969, 25–26). He told Carl Seelig on the subject, "I did not grow up in the tradition of Kant, but only at a late point I understood the valuable point that—next to errors that are obvious today—lies in his teachings. It in contained in the statement: 'Reality is not given to us, but it is posed to us in the manner of a riddle; that is, there exists a conceptual construction for the perception of the inter-personal, whose authority is grounded solely in successful performance'" (Seelig 1954, 135).

It is not hard to see how Einstein's theory of relativity challenged Kantian philosophy, particularly its core concept of the a priori forms of space and time. Consequently, a huge debate developed among Neo-Kantian philosophers about the implications of Einstein's theory. (See the thorough documentation in Hentschel 1990.) One example was Ilse Schneider's (1921) attempt at reconciliation.[10] "If understood correctly, Kant's transcendental idealism does not at all contradict Einstein's physics and its epistemological results" (Schneider 1921, 64). Schneider emphasized that Kant's forms of space and time were really *psychological* aprioris, whereas modern physics examined the physical realities of space and time.

Einstein himself was less convinced of such compatibility. "If one does not want to claim that the theory of relativity contradicts reason, one cannot maintain Kant's system of a priori concepts and norms," he wrote in a book review (1924, 1688).[11] Although Einstein did not altogether ignore the rapidly growing literature on the Kant-Einstein issue, his instincts were to stay at arms' length. In his opinion, little was to be gained from sterile philosophical disputations about this matter, obfuscated by a great variety of Neo-Kantian interpretations of the philosopher's work. In addition, Einstein was generally skeptical about philosophy, especially about that of the speculative kind. He appreciated its seductive powers but considered it lacking a solid foundation. "Is not the whole philosophy as if written in honey? When you look at it, everything seems marvelous, but when you look at it again, everything is gone. Only the ooze remains" (Rosenthal-Schneider 1980, 90).

In a letter to the philosopher Hans Vaihinger (May 3, 1919), Einstein tried to keep his distance from Kantian philosophy—diplomatically emphasizing that he was not a professional philosopher: "The statement that I promised the 'Kantstudien' a paper is erroneous. I know much too little of philosophy to participate actively in it; I am already glad if I can keep up receptively with the works of the experts in this discipline. I only promised that I would give explanations, in oral and written form, about things that pertain to my specific discipline and are of interest to the philosopher. This is the only way in which I can perhaps be of service to philosophy. Cobbler, stick to your last!" (Einstein Archive #74264).

To Ilse Schneider, Einstein expressed his sarcastic opinion of the Kantian epigones: "Kant is like a country road with lots of mile markers. Then come the little doggies, and each makes its own deposit at the mile markers" (Rosenthal-Schneider 1980, 90). Seelig reported that, after giving a talk in Paris, Einstein was asked whether his theories conflicted with those of Kant (1954, 96). He replied, "This is hard to say. Every philosopher has his own Kant." In a somewhat resigned mood, Einstein once wrote to Paul Ehrenfest (October 24, 1916), "Hume has made a really huge impression on me. Compared to him, Kant seems rather weak, but in the interest of saving time, I have dissuaded myself from arguing this thesis" (Einstein Archive #9388.2).

Whereas Einstein felt that every philosopher had "his own Kant," one might conversely also surmise that every philosopher had "his own Einstein." Schneider, a Kantian herself, unflinchingly believed that Einstein greatly admired Kant. When Einstein sent her a postcard in which he denigrated Kant's theory by likening the philosopher's concept of time to the emperor's new clothes, Schneider felt he "just wanted to tease me" (Rosenthal-Schneider 1980, 83).

We saw, in several of the above quotes, that Einstein compared Hume favorably to Kant. In marked contrast to his ambivalence toward Kant, Einstein had a deep and enduring respect for the British philosopher's work. He encountered Hume's *Treatise of Human Nature* in the Olympia-Akademie. Maurice Solovine remembered that the Akademie members "devoted weeks to the discussion of David Hume's eminently penetrating criticism of conceptions of substance and causality" (1987, 9). Einstein often acknowledged his intellectual debt to Hume. In addition to the already quoted passages, here is another one from Einstein's "Autobiographisches," where he once again juxtaposed Hume and Kant, preferring the former to the latter: "Hume clearly understood that certain concepts, e.g., that of causality, cannot be deduced by logical methods from the material of experience. Kant, who was convinced of the inescapability of certain concepts, considered them—as he chose them—necessary premises of all thinking, and distinguished them from concepts of empirical origin. I am convinced, however, that this distinction is fallacious, or, put differently, does not fit the problem in an unforced way" (1979a [1949], 5). For Einstein, Hume's philosophy complemented Mach's and Avenarius's works in their emphasis on concrete empirical observation, and in their critique of a logical deduction of concepts, such as causality and, one might add with emphasis, time and space.

At this point, we should note that Einstein's interest in Kant focused on the *Kritik der reinen Vernunft* and that the *Kritik der praktischen Vernunft* hardly appeared to have affected his outlook. By restricting science to the realm of the factual, Einstein remained somewhat distant from the tradition of Wissenschaft that claimed the domain of ethics for "practical reason" and to which the more radical proponents of a *scientific Weltanschauung*, both on the Marxist and non-Marxist sides, had an affinity. Although Einstein occasionally made forays into the ethical and political domains, he clearly did not consider his statements in those areas on par with his scientific work on the physical Weltbild. That is, Einstein did not seriously believe in a scientific Weltanschauung, and for that he came under fire from Lenin (1927), who correctly perceived that Einstein was not an ally on the side of a comprehensive scientific Weltanschauung, of the kind the communists espoused.

Einstein certainly had a scientific Weltanschauung in the sense that science gave his life its deepest meaning, but, in contrast to the most extreme proponents of that Weltanschauung, he also recognized certain limitations. In his brief essay "The Laws of Science and the Laws of Ethics," Einstein acknowledged that the foundation of ethics had to lie outside of the realm of science—scientific facts cannot bring forth ethical directives—and that, in other words, a comprehensive scientific Weltanschauung could not be sustained (1979c, 53–55). Yet, he continued, it is possible to apply the general rules of rational argument to the realm of ethics. Once certain ethical axioms have been accepted, other propositions can be derived from them, and, conversely, individual ethical propositions can be logically reduced to those axioms. As an example, he presented a short logical syllogism about lying: Lying destroys trust in other people's statements. Without such trust, societal cooperation is obstructed. Societal collaboration is necessary to make the life of humans possible or at least bearable. Hence, the prohibition of lying can be reduced to the basic ethical axioms that human life should be preserved or that pain and suffering should be minimized. This is also a conspicuous example of how a superficially rational deduction can go totally astray, if detached from empirical facts. Doing very detailed studies of human interactions, sociologists have found that people routinely apply a rich arsenal of lies, deceit, and dissimulation to make human relationships bearable. From this perspective, lying, in at least some of its manifestations, appears to be the glue that holds society together (though there are, of course, tipping points).[12]

Coming from Hume's skepticism, Einstein knew that, strictly speaking, there could be no rational proofs for rationality. He also realized that several areas of life lie beyond the purview of science and that somewhere in those areas are the wellsprings from which the deepest impulses for devoting one's life to science flow. In this sense, Einstein recognized religion—abstractly understood—as the root element of his Weltanschauung. "A religious person is devout in the sense that he has no doubt about the significance and loftiness of those superpersonal objects and goals which neither require nor are capable of rational foundation" (Einstein 1954, 54).

For Einstein, religion articulated itself on the philosophical plane, and we now turn to the two philosophers who had the strongest impact on Einstein's Weltanschauung. I propose that, to a large part, his mature Weltanschauung was a synthesis of negative impulses from Schopenhauer and positive impulses from Spinoza. The following sections will elaborate and support this proposition.

8.3. SCHOPENHAUER

Schopenhauer was not on the Olympia-Akademie reading list, but Einstein had become acquainted with the philosopher even before the Akademie convened in 1902 (Solovine 1987, 8–9). According to Michelmore and Hermann, Einstein first encountered Schopenhauer's writings in his teens (Michelmore 1962, 29–30; Hermann 1994, 24). Einstein's son-in-law reported that Einstein read Schopenhauer, along with Kant and Hume, while a student at the Polytechnic Institute in Zürich (Reiser 1930, 55). And in September 1901, in a letter to Marcel Grossmann, Einstein himself mentioned reading the philosopher: "Have you had Schopenhauer's Aphorismen zur Lebensweisheit in your hands yet? It is a part of Parerga & Paralipomena, and I liked it a lot" (CP, 1: 316).

Whereas Einstein was undeniably a dedicated follower—a self-proclaimed "disciple"—of Spinoza, his relationship to Schopenhauer's philosophy was less clear (Hoffmann and Dukas 1972, 94). Without doubt, Einstein appreciated Schopenhauer, but scholars have been divided on a major question—did Einstein subscribe to any substantive points of Schopenhauer's work, or did he read it just for its stylistic and aesthetic brilliance?

Philipp Frank was among those who believe that Einstein did not at all take the substance of Schopenhauer's philosophy seriously. "One likes to read some philosophers, because they make more or less superficial and opaque statements about all sorts of things in a beautiful language—statements that often create a sentiment like beautiful music and give rise to pleasant reveries about the world. For Einstein, Schopenhauer, first of all, was such an author, whom he liked to read without taking his views seriously at all. Philosophers like Nietzsche belonged to the same group. Einstein read them, as he sometimes put it, for 'edification,' just like others listen to a sermon in church."[13] This is supported by a passage in a letter Einstein wrote to his son Eduard: "I too find the splendid style [of Schopenhauer] is worth far more than the actual contents" (quoted in Highfield and Carter 1993, 230). Similarly, Moszkowski observed, "[Einstein] holds Schopenhauer and Nietzsche in very high regard as writers, as masters of language and creators of evocative thoughts; he values them for literary height and denies philosophical depth. As to Nietzsche, whom, by the way, he calls too glitzy, an ethical resistance, too, is certainly aroused in him against the prophet of the Herrenmoral [morality of the Master], which harshly contradicts his own view on human relationships" (1922, 234).

One cannot doubt that Einstein read some philosophers in a light-hearted vein. Nietzsche, perhaps, belongs in this group—and definitely Spengler, of whom Einstein wrote to Born, "Spengler did not spare me either. One sometimes enjoys him putting ideas in one's mind in the evening, and one smiles about it in the morning" (Einstein, Born, and Born 1969, 44). And, to be sure, an occasional quote of a philosopher's bon mot does not make someone a follower of that philosopher. For instance, Einstein said at a New York banquet in late 1933, commemorating the hundredth anniversary of Alfred Nobel's birth, that he agreed with Schopenhauer's dictum that willpower and intelligence were antagonistic (Brian 1996, 255). This, in isolation, might have been nothing more than a clever swipe at the new German regime that glorified willpower.

However, following Herneck and Howard, I think that Schopenhauer had a more serious influence on Einstein, and the following will document substantive affinities between Schopenhauer and the physicist (Herneck 1976, 199–210; Howard 1997). Einstein's references to Schopenhauer were numerous—and more than casual in nature. When, late in his life, Einstein was asked what literature he had read as a young man, his reply accorded Schopenhauer a prominent place: "As a young man, and also later, I had little interest in poetic literature and novels. . . . Rather, I preferred books of weltanschaulichen content, and especially philosophical books. Schopenhauer, Hume, Mach, partly Kant, Plato, and Aristotle" (Seelig 1954, 134). Moreover, according to Einstein's son-in-law, Schopenhauer's portrait, alongside pictures of Faraday and Maxwell, graced the walls of Einstein's Berlin study (Reiser 1930, 194).[14] Konrad Wachsmann observed that Einstein often read "already well-worn" Schopenhauer volumes in both his Berlin home and his Caputh cottage (1990, 243). And a friend of Einstein's, the psychiatrist Otto Juliusburger, was an expert not only on Spinoza but also on Schopenhauer (Dukas and Hoffmann 1979, 79).

In fact, one can identify several resonances between Schopenhauer and Einstein. Both rejected conventional religion, and both distanced themselves from the convention-dominated existence of most humans. Both were opposed to becoming public figures (Hermann 1994, 102). Schopenhauer was notorious for being a misogynist; he also favored polygamy. Hermann traced misogynistic elements in Einstein's beliefs about women (ibid., 301). Furthermore, Einstein thought that monogamy was unnatural and acted accordingly (Highfield and Carter 1993, 210). Schopenhauer was part of the cultural tradition of rebellion to which Einstein cleaved. With Schopenhauer, Einstein shared a defiant attitude toward authorities. Both believed in having the courage of one's convictions, even if they flew in the face of received wisdom. Einstein also found Schopenhauer's penetrating skepticism attractive—a stance that, at a more epistemological level, was mirrored by that of Mach and Hume. Schopenhauer's denunciation of patriotism in science as a "schmutzigen Gesellen [dirty fellow]" that needed to be thrown out (Hermann 1994, 153) found its parallel in Einstein's untiring opposition to any nationalist tendencies in science.

In his allegiance to the principle of causality, and especially in his rejection of free will, Einstein drew on Schopenhauer for support. He said in a voice recording made in 1932 for the *Deutsche Liga für Menschenrechte* (German League for Human Rights), "I do not believe in free will. Schopenhauer's saying—Humans can do what they want, but they cannot will what they want—has accompanied me through all situations of life and has reconciled me with the actions of humans, even if they are rather hurtful to me. The realization that the will is not free has protected me from taking myself and my fellow humans all too seriously as acting and judging individuals, and from losing my good humor."[15]

The most important resonance, for our purpose, is an almost existentialist ennui at the human condition that, as Herneck showed, the physicist shared with the philosopher (1976, 204–205). Einstein said, for instance, "Our situation on this earth appears strange. Involuntarily and uninvited, each of us shows up for a brief stay, without knowing cause or purpose" (quoted in ibid., 204). His autobiographical notes contained a variation on the same theme: "As a quite precocious young person, I vividly realized the vanity of hopes and strivings that relentlessly drives most people through their lives. I also soon saw the cruelty of this scene, which in those days was hidden more carefully than today by hypocrisy and flashy words. Everyone was condemned by the existence of his stomach to participate in this folly. The stomach could be satisfied by such participation, but not the human being as a thinking and feeling entity" (Einstein 1979a [1949], 1). With Hedwig Born, he agreed "that the liberation from the fetters of the I is the only way to a satisfying human world" (Seelig 1956, 39). And in 1945, thanking Hermann Broch for a copy of his *Vergil*, Einstein wrote, "I am fascinated by your *Vergil*—and am steadfastly resisting him. The book shows me clearly what I fled from when I sold myself body and soul to Science—the flight from the I and WE to the IT" (Hoffmann and Dukas 1972, 354).

Returning to Einstein's speech on the occasion of Max Planck's sixtieth birthday in 1918, we find him agreeing with Schopenhauer on the deepest motive for science: "First, I believe with Schopenhauer that one of the strongest motives that lead to the arts and sciences is the flight from everyday life with its painful coarseness and desperate desolation, away from the fetters of one's own perpetually shifting desires" (Einstein 1955a, 108). Yet this negative root of science— escape from a miserable human existence—has a positive twin—the majestic beauty of the universe. A Spinoza-inspired view of the universe rescued Einstein from a pessimistic and nihilistic stance. For him, rational understanding of the universe held the key for salvation, even the promise of paradise. "The intellectual comprehension of this extra-personal world within the range of our possibilities has been— partly consciously, partly sub-consciously—my highest goal. . . . The path to this paradise was not as comfortable and alluring as the path to the religious paradise; but it has proved dependable, and I have never regretted choosing it" (Einstein 1979a [1949], 2).[16]

The following passage in Einstein's "Religion and Science" again demonstrates the twinning of the positive and negative roots of science in his thinking: "The individual feels the futility of human desires and aims and the sublimity and marvelous order which reveal themselves both in nature and in the world of thought. Individual existence impresses him as a sort of prison and he wants to experience the universe as a single significant whole" (Einstein 1954, 47–48).

Finally, I should note that Einstein was also critical of Schopenhauer. Moszkowski reported that Einstein criticized Schopenhauer's attack on Newton, which derogated Newton's accomplishment by overemphasizing Robert Hooke's (1922, 53). Einstein thought that the root of Schopenhauer's misguided attack was his nonmathematical way of thinking. Thus, Einstein's critique centered on a deviation from the scientific method. As always, he considered such a deviation an illegitimate shortcut at best: there could be no substitute for science.

8.4. SPINOZA

Albert Einstein's deep admiration for Spinoza is well known.[17] Among the philosophers, he told Peter Bucky, Spinoza was his "favorite of all" (1992, 112). Here are just three additional examples selected from Einstein's numerous homages to the philosopher. In a letter written in 1929, he called himself a "disciple" of Spinoza (Hoffmann and Dukas 1972, 94). He also wrote to Carl Seelig, "I have always been close to Spinoza's point of view, and I have always admired this man and his teachings" (Seelig 1954, 187). Finally, Einstein also used to quote, from memory, Schleiermacher's praise of the philosopher (Pesic 1996, 195).[18] This might indicate that, to some extent, Schleiermacher mediated Spinoza's influence on Einstein. As we have seen, Schleiermacher considered himself a follower of Spinoza when he developed his idea of Weltanschauung (which, in turn, was an important milestone in the evolution of the concept). The Spinoza-Schleiermacher connection thus may be one of the strands in the mighty tie that bound Einstein to the goal of a Weltanschauung.

Einstein became acquainted with the philosopher's writings early in life. Peter Michelmore even claimed that Einstein was only a teenager when he read Spinoza, along with a number of other philosophers (1962, 29–30). Einstein's first encounter with Spinoza's philosophy, for which we have a firsthand witness, was in 1902, when he was in his early twenties, at the "Olympia-Akademie." From Maurice Solovine, a fellow member of this philosophical discussion circle, we know that the reading list of the Akademie included, among many other scientific and philosophical works, Spinoza's *Ethik* (1987, 8–9). In these days, when Einstein suffered economic hardship and insecurity (before he obtained a position at the Bern patent office), Einstein might well have taken solace in Spinoza's shunning of conventional career success. Nonetheless, there is evidence that Spinoza came to dominate Einstein's philosophical thinking only at a later stage.

More than a decade later, in 1915, Einstein's situation was worlds apart from that of the indigent job seeker in Bern. He had won his colleagues' admiration for his trailblazing work and had moved to Berlin to take a prestigious and well-endowed position at a very different kind of Akademie, the Prussian Academy of Sciences. And he was about to complete his most glorious contribution to physics, the general theory of relativity. Yet at this point, he turned again to Spinoza's *Ethik*. On September 3, 1915, he wrote to Elsa, who was to become his second wife, "I have read here almost the whole *Ethics* by Spinoza, much with great admiration. Kraft had very good sense to bring this profound work to my attention. I believe it will have a lasting effect on me" (Einstein Archive #73050).[19]

Interestingly, Einstein sounded as if he had forgotten that he had read Spinoza at the Olympia-Akademie thirteen years earlier. Apparently, the *Ethik* had not stood out, in Einstein's mind, from the varied and rich reading list of the Akademie. Hume, first of all, seemed to have made a much stronger impression on the young Einstein. Now, however, Einstein signaled the beginning of a deep admiration for Spinoza.

In 1920, Einstein composed a poem about Spinoza, which begins:

Wie lieb ich diesen edlen Mann
Mehr als ich mit Worten sagen kann.
Doch fuercht ich, dass er bleibt allein
Mit seinem strahlenden Heiligenschein.

Oh, how I love this noble soul
More than words can e'er extol.
I fear he'll remain alone
With shining halo of his own. (quoted and translated in Sayen 1985, 3)

Einstein went on to speak and write about Spinoza frequently, always in admiring tones. In a 1946 letter, for instance, Einstein referred to Spinoza as "one of the deepest and purest souls our Jewish people has produced" (Hoffmann and Dukas 1972, 95). Occasionally, he also made public statements about the philosopher. He provided the introduction to his son-in-law's Spinoza biography (Kayser 1946, ix–xi), the foreword to a Spinoza dictionary (Runes 1951, iii–iv), a letter to a Spinoza *Festschrift* (Hessing 1933,

221), and a message for the Spinoza Institute of America's celebration of the tercentenary of the philosopher's birthday in 1932 (*New York Times*, November 24, 1932). His interest in Spinoza may have been bolstered in discussions with his friend, the psychiatrist Otto Juliusburger, who, as we heard earlier, was quite an expert on Spinoza (and Schopenhauer) (Dukas and Hoffmann 1979, 79).

One of Einstein's most spectacular professions of allegiance to the philosopher took the form of a short telegram. In 1929, Boston Cardinal O'Connell branded Einstein's theory of relativity as "befogged speculation producing universal doubt about God and His Creation" and "cloaking the ghastly apparition of atheism" (*New York Times*, April 25, 1929). To defuse the atheism charge, New York Rabbi Herbert S. Goldstein asked Einstein per telegram whether he believed in God—paid reply fifty words (Seelig 1954, 187). In response, Einstein needed but twenty-five words (in the German original) to state his belief succinctly: "Ich glaube an Spinozas Gott, der sich in gesetzlicher Harmonie des Seienden offenbart, nicht an Gott, der Sich mit Schicksalen und Handlungen der Menschen abgibt" (I believe in Spinoza's God who reveals himself in the orderly harmony of what exists, not in a God who concerns himself with fates and actions of human beings) (*New York Times*, April 25, 1929). The rabbi cited this as evidence that Einstein was not an atheist, and even went as far as to declare that "Einstein's theory if carried to its logical conclusion would bring to mankind a scientific formula for monotheism." Einstein himself remained aloof. "Informed of Cardinal O'Connell's assertion, Professor Einstein later said, speaking through his wife, that the attack left him 'cold' and was 'devoid of interest.' He said he was wholly disinclined to enter into a controversy with the Cardinal" (ibid.).

Eighteen years later, in 1947, when again asked to sum up his views on the belief in a Supreme Being, he restated his Spinozist position: "My views are near those of Spinoza: admiration for the beauty of and belief in the logical simplicity of the order and harmony which we can grasp humbly and only imperfectly. I believe that we have to content ourselves with our imperfect knowledge and understanding and treat values and moral obligations as a purely human problem—the most important of all human problems" (Hoffmann and Dukas 1972, 95).

Although Einstein acquainted himself with Spinoza's philosophy in his early years, the preceding material thus indicates that his interest in this philosopher grew markedly more intense over the span of his life. Lewis Feuer correctly observed that "a shift took place in his middle years which we might describe as from Hume to Spinoza. His references to Spinoza became more numerous than and finally supplanted those of empiricist critics" (1974, 78). Now we will examine, in detail, the aspects of Spinoza's life and work that impressed Einstein—and it was clearly *both* life and philosophy that Einstein revered, as these two aspects formed a consistent whole in Spinoza.

Einstein greatly admired Spinoza for choosing independence of mind over things that most individuals strive very hard to obtain, such as fame and career success. For instance, Spinoza rejected the offer of an academic career and, instead, preferred an outwardly inconspicuous and humble life that afforded him the freedom to pursue his quest for the truth. Earlier, Spinoza's independent-mindedness had set him on a collision course with religious authority, which had caused the Amsterdam Jewish community formally to expel him for his views. This may have further endeared him to Einstein, who, after his brief but intense period of exuberant religiosity that ended at the age of twelve, was deeply skeptical of organized religion and conventional religious beliefs.

Toward the end of his life, Einstein may even have had the nagging feeling that Spinoza had been better able to remain true to his convictions than had been Einstein himself, who had entered the top echelons of the academic establishment and, though entirely against his will, had turned into an icon of popular culture. In 1954, Einstein said in a letter to *The Reporter* (November 18, 1954): "If I would be a young man again and had to decide how to make my living, I would not try to become a scientist or scholar or teacher. I would rather choose to be a plumber or a peddler in the hope to find that modest degree of independence still available under present circumstances." When an upset reader named Arthur Taub wrote to Einstein, protesting this apparent desertion from science,

Einstein explained his stance, which was specifically directed against what Einstein considered an oppression of intellectuals in the McCarthy era. "I wanted to suggest," Einstein replied, "that the practices of those ignoramuses who use their public positions of power to tyrannize over professional intellectuals must not be accepted by intellectuals without a struggle. Spinoza followed this rule when he turned down a professorship at Heidelberg and (unlike Hegel) decided to earn his living in a way that would not force him to mortgage his freedom" (*The Reporter*, May 5, 1955).

Determinism, including the rejection of free will, was a plank of Spinoza's philosophy that Einstein particularly cherished. In his 1932 contribution to the Spinoza *Festschrift*, Einstein praised the philosopher as "the first to apply with true consistency to human thought, feeling, and action, the idea of the deterministic constraint of all that occurs."[20] Spinoza's deterministic stance resonated with the thema of deterministic causality that played a major role in Einstein's physics. In his later years, when Einstein's stubborn efforts to rescue determinism in physics did not progress, he seemed to have drawn encouragement from Spinoza's sticking to his position, even in the absence of sufficient supporting evidence. Drawing a parallel between Spinoza's and his own times, Einstein noted: "[Spinoza] was utterly convinced of the causal dependence of all phenomena, at a time when the success accompanying the efforts to achieve a knowledge of the causal relationship of natural phenomena was still quite modest" (in Kayser 1946, xi).

In 1937, Niels Bohr and Einstein met to discuss their long-standing disagreement on questions of determinism and got, as Bohr recalled, into "a humorous argument . . . about which side Spinoza would have taken if he had witnessed the current developments" (1979, 147). This whimsical debate had at its bottom a serious philosophical issue. As Peter Pesic (1996, 2000, 2002) pointed out, Spinoza's position on determinism could have been read differently from how Einstein interpreted it—in a way that would allow for the lack of determinacy in quantum theory. According to Pesic, the question of individuation is crucial here. He argued that Spinoza dissolved the distinguishability of individuals within the whole of nature (*deus sive natura*) and that the philosopher's stance was consistent with the indistinguishability of quantum particles, and hence with the indeterminacy of quantum theory. Einstein, however, insisted on the individuality of particles and on determinism in the quantum domain, for which he used his own interpretation of Spinoza as support.

Einstein realized that determinism could be a double-edged sword, as it destroys childlike feelings of security and consolation. "The habit of casual [should probably read: causal] interpretation of all phenomena, including those in the psychic and social spheres, has deprived the more wide-awake intellectual of the feeling of security and of those consolations which traditional religion, founded on authority, offered to earlier generations. It is a kind of banishment from a paradise of childlike innocence" (in Kayser 1946, ix). In Einstein's view, Spinoza's rationalism showed how to live after the disenchantment. The physicist saw a kind of feedback loop in which rational understanding keeps the power of emotions in check, and the control of emotions enables clear and creative thinking. One the one hand, he said, "In the study of this causal relationship [i.e., the causal dependence of all phenomena] he [Spinoza] saw a remedy for fear, hate and bitterness, the only remedy to which a genuinely spiritual man can have recourse" (in ibid., xi).[21] On the other hand, he told William Hermanns in a 1930 conversation, "For me [Spinoza] is the ideal example of the cosmic man. He worked as an obscure diamond cutter, disdaining fame and a place at the table of the great. He tells us of the importance of understanding our emotions and suggests what causes them. Man will never be free until he is able to direct his emotions to think clearly. Only then can he control his environment and preserve his energy for creative work" (Hermanns 1983, 26).

Einstein found the key to attaining the Schopenhauerian goal of liberating himself from the "merely personal" of his existence in grasping—understanding and admiring—the objective outside world. The expanded version of the passage in "Autobiographisches" quoted above clearly drives home this point:

It is clear to me that the . . . lost religious paradise of my youth was a first attempt to liberate myself from the fetters of the 'merely personal,' from an existence that was dominated by desires, hopes, and primitive feelings. Out there, there was this big world, which exists independently of us humans and stands in front of us like a great eternal riddle, which is at least in part accessible to our perception and thinking. Its examination was like a liberation, and I soon noticed that quite a few of those whom I had learnt to respect and admire had found inner freedom and security in their devoted preoccupation with it. The intellectual comprehension of this extrapersonal world, within the limits of the capacities that are available to us, appeared to me, partly consciously and partly subconsciously, as the highest goal. Present and past individuals of similar convictions, as well as the insights they attained, have been my friends forever. The path to this paradise was not as comfortable and enticing as the path to the religious paradise; but it has proved dependable, and I have never regretted choosing it. (Einstein 1979a, 2)

It is obvious that Einstein, though he did not mention names, counted Spinoza prominently among those kindred spirits and "friends forever," those members of the supratemporal community of geniuses with which Einstein identified.

Einstein's God was, as he once put it, "not Jahwe or Jupiter but Spinoza's immanent God" (Hoffmann and Dukas 1972, 195). For him, as for Spinoza, nature and God coincided. David Reichinstein saw in Einstein a tendency "toward religious mysticism," and Max Jammer observed an apparent oscillation between a more rationalist Spinozism and a more mystic point of view (Reichinstein 1935, 204; Jammer 1995, 33). One can indeed find some passages that appear to position Einstein's pantheism close to mysticism, for instance, when Einstein spoke of the "eternal mystery of nature" (1929, 6), or when he exclaimed,

The greatest beauty that we can experience is the mysterious. It is the fundamental feeling that is at the cradle of true art and true science. He who does not know it and is unable to be astonished and amazed, is so-to-speak dead, and his eye has been blinded. The experience of the mysterious—even though mixed with fear—also created religion. Knowledge of the existence of that which we cannot penetrate, of the manifestations of deepest reason and most brilliant beauty, which our reason can grasp only in its most primitive forms—this knowledge and sentiment is true religiosity. In this and only this sense, I belong to the deeply religious persons. (Einstein 1955a, 9–10)

What Einstein said about Max Planck was clearly true about himself, too: "The state of mind which enables a man to do work of this kind is akin to that of the religious worshipper or the lover" (1954, 222).

Nonetheless, Einstein scoffed at the label "mystic": "What I see in Nature is a magnificent structure that we can comprehend only very imperfectly, and that must fill a thinking person with a feeling of 'humility.' This is a genuinely religious feeling that has nothing to do with mysticism" (Dukas and Hoffmann 1979, 39). And there are additional quotes that indicate a more rationalist stance: "I have no better term than the term 'religious' for this trust that reality is rational and at least somewhat accessible to human reason. Where this feeling is absent, science degenerates to mindless empiricism" (Einstein 1987, 118). And, "To [the sphere of religion] there also belongs the faith in the possibility that the regulations valid for the world of existence are rational, that is, comprehensible to reason. I cannot conceive of a genuine scientist without that profound faith" (Einstein 1954, 55).

Einstein marveled at the fact that the structure of the universe itself was rational rather than chaotic. In a letter to his old friend Solovine (March 30, 1952), he wrote,

You find it surprising that I feel that the intelligibility of the world (as far as we are entitled to speak of such) is a mystery or eternal secret. Well, a priori, one should expect a chaotic world, which in no way could be comprehended by reasoning. One could (even *should*) expect that the world would prove orderly only to the extent that we impose order. It would be a kind of order like the alphabetical order of the words of a language. By contrast, the kind of order that for instance was created by Newton's theory of gravity is of an entirely different character. Even though the axioms of the theory are stated by a human being, the success of such a project presupposes a high-grade order of the objective world, which to expect a priori one had no right. Here lies the 'mystery,' which only intensifies with the advance of our knowledge. (Einstein 1987, 130)

The question of whether or not Einstein was a mystic parallels a similar conundrum about Spinoza, who variously has been seen as a mystic or as a rationalist. Whereas the Romantic "Spinozists," such as Schleiermacher and Schelling, gave Spinoza a Romantic spin, Einstein clearly appreciated the underlying rationalism in this philosophy. Ultimately, the question of Einstein's mysticism turns on his understanding of the concept of mystery that he used so often. Is his mystery something entirely alien to reason, or is it something hard or even impossible to understand, but still of a fundamentally rational structure? To Einstein, the blueprint of the world was immensely complex and difficult to decipher, but it was not irrational or arbitrary. One of Einstein's much-quoted adages was: "Raffiniert ist der Herrgott, aber boshaft ist er nicht (The Lord is slick, but He ain't mean.)."[22]

To a large part, Einstein's mystery owes its existence to the limited rational capacity of humans. The religious sentiments flow from having a glimpse but not full insight into the rational structure of the universe: "We see a universe marvelously arranged and obeying certain laws, but only dimly understand these laws."[23] Because understanding is at the same time possible and limited, scientists get "hooked" on the exploration of nature and stay committed to their quest for the scientific Weltbild, even when they realize that the ultimate goal is elusive and that theirs is a journey without end, or an "endless frontier," as Vannevar Bush (1945) termed it. In other words, it is the combination of intermittently and unpredictably occurring breakthroughs (with their intense momentary gratification) and the elusiveness of the ultimate goal that creates a motivational structure akin to that of an inextinguishable obsession.

We can now conclude our discussion of Einstein's "mysticism" by asking whether and to what extent Einstein might have been a Romantic. Indeed, a number of similarities between Einstein and the Romantics were quite conspicuous—grandeur of vision, passion in its pursuit, embrace of tragic failure, and rejection of social convention. The Romantics and Einstein shared their ultimate goals, their deep interest in a Weltanschauung and the quest for a Weltbild. Their attitude toward pursing these goals was also similar. Whereras the Classics, in particular Seneca and the Stoics, had considered peace of mind the ultimate achievement, and had extolled the mind at rest in itself, contented and measured, the Romantics were fond of unbounded, wistful passion. In Einstein's life, a fervent, almost fanatical quest for scientific breakthroughs continued literally onto his deathbed. The Romantics, furthermore, were fully aware that their passion might lead to ultimate failure, but they found that potential for a tragic crash strangely fascinating. According to Isaiah Berlin's perceptive analysis of Romanticism in *The Crooked Timber of Humanity*, the Romantics were interested in "a search after means of expressing an unappeasable yearning for unattainable goals" and felt "an irresistible gravitation towards the unattainable centre of the universe" (Berlin 1992, 92, 93). Here another communality with Einstein becomes evident. He, too, was driven by an unrelenting desire for a comprehensive Weltbild—to remove another one of Isis's veils—in the clear knowledge that his pursuit could never succeed fully. Einstein thus embraced an ultimately tragic project and cherished the nobility of futility. Significantly, *Don Quichotte* was one of his favorite books. The posture of defiance that was an element in Einstein's Weltanschauung (as in that of many "priests of Isis") was all the more extreme because—in contrast with most religious figures—the scientist priests stuck to their elusive goal without hope for transcendence or for rewards in an afterlife. Finally, revering the maverick genius, the Romantics perpetually defied social convention and struggled against the oppressive forces of philistine society—and so did Einstein.

Yet as soon as one looks closer at the methods of Einstein's thinking, it becomes obvious that they are embedded in an Enlightenment tradition that, in many respects, was opposed to Romantic thinking: reason vs. will and sentiment, universalism vs. particularism and relativism, internationalism vs. nationalism, determinism vs. free will. Einstein was convinced that the world existed irrespective of any human consciousness and that scientific methods, rather than idealistic speculation or mystical meditation were necessary to understand it. Conversely, the Romantics would have scoffed at his empirical and rational methodology. For them, the creative synthesis was an act of special vision—"beyond . . . the framework of normal thought and sober reasoning"—whereas Einstein's project of a scientific Weltbild remained essentially rationalist and based on empirical facts (Berlin 1992, 93). He adamantly rejected the admixture of mystical elements at this level. A key quote is Einstein's characterization of Romanticism as an illegitimate shortcut: "Romanticism as a kind of illegitimate shortcut to arrive relatively cheaply at a deeper understanding of art" (Seelig 1954, 135).[24] His side interest in technology may be one instantiation of his general insistence that theory—if it is to be accepted as valid—must, in principle, stand the test of its successful application in the real world. He might thus have considered it healthy for a theorist who usually works at the highest levels of abstraction to deal with actual machinery. Einstein confined the element of mysticism (in the meaning explained previously) to the metalevel of his Weltanschauung. In other words, mystic elements of awe and wonder about the world joined the sense of inadequacy of the "mere-personal" in providing the subjective *impulse* for expanding the rational and scientific understanding of the world, that is, for creating a more comprehensive Weltbild. Moreover, the (albeit partial) success of this rationalist project led to a subjective gratification and *reinforcement* that also expressed itself in increased mystic awe. In sum, Einstein shared with the Romantics the grand goals of Weltbild and Weltanschauung, but, in contrast with them, he insisted on rational and empirical means to reach those goals.

At the deepest level, Einstein's Weltanschauung, with the quest for the scientific Weltbild as its centerpiece, rested on a religious experience that played itself out in philosophical and scientific terms. Applying William James's concept of the religious experience, we now realize that Einstein's rebelliousness can be understood as the outward expression of the first religious stage: when one realizes that the status quo is deeply flawed, and when one becomes increasingly uneasy, irritated, and even obsessed, while searching for a solution. Einstein's religious seeking was driven by his discomfort with the existing physical Weltbild. Others who were also aware of problems within the physical Weltbild of the time may have considered them minor flaws. Not so Einstein! He was preoccupied, almost infuriated, by what he considered unacceptable shortcomings, and he singlemindedly devoted his whole being to finding the remedy. The passion he developed in this quest perhaps resembled that special kind of love, *amor Dei intellectualis*, which we discussed previously. His major breakthrough, in the form of the general theory of relativity, was akin to entering William James's stage two, a religious conversion experience. Now Einstein glimpsed the beauty of the cosmos and saw it increasingly through Spinoza's eyes. Here, however, we encounter an important difference between the religious experience described by James and Einstein's experience. The second stage, the state of grace that manifests itself in an enchanted serenity against which the common problems of lesser mortals fade into insignificance was not permanent but merely fleeting. Einstein was very aware that his accomplishment, amazing as it was, had only reached a way point and that more work remained to be done, which he tried to do for the rest of his life—devising a general field theory—albeit without success. One of his statements quoted previously revealed that the deepest source of his religion was his reverence for "something exceedingly fine, intangible, and inexplicable; the awe of this power beyond our grasp is my religion; in this sense I am indeed religious" (in Kessler 1982, 550). He clearly realized that the complete scientific Weltbild was beyond the grasp of human cognition. Einstein's religious experience epitomizes a chief characteristic of the religion of science (see above): This religion consciously forgoes ultimate fulfillment and thus remains strangely ascetic, or, perhaps one might say, tragic.

Einstein's religious journey, at an intensely personal level, illustrates the religiosity that, at the collective level, was projected by the German scientists who presented themselves as Priests of Isis. When Medizinalrat Julius Baumgärtner opened the fifty-second meeting of the association of German scientists and physicians in Baden-Baden, he exclaimed: "He who cannot subordinate his way of thinking and feeling to the prescribed statutes of a church, may he take his refuge under the banner of research, and may he ascend the arduous path, which allows only slow progress, but which at every step reveals insights into new splendors and into a cornucopia of wonders. Not through myth or ecclesiastical tradition, but through observation during the course of research do we recognize the majesty and infinity of creation" (Schipperges 1976, 54–55). These words, uttered in 1879, were a fitting epigraph for the life of a baby born in the same year some one-hundred miles to the east in another southern German town: Albert Einstein. In this chapter, we have seen that, although he was not at all religious in any conventional sense, Einstein's unquenchable desire to discover the laws of the universe had definite religious undertones. The Priests of Isis would have easily recognized him as one of their own.

NOTES

1. Accordingly, the sign of the macrocosm might well be the one on the title page of Nostradamus's (1943 [1555]) *Prophecies*: a globe formed of lines of longitude and latitude, with a solid core and with a band of pictures surrounding it.

2. On the Jewish mysticism of the Kabbalah, see Epstein 1998.

3. The fact that the two last-mentioned books both were published in Frankfurt, Goethe's hometown, may suggest that a certain interest in this kind of scholarship existed locally, and that, to some extent, impulses from this environment might have influenced young Goethe's attitudes. Goethe's interest in Egyptian-inspired esoterics peaked in his youth; in his later years, he became increasingly critical (Hornung 2001, 128–29).

4. Cyrus Hamlin (1998) pointed out that, in Goethe's youth, a school of painters existed in Frankfurt that was heavily influenced by Rembrandt. Goethe owned a copy of the etching in question. It is not clear, however, whether the sign shown in Rembrandt's etching should be considered the sign of the macrocosm or a formula conjuring up the Erdgeist. Hamlin supported the latter version.

5. It is a matter of controversy if the etching indeed was intended to depict Dr. Faustus. See, for instance, Behling 1964; Carstensen and Henningsen 1988; Waal 1964.

6. This is the translation given in Bojanowski. In his earlier article, Bojanowski (1938, 530) translated: "Das Häufige magst du berühren; verborgen bleibt Dir die himmlische Liebe (You may touch that which is frequent, heavenly love remains hidden)" (1956, 527). For different interpretations, see Rotermund 1957; Waal 1964.

7. In the later part of the fifteenth century, Florence became a center of Platonic philosophy. These thinkers identified Love as the prime motive force for knowledge and redemption (Churton 1997, 103–104).

8. Letter to Hans Mühsam, March 30, 1954; quoted in Calaprice 1996, 158.

9. *New York Times Magazine*, November 9, 1930, pp. 1–4, reprinted in Einstein 1954, 46–49.

10. Ilse Schneider was a student of philosophy and physics at Berlin University.

11. A letter to Moritz Schlick, in which Einstein commended Schlick's essay on the philosophical implications of relativity, also reflected Einstein's critique of Kant's a priori forms and categories (December 14, 1915): "The relationship of the theory of relativity to Lorentz's theory is described in an excellent manner, its relationship to the teachings of Kant and his successors [is described] masterfully indeed. The trust in the 'apodeictic certainty' of the 'synthetic a priori judgments' is severely shaken by the realization of the invalidity of even a single one of these judgments" (Einstein Archive #21610).

12. See especially Erving Goffman's (1959) work. Using a very different approach, sociobiologists come to the same result. As E. O. Wilson put it, "Deception and hypocrisy are neither absolute evils that virtuous men suppress to a minimum level nor residual animal traits waiting to be erased by further social evolution. They are very human devices for conducting the complex daily business of social life. . . . Complete honesty on all sides is not the answer. The old primate frankness would destroy the delicate fabric of social life that has built up in human populations beyond the limits of the immediate clan" (1980, 277).

13. Frank 1949a, 90; also see Brian 1996, 44.

14. Regarding the pictures in Einstein's study, see the earlier endnote (p. 45 n. 33).

15. In Herneck 1976, 199; see an almost identical passage in Einstein 1955a, 7. Einstein's denial of the existence of a free will did not totally coincide with Schopenhauer's views on the free will, which were rather complex. The will, a central category in Schopenhauer's philosophy, was indeed free in its pure form, although it was not free in its empirical manifestations (that is, for

instance, as psychological motives that were determined by causality). Using a Vedic mataphor, Schopenhauer argued that space and time constituted a veil, the veil of Maya, that concealed the unity of the world as will and brought about the humans' perception of dispersed and separate manifestations of the will. On the concept of spatiotemporal separability (also called *principium individuationis*) and its philosophical history, see Howard 1997.

16. Here religion refers to the belief in Judaism he held as a child.

17. On Einstein and Spinoza, see Degen 1989, Paty 1986, and Pesic 1996.

18. "The infinite was his beginning and his end, the universe his only and everlasting love. In holy innocence and deep humility he beheld himself mirrored in the eternal world and perceived how he was its most amiable mirror, wherefore he stands there alone and unequalled, a master of his art but sublime above the profane rabble, a peerless beacon forever" (quoted in Pesic 1996, 195; original in Schleiermacher 1912, 40). Einstein's library contained one publication by Schleiermacher, a special print of his speech "Über die Bildung zur Religion" (Through Bildung to Religion), which was issued in 1916. This, however, was not the source of the quotation. The passage is from a different speech with the title "Über das Wesen der Religion" (On the Nature of Religion).

19. In light of the clear evidence that Einstein read Spinoza's *Ethics* at the Olympia-Akademie and again in 1915, I cannot follow Peter Degen's suggestion that there is "no indication that Einstein was thoroughly familiar with Spinoza's original works" (1989, 143).

20. Hessing 1933, 221; translation from Hoffmann and Dukas 1972, 95.

21. Similarly, Einstein stated on occasion of the three-hundredth anniversary of Spinoza's birth, "The recognition of the causal interrelationships of human behavior should lift our actions to a higher plane of conduct, which should not be subject to the irrational reactions of blind emotion" (*New York Times*, November 24, 1932, p. 27).

22. See, for instance, Pais, who also used the first part of the phrase in the title of his book on Einstein (1982, vi). For an account of how Einstein coined the phrase, see Sayen, who also noted that Einstein once told his assistant Valentine Bargmann, "I have second thoughts, maybe God is malicious" (1985, 50–51).

23. Interview with George Sylvester Viereck (Brian 1996, 186).

24. Furthermore, Einstein was rather indifferent toward Freud's theory. This theory's emphasis on speculative interpretations of the subconscious, to him, may have seemed like another illegitimate exit from the rationalist project of understanding the world.

9.

EINSTEIN AND AMERICA

The Priest and the Tinkerers

"Two kinds of people go into physics: those who have trouble with their car, and those who have trouble with God," the great physicist I. I. Rabi once told Gerald Holton. The deep insight lying at the bottom of Rabi's lighthearted remark serves well to frame our concluding discussion that links German science with American science. For it can easily be extended to apply not only to the motivations of individual scientists but also to the legitimation of national science systems. To see physics—or science in general—as religion, as the search for God, was the essence of the nineteenth-century German science tradition striving for a scientific Weltbild and Weltanschauung, as we have discussed it. Although the practical importance of science for military and civilian technology was not overlooked, scientists—as part of the Wissenschaftler group—were primarily commissioned by society to contribute to the collective identity of the Germans; they had the function of Kulturträger who defined and supported the nation. In the absence of a unified German state, of an advanced economy, and of a democratic society, Kultur, in its specific German definition, made the nation through much of the nineteenth century.[1] Kultur was a major source of collective identity and pride, both internally and externally (vis-à-vis other nations). Wissenschaft was seen as a major contributor to Bildung and Kultur, via the central concepts of Weltanschauung and Weltbild, and science was a part, if perhaps a somewhat marginal part, of Wissenschaft. The scientists were junior members of an elite of *Dichter und Denker*, who, as foremost Kulturträger, were priests in the realm of pure knowledge rather than mechanics—priests not only in the sense of the science-internal metaphor of the priesthood of Isis but also in the societal sense of meaning-producers who sustain the national identity. This role of the Wissenschaftler, and thus of the scientist, was institutionally embodied in the Humboldtian research university and socially defended by the Bildungsbürgertum, who legitimized their elevated status by pointing to their nonutilitarian Bildung. There was a pervasive popular deference to the mission of the Kulturträger who dedicated their lives to pure Wissenschaft. The German university could be considered "the greatest expression of the publicly supported and approved version of the theoretical life" (Bloom 1987, 322–23).

The preceding chapters also portrayed Einstein, arguably the most successful scientist of the epoch, as the quintessential embodiment of the Weltbild-searching tradition of German science. Einstein's transit from Berlin to Princeton symbolizes the twentieth-century move of German-style science to America. If the German archetype of the scientist was the priest, the American archetype was Rabi's car mechanic, or tinkerer: someone who focuses on narrowly circumscribed problems in an empirical trial-and-error mode, someone who works on devising and improving practically useful things. That American science tends to be dominated by "car mechanics" is a time-honored hypothesis, going back, as we have seen, to de Tocqueville's classic analysis of the utilitarian penchant of American science.[2] We now survey the influx into America of scholars and scientists from the Central European tradition.[3]

One of the first measures of the National Socialist government was to purge the civil service—which included the universities—of those whom it identified as Jews.[4] Germany thus deprived itself of many of its most productive Kulturträger, and the United States was among the main beneficiaries of the mass exodus of scholars and scientists. The large-scale empirical study by Davie and his collaborators identified 707 refugees who had formerly been professors (1947, 315). Of those, 523 succeeded in finding some academic position in the United States.[5] An additional thirty-six refugees who had not been professors in Central Europe became professors or assistant professors in the United States by 1947 (ibid., 316). Einstein and the small number of world-famous scholars and intellectuals among the immigrants faced relatively minor difficulties in securing suitable positions because many American institutions felt honored and were eager to have them in their midst. The problems were greater for the much larger group of refugee scientists and intellectuals who had less-than-super-star reputations in America. Yet assistance from many sources eased the transition of the European scientists and scholars into the American system.

A major haven for social scientists was the University in Exile of the New School for Social Research in New York, which was backed by Alvin Johnson; the entire faculty of the University in Exile consisted of Europeans.[6] The Emergency Committee in Aid of Displaced German Scholars (later called Emergency Committee in Aid of Displaced Foreign Scholars), which existed from 1933 to 1945, helped refugee scholars find academic positions in the United States and partly covered their salaries (Duggon and Drury 1948). During its lifetime, the Emergency Committee assisted a total of 335 scholars. In similar ways, the Oberlaender Trust of Philadelphia supported about 330 scholars, and the American Friends Service Committee, approximately 200. The American Christian Committee for Refugees started the Refugee Scholar Fund in 1943. In 1945, it created the American Committee for Émigré Scholars, Writers and Artists, which, in a more limited way, carried on the work of the folding Emergency Committee in Aid of Displaced Foreign Scholars. The Rockefeller Foundation was also active in aiding refugee scholars (Raven and Krohn 2000, xxv).

The group of scholars and scientists who had fled National Socialist or Fascist rule in Europe gave remarkable impulses to the American mind and to American science.[7] An eastern college president wrote, "There is no question that the presence of these foreign scholars has enriched American education. It has provided a great diversity of points of view, and it has brought to us many men of distinction" (quoted in Davie 1947, 314). As H. Stuart Hughes rightly pointed out, many intellectuals and scientists arriving from Germany and other European countries belonged, in terms of their Weltanschauung, to the most modern, cosmopolitan, and critical of the Kulturträger (1975, 18–19). Only few of them still felt attached to the (in most cases) Jewish religion of their ancestors, which they tended to consider a pre-emancipation relic. Instead, Kultur—first and foremost, the quest for Weltbild and Weltanschauung—was typically the most treasured possession that they carried in their baggage, and Kultur was, in their eyes, the most important contribution they could made to American science and American society.[8]

We now have to address what at first glance might seem like a counterclaim against the notion that America derived intellectual benefits from this wave of immigration, although it is more about the allegedly pernicious influence on America of German Wissenschaft in general than about the impact of German science. In his famous book *The Closing of the American Mind*, Allan Bloom (1987) not only decried the current state of American higher education but also identified the cause of its decline: According to his analysis, Continental European intellectual traditions, such as relativism and nihilism, have been seeping in and have been corroding American education. Bloom's broad category of Continental influences included more than merely German traditions; especially in recent years, Paris had become a hub for this kind of thought. But German traditions made up a prominent part of those influences (and they also inspired the recent French offshoot, as Bloom noted). His book dealt in detail with savants like Hegel, Marx, Nietzsche, Freud, Weber, and Heidegger. These thinkers were no tinkerers; they were the high priests of an alien intellectual religion, which nonetheless seemed to fill a void in America. "Fancy German philosophic talk fascinates us and takes the place of the really serious things" (Bloom 1987, 380). In Bloom's scenario, a naive American public was beguiled by the unfamiliar grand theories that provided Weltbild and Weltanschauung, or alternatively sophisticated critiques thereof, but it also misunderstood and vulgarized these theories. For him, the problem with the ideas disseminated by "our German missionaries" was that "we chose a system of thought that, like some wines, does not travel; we chose a way of looking at things that could never be ours and had as its starting point dislike of us and our goals" (ibid., 156, 153). In their new American environment, the German traditions were transformed into "a Disneyland version of the Weimar Republic for the whole family" (ibid., 147).

Bloom was not centrally concerned with science, but he did criticize a certain reductionism he observed among contemporary American scientists (ibid., 345). Nowadays, he noted, most scientists are engulfed in their narrow area of specialization and have lost touch with the other branches of knowledge. The great scientists of old, according to Bloom, had a broader horizon. They "were in general cultivated men who had some experience of, and real admiration for, the other parts of learning" (ibid., 350). But this very group of great scientists included prominently those who had been trained in the spirit of German science. Bloom thus in fact deplored the ascendancy of Rabi's car mechanics and denounced the lack of a synthetic vision in American science—the very kind of vision to which the German tradition of the scientific Weltbild aspired.[9] If anything, the immigrant scientists would serve as a counterweight to the defects observed by Bloom. We now look more closely at the refugee scientists' experience and influence.

In addition to Einstein, the group of refugee scientists included others of exceptional talent. Davie counted nine Nobel Prize winners among the immigrants (1947, 322). The Central European scientists brought their knowledge and skills to many academic fields in America, with their greatest impact, numerically speaking, being on physics and the medical sciences. Among those 707 former university professors mentioned earlier, we find seventy-one physicists and an equal number of professors of medicine, sixty-three chemists, sixty economists, fifty-three mathematicians, twenty-five psychologists, nineteen sociologists, fifteen biologists, nine political scientists, seven engineering professors, and small numbers in various other disciplines (ibid., 315).

In the eyes of many Americans, the refugee scientists' most important and valued contribution was that to war-related projects. In the immediate aftermath of World War II, Davie wrote, "One of the strongest and most tangible of all contributions which the refugee scholar made to his new country was in connection with the war effort. Here the scientific gifts of the refugees were used to the fullest possible extent in the development and testing of implements of war, while the art historians mapped monuments and historic sites in areas to be bombed and economists and technicians indicated strategic points for aircraft penetration" (ibid., 312–13). At least eight physicists and one chemist of the refugee scientists identified by Davie were known to have been directly involved in atomic energy research.

Einstein himself felt highly ambivalent about this facet of European influence. Although his famous letter to President Roosevelt had helped launch the Manhattan Project, he soon became an outspoken critic of nuclear weapons. By 1946, he served as chairman of the Emergency Committee of Atomic Scientists, which opposed the government's agenda for the military use of nuclear technology (Sonnert and Holton 2002, 151–52). In terms of scientific impact, Einstein, the most prominent among the German scientist-priests moving to America, had relatively little influence on the tinkerers. His obstinately pursued research program of devising a general field theory remained barren, and physicists in America (as elsewhere) took his work less and less seriously. Einstein felt disrespected and isolated—so much so that he complained to a friend that, at Princeton, "they regard me as the village idiot" (quoted in Holton 1998, 4).

Other scientists from Central Europe, by contrast, became extraordinarily productive in their new environment. For a case example, we now turn to the transplantation of the members of the Vienna Circle (with whom Einstein shared important intellectual roots) to America, which has been documented in some detail (Galison 1998; Holton 1993a, 1–55; Holton 1993b). As we have seen, these scholars and scientists were somewhat atypical in the tradition of Wissenschaft. Following Mach's example, they radically rejected metaphysical thinking and abstract ideas of unity. In their view, the goal of the unity of a scientific Weltanschauung, to which they wholeheartedly subscribed, should be accomplished through method, not through abstract concepts. Or, as Gerald Holton put it, unification was to be achieved not through metaphysics but through the elimination of metaphysics, thus standing Hegel on his head yet another way (1992b, 62).

The Viennese brand of Weltanschauung had early on gained some distinguished adherents in America, among them Paul Carus, William James, B. F. Skinner, W. V. Quine, and P. W. Bridgman. These connections eased the reception of the displaced Viennese Circle scholars. Many found themselves in the environs of Cambridge, Massachusetts, and eagerly participated in numerous applied research projects that went on at Harvard University and MIT as part of the war effort. But they also quickly reconstituted a forum for scientific exchange to continue the program of the Viennese Circle.

At Harvard University, Philipp Frank organized the pivotal Inter-Scientific Discussion Group in 1944, which, in 1947, turned into the Institute for the Unity of Science, after having received three years' funding from the Rockefeller Foundation. This institution organized meetings and provided the framework in which exiled and indigenous scientists could exchange their ideas.[10] Their discussions provided the fertile soil for interdisciplinary research and for the development of new fields, such as information theory, computing, systems analysis, game theory, and cybernetics, which would shape the course of science for decades to come.

A rather marginal movement in the Central European intellectual landscape, the Vienna Circle found America congenial. It perhaps helped that the Vienna Circle had been one of the less priestlike groups to begin with, not only in the sense of being suspicious of metaphysics, but also by harboring a certain appreciation of applied science. These scholars by and large agreed on a "scientific humanism," as Rudolf Carnap called it (1963, 83), that also had extrascientific—social and political—goals, and thus took the shape of a Weltanschauung (Holton 1993b, 58). Philipp Frank emphasized the "great relevance for the social and cultural life" of their efforts (1949b, 34), and the Vienna Circle formed productive connections with movements like the Bauhaus school of architecture (Galison 1990). Once in America, the Vienna Circle scholars focused on a more pragmatic and applied version of their ideal of the unity of science and of their scientific Weltanschauung (Galison 1998; Holton 1993b). They focused on pursuing their quest for God through and in "fixing cars."

The organization of society according to reasonable and efficient— or scientific—principles has been an American penchant of long standing. Supplanting Platonic philosopher-kings with scientist-kings, Alfred North Whitehead, in *Science and the Modern World*, proclaimed that the men of science were "ultimately the rulers of the world" (1925, 292).[11] Fed by this and other sources, efforts to advance the technocratic management of societal affairs burgeoned. The scientist-kings were to be car mechanics writ large, and they could count on the sympathy and the support of the transplanted Vienna Circle. In *The Rise of Scientific Philosophy* (1951), Hans Reichenbach (who had belonged to the Berlin counterpart of the Vienna Circle and had landed on the West Coast) propounded a Weltanschauung along the lines of analytic philosophy: The decision for democracy was ethical, but once this decision had been made, all secondary questions then became scientific, to be decided on the basis of empirical evidence. This was the application to societal practice of the scathing critiques of the metaphysical and untestable statements pervading many Weltanschauungen (especially Marxism)—critiques that had regularly been emerging from associates of the Vienna group, from Mach through Popper. Based on these ideas, American scholars forged a powerful technocratic vision and advocated the comprehensive application of empirical knowledge to societal problems. The title of Daniel Bell's book *End of Ideology* (1960) became emblematic of the rejection of the ideologue in favor of the scientist-technocrat. In this sense, Central Europe provided not only the Bloomian intellectual seducers but also the antidote, a science that denounces the nonempirical flights of certain Weltanschauungen but at the same time does not reduce itself to small-scale "tinkering."

The group of refugee scientists as a whole appears to have had considerable structural and long-term effects on the American science system by helping increase the emphasis on basic science and the "ultimate questions." The distinguished chemist Eugene Rabinowitch once said, "The greatest role of the European-born scientists was to change the American concept of science—it used to mean invention; it seemed impossible to convince the congressmen that basic science is important"(quoted in Fermi 1971, 315). But then the refugee scientists and, first of all, the decisive contributions of basic science to the war effort made a difference. At the end of World War II, Vannevar Bush's (1945) famous report *Science, the Endless Frontier* and other contributions by visionary science policy makers inaugurated massive federal investments in basic science and thus brought about a "Golden Age" for American basic research. We must keep in mind, however, that the compact between science and American society that made possible the unprecedented growth in federally funded basic research did not rest on the citizenry's passion for basic science itself. In the view of most nonscientists and science policy makers, it rested on the promise that, through unpredictable spin-offs, basic science would lead to useful advances in the domain of the "car mechanics": For instance, basic science was trusted to deliver in the technological arms race of the Cold War, as it had in World War II. In other words, it was not supported as a quasi-religious quest for understanding the cosmos but as an indirect, yet eventually effective, way of generating practical benefits. The successful spin-offs that indeed materialized provided the "halo" for basic science in general that made the expenditure of tax money on it palatable.

Nonetheless, the influence of the German science tradition has not gone unchallenged. Basic science has remained under periodically rising pressure to justify itself to the public. Although in certain fields and subfields (e.g., string theory), the quest for synthesis so strenuously pursued by Einstein is still alive and well, there is a tendency for science to revert to what might be called a Tocquevillian baseline favoring the "car mechanics." Several factors appear to contribute to this tendency. First, there is the argument that the shift toward short-range "tinkering" is necessary because contemporary science has surpassed the synthetic capacity of the individual scientist—an argument to which Einstein gave short shrift. He was known to heap scorn on those of his colleagues who gave up on the quest for synthesis and were content boring holes into the wood where it was thinnest, as he sneered (Frank 1979, 23). Second, the leading scientific themata may be shifting from unity to diversity.[12] Whereas Einstein and his cohorts were positively irked by diverse facts or theories for which they knew no unifying theoretical framework, contemporary scientists seem less disturbed. Perhaps diversity has become as aesthetically pleasing to them as unity was to Einstein. Third, the current strength of the "car mechanics" may be boosted by talk of the "end of science": The fundamental scientific questions have already been answered, the proponents say, and all that remains are some minor improvements of the tinkering kind (although such a pronouncement may be just as premature as its predecessors have turned out to be).[13]

What we have here, then, in contemporary American science, is not a wholesale acceptance of the science tradition of Central Europe but the cross-fertilization of national traditions. The "tinkering" tradition of American science has benefited greatly from the infusion of the more "priestly" European, and especially German, science tradition. The resulting mixture has proven enormously potent and, given wise policy guidance, will be even stronger in the future.[14]

NOTES

1. The political unification of Germany into the "Second Reich" (under Prussian hegemony) came about only in 1871. Toward the end of the century, Germany also caught up economically with the leading nations.

2. Let me reiterate: These two national science styles must be understood merely as tendencies; no homogeneous and pure national traditions of science have existed in the era under consideration because there have

been substantial variations within each of them, as well as overlaps and exchanges between them (as we shall see). And even some of those who got into science as car mechanics became interested in the bigger questions as they grew older.

3. Among the scholars who have written on this topic are, for instance, Ash and Söllner 1996; Coser 1984; Fermi 1971; Fleming and Bailyn 1969; Heilbut 1983; Hughes 1975; Jackman and Borden 1983; Kent 1953; Möller 1984; Stadler 1987, 1988; Strauss, Fischer, Hoffmann, and Söllner 1991.

4. To give but one highly significant example, the world-renowned Institute of Mathematics at the University of Göttingen was virtually depleted of many of its leading members. Among the former institute members who came to the United States were Richard Courant, the former director, as well as Otto Neugebauer, Emmy Noether, William Prager, Hans Lewy, and Hermann Weyl (Fermi 1971, 286).

5. These numbers include all refugees from National Socialism. The former Germans and Austrians can be estimated, according to other data from Davie, to comprise three-quarters of those who found an academic job in America, or close to four hundred (1947, 301). Further note that Davie also mentioned 424 "nonuniversity scholars and other scientists" (ibid., 320–21).

6. See Davie 1947, 109; Krohn 1993; Luckmann 1981. The so-called Frankfurt School of unorthodox Marxist intellectuals and social scientists moved its *Institut für Sozialforschung* to New York City and, thanks to its endowment, was relatively self-sufficient.

7. This wave of scientific migration in the 1930s and 1940s was not the first instance of an interaction between the national scientific traditions. The German model has, without doubt, had an impact of long standing. Because Germany was leading in most scientific disciplines during the late nineteenth and early twentieth centuries, it markedly influenced the scientific enterprise of other nations, such as Britain and the United States. Johns Hopkins University was modeled after the German research university, and many of the leading American scientists of that era received part of their training at German-speaking universities. At a more general level, the tradition of the American liberal arts education has drawn, to a considerable extent, on the German ideal of Bildung.

8. In his essay "The Intellectual Preeminence of Jews in Modern Europe," the sociologist Thorstein Veblen examined why people of Jewish ancestry excelled in intellectual pursuits (1934 [1919], 219–31). He saw the

cause for the Jews' overproportional contributions in their assimilation, thorough yet incomplete, into modern Gentile culture. "[T]his intellectual pre-eminence of the Jews has come into bearing within the gentile community of peoples, not from the outside; . . . the men who have been its bearers have been men immersed in this gentile culture in which they have played their part of guidance and incitement, not bearers of a compelling message from afar or proselyters of enlightenment conjuring with a ready formula worked out in the ghetto and carried over into the gentile community for its mental regeneration" (ibid., 224). According to Veblen, their inquiring and skeptical mind, a requisite for scientific achievement, sprang from their rejection of traditional Judaism yet also remained unfettered by the comfortable security of Gentile conventions that dulled the inquisitiveness of many of their Gentile cohorts. Veblen's sociological analysis of the Jewish intellectuals thus was a variation of Georg Simmel's classic concept of the "stranger," which emphasized the intellectual benefits that can accrue from that status (1964 [1908], 402–408). In American sociology, the "marginal man" was a somewhat similar concept, pioneered by Robert E. Park (1928; see Park and Burgess 1921) and elaborated by Everett Stonequist (1937). The notion of the stranger was further explored by Simmel student Siegfried Kracauer (1969). Werner Sombart (1911) hypothesized that their marginality would benefit Jews in their business pursuits. On Veblen's hypothesis, also see Feuer 1963, 297–318.

In his book *Science, Jews, and Secular Culture*, David Hollinger explored "the role of Jewish intellectuals in the process by which the cultural program of Christianity loses some of its public standing" (1996, xi; also see 2002). The rise of the Jewish intellectual in what used to be a predominantly WASP domain led, according to Hollinger, to the secularization of the Christian vestiges and to the promotion of cosmopolitanism. It also gave American science a boost that propelled it from provincialism to the forefront of world science. Before the ascendancy of the Jewish intellectuals, secularism had been weaker in America than in the developed European countries. As we pointed out, German Protestantism in particular modernized itself by merging into a religion of Kultur. Our topic only partly overlaps with Hollinger's, of course. Most of the Jewish intellectuals whom Hollinger mentions had their roots in the Jewish immigration from Russia and Eastern Europe, some in an earlier nineteenth-century immigration from Germany, evidence of which can be found, for instance, in the chronics of Jebenhausen, the ancestral home of Einstein's maternal family. Nonetheless, the intellectuals and scientists who arrived as refugees from

National Socialism certainly embodied the assimilated, critical, secular, and cosmopolitan mind-set to an unusually high degree.

9. Here lies a poignant irony: It was precisely what Bloom and others considered the most dubious fruits of the German intellectual tradition that they found flourishing most visibly in America. The most flashy success belonged to Marx's, Freud's, and Nietzsche's grandiose Weltanschauungen, which were only tenuously tied to empirical evidence. As we have shown in the case of Einstein, the exponents of the scientific Weltbild and Weltanschauung—the scientists on a quest for God, in Rabi's terminology—insisted that any grand scheme of Weltbild had also to be empirical, had to be consistent with the data. Their project was to keep the goal of Weltbild and Weltanschauung, which permeated wider German Wissenschaft and Kultur, grounded in empirical contents. Such an endeavor was rather less glamorous than the bolder ones that caught the attention of the general public on the Continent and in America. In terms of popularity among American intellectuals, the scientists could not rival Freud and Marx, the latter of whom also gained a mass following in the former Communist Block and even in the Third World.

10. Analogous meetings took place in Chicago (under Morris), Los Angeles (Reichenbach), Minneapolis (Feigl), Berkeley (Lenzen), and Princeton (Hempel); see Holton 1993b, 63.

11. This was, of course, an old core idea of the proponents of a scientific Weltanschauung, such as Ludwig Büchner.

12. For an early expression of this view, see Adams 1918.

13. See Glass 1974 and Horgan 1996. For an opposing view, see Maddox 1998. Recall how, at the end of the nineteenth century, the notion of physics as a closed science evaporated in a flash of dazzling discoveries, not the least those made by Einstein himself.

14. In *Ivory Bridges: Connecting Science and Society*, Gerald Holton and I argue for a reform in the current federal science policy so as to better accomplish the twin goals of advancing our knowledge of nature and bringing tangible benefits to society (Sonnert and Holton 2002; also see Branscomb, Holton, and Sonnert 2001 and Holton and Sonnert 1999). Since World War II, science policy has been rooted in the conceptual dichotomy of basic science (we call it "Newtonian") and applied ("Baconian") science. (On the limitations of the conceptual dichotomy, see Sonnert and Brooks 2001; for a case study of how the actual process of scientific discovery transcends that dichotomy, see Holton, Chang, and Jurkowitz 1996.) We suggest complementing these two modes of research

with a third, which we call "Jeffersonian science." It is defined as basic research in an area of identified societal need, that is, in an area of basic scientific ignorance that appears to lie at the heart of a social problem. (To some extent, Jeffersonian science is similar to what has been described as "Mode 2" of knowledge production [Nowotny, Scott, and Gibbons 2001].) In our opinion, an integrated science policy that is based on an explicitly tripartite rationale (Newtonian-Baconian-Jeffersonian) will be a decisive step forward. The addition of a Jeffersonian component to a balanced federal science policy will, in our view, ensure the continuing flourishing of basic science even in a basically Tocquevillian cultural environment that cares primarily for practical benefits, and it will lead to quicker and better results, compared with a science policy that operates only with the concepts of basic and applied research.

CONCLUSION

Einstein loved *Gedankenexperimente* (thought experiments),
which he used with great success at several crucial junctures in
the development of his physical Weltbild. (See Wagner 1970,
100–16.) At the conclusion of this study, let us now conduct our
own *Gedankenexperiment* with Einstein: If he had read this book,
what would he have said? I suppose that his reaction in Europe
would have differed from his reaction in America. During his stay in
Berlin, for instance, he may well have commented wryly, "It's all
quite obvious to me," but, in Princeton, he might have been more
interested. One of the defining characteristics of culture is that it is
taken for granted: The very things that appear obvious (or perhaps
even trivial) constitute much of the core of a culture. Cultural self-
awareness tends to be heightened when one is confronted with a
strange culture and its different take on supposedly self-evident pre-
sumptions. Therefore, I suggest, Einstein in America might have
found this book much more relevant than Einstein in Europe would

have. Conversely, I hope that the in-depth presentation of the nine-teenth-century Central European culture of science will contribute to sharpening the self-awareness of the scientific culture in America at the start of the twenty-first century.

Let us now briefly retrace the journey on which this book has taken us. Our point of departure was Einstein's popular rebel image. To challenge and amend this rather one-sided perception was a main purpose of the book. The rebel image is, of course, not entirely unfounded. In fact, we began by viewing the body of evidence that portrayed Einstein as a rebel in various aspects of his life. Then we confronted that representation with a list of counterevidence, showing the limitations to Einstein's rebelliousness and emphasizing his traditionalist side. At that point, Einstein may well have appeared as a contradictory enigma that would be hopeless to figure out. But an extensive tour through the cultural landscape of nineteenth-century Germany enabled us to arrive, at the end, with a more thorough understanding of the complex nature of the famous physicist.

Our tour began with the concept of Kultur itself and its companion concepts Bildung and Zivilisation. We realized how specific historical and socioeconomic formations in the German-speaking areas of Central Europe encouraged certain cultural ideas that expressed themselves most significantly in the concept of Kultur itself. At the very pinnacle of Kultur, we found the concepts of Weltanschauung and Weltbild. Especially the former concept underwent a dramatic evolution, which we traced in detail. Soon after it originated in Kant's transcendental theory of perception, it acquired the additional meaning of a rational synthetic view of everything, under the tutelage of idealistic philosophers and theologians. Then, however, this concept collapsed under its own weight and gave way to a relativism that, among some radical exponents, became so rampant that it not only rejected the idea of an all-encompassing rational Weltanschauung but also abandoned the narrower notion of factual objectivity.[1] The preoccupation with Weltanschauung persisted, but now elements of choice, decision, and ultimately power entered the notion. The terminal point was reached in the National Socialist Weltanschauung with its far-reaching epistemological relativism.

After visiting the major landmarks in the conceptual landscape of Kultur, we examined how science related to it and how German science was uniquely shaped by its cultural environment. Because German science, generally speaking, strove for a place in the treasured Kultur, it adopted certain characteristic features in the process—German science became more nonutilitarian and focused on the ultimate questions. It moved close to religion, and many German scientists viewed themselves as some kind of priests. These features of German science became most apparent when looking at how enthusiastically it took on the cultural aspirations of Weltanschauung and Weltbild, and how deeply it committed itself to the quest for a comprehensive synthesis. Although only a minority of German scientists upheld the original idea of a comprehensive scientific Weltanschauung that covered all areas of life (and ran into some of the same problems that plagued the attempts at an all-encompassing Weltanschauung in general), most scientists concentrated their energies on the construction of a scientific Weltbild, with enormous productivity and success.

Here we found the key to solving our original conundrum, and we could return to Albert Einstein in hopes of gaining a better understanding of the seeming contradictions. We first passed through the way station of a sociological account showing that the mobility processes propelling the German Jews into modern society tended to foster a particularly fervent allegiance to the ideals of Kultur, Weltanschauung, and Weltbild. Then we were ready to demonstrate how strongly Einstein's physical research program was shaped by pivotal cultural goals of German society and of the German science community, especially by the quest for a scientific Weltbild. Einstein thus revealed himself not as a countercultural rebel but as a person deeply indebted to cultural tradition. He was crucially influenced, in his general outlook on life and in his physics, by the surrounding culture of the nineteenth century, in which he grew up and was socialized. To an exceptionally strong and rigorous extent, he identified with the major goals of Kultur, as encapsulated by the quest for Weltanschauung and Weltbild. His position vis-à-vis Kultur also made possible a highly productive rebellion. Einstein was not merely a cultural traditionalist; he operated at what one might consider an optimal distance from the *actual* state of Kultur and rebelled against it insofar as he found it wanting, as measured by its own core *ideals*. His was the rebellion of the idealistic radical *in the name of Kultur* against its perceived failings and shortcomings. He enjoyed obliterating the pragmatic compromises the more complacent and pedestrian contemporaries—philistines, as he loved to belittle them—liked to make. Einstein's cultural environment, in turn, understood, tolerated, and to some extent supported his radical quest because it served Kultur's highest goals. Kultur was wide enough to allow for the type of the culturally legitimate rebel.

We then turned to Einstein's Weltanschauung to understand the psychological motivating force behind his quest for the Weltbild. Although Einstein rejected organized religion, we found him to be a deeply religious person (in a wide sense of the term). Even while he realized that a full understanding of Nature exceeded human capabilities, the glimpses of Nature he caught inspired him with a profound sense of awe and humility that drove his ceaseless (though ultimately futile) pursuit of the scientific Weltbild. Here again, Einstein showed himself as a true creature of Kultur—epitomizing the German scientist-priest.

Finally, we observed the transit to the United States of some strands of the German Kultur (those for which Einstein stood). The mass exodus of scientists and other Kulturträger during National Socialist rule symbolized a demise of Kultur, but it also set off a fertile synthesis, blending the Kultur-inspired science tradition with the quite different American one.

Throughout this book, our analysis focused on two levels—the individual level (Einstein) and the collective level (Kultur)—and explored their interaction. To an amazing degree, cultural trends became focused and crystallized in one person: Einstein personified the Goethean drive to unity and the quest for a scientific Weltbild. Kultur was certainly no monolith; within a shared and stable framework, its specific content was under debate, often controversial. We have traced the enormous shifts in meaning that the concept of

Weltanschauung underwent in the course of one century and a half; we have seen how it splintered up into diverse and mutually hostile strands. We have observed how Einstein's Weltanschauung and Weltbild were determined by ideas rooted in the mid-nineteenth century or earlier; the later developments of those concepts did not make a comparable impression. Einstein's cultural outlook thus differed from that of others, and an example of this divergence could be found in the area of the physical Weltbild. The issue that most decidedly separated Einstein from many of his colleagues was the problem of causality and determinism in quantum mechanics. Here he did not budge but remained true to his deep-seated instincts rooted in the early and mid-nineteenth century, whereas other, mostly younger, physicists had less trouble jettisoning the classical concepts of determinism and causality. On this core question of modern physics, we found Einstein vociferously taking the traditionalist position. At a more general, and much more dramatic, level, Einstein's Kultur collided with the Kultur that became ascendant after 1933. What Einstein cherished in Kultur and what the German government propagated as Kultur at that time were totally opposite, and Einstein suffered vilification and expulsion as one of the prime enemies of Kultur of the National Socialist variety.

Why is it a worthwhile pursuit for a sociologist of science to examine the influence of major elements of Kultur on Einstein? Primarily because it provides an illuminating case study on the important question to what extent scientific progress is driven by forces external to science. The concept of thematic analysis in the history of science is central in this undertaking (Holton 1973). The concept proposes that the research programs of some scientists cannot be wholly explained by problems and questions somehow imminent in the current state of their discipline. Those scientists are considered to be deeply committed to certain themata, which they pursue in their research relatively independently of whatever are the latest research results in their field. Einstein's case serves to corroborate that central premise of thematic analysis. How Einstein conceived and pursued his research program: neither could be fully explained by issues that would present themselves or rise automatically from experimental results. Moreover, Einstein himself repeatedly emphasized the speculative and constructive nature of scientific thought. He was convinced that theories did not emerge from empirical facts in any necessary way. "A theory can be tested by experience," he said, "but there is no way from experience to the formation of a theory" (Einstein 1979a, 33). He thought that intuitive leaps played a major role here and that these leaps were guided by what he called the researcher's freely choosable categories, and what Holton called themata (Einstein 1979b, 500). Indeed, Einstein's research program, which addressed itself to the precise formulations of major physical laws that had been known for a long time (e.g., Newtonian mechanics), or had more recently become common knowledge among physicists (e.g., Maxwell's electrodynamics), was driven by his allegiance to certain enduring themata (Einstein 1979b, 500). And those themata—unity and synthesis first among them—resonated with the surrounding Kultur.

Of course, Kultur cannot completely explain Einstein. Let us emphasize again that additional factors—physics-internal, biological, psychological, sociological—must also be considered to gain a full understanding of a singular genius who stood out from his cultural surroundings. But these additional factors were reinforced and validated by the cultural environment in which Einstein grew up and lived for most of his life, and which left an indelible mark on his outlook. The following example may illustrate the interplay of cultural and noncultural factors that shaped the scientist Albert Einstein. Late in life, Einstein still recalled that, at the young age of four or five, he was shown a compass by his father. "I still remember—or believe to remember—that this experience has made a deep and lasting impression on me. There had to be something behind the things that was deeply hidden" (Einstein 1979a, 3). The amazement about the "miracle" of the compass—the sheer joy of it—was one of those seminal moments that might trigger a lifelong fascination with science (ibid.). As we know, the theoretical exploration of fields, similar to the magnetic field of the compass, became one of the central planks of Einstein's physical research program, up to the (unsuccessful) search for a general field theory.

I do not suggest that young Albert's delight with encountering the "miraculous" workings of nature was itself culturally determined. However, that delight was certainly not enough to explain his further involvement in science. There was no causal chain that as if by logical necessity linked the original fascination with the compass to the pursuit of field theories. Other possibilities would have been conceivable: an interest in using the compass in practical applications, in improving the instrument, in exploring specific aspects of magnetism, and so forth. Here is one of the points where the cultural influence comes in—giving form to the sheer enthrallment of an encounter with nature and channeling it into certain directions. A fascination with the compass could, in a different cultural environment, well have led to the career of a science tinkerer; in Kultur, it may have helped to lead Einstein to the career of a science priest.

Being part of an epic transplantation of Kultur into a different environment, Einstein spent roughly the last two decades of his life in America. There, Einstein steadfastly clung to his research program, although it generated next to no interest in the physics community and lacked success. He also publicly stuck to his political convictions, although they were wildly unpopular (especially in the McCarthy era). When Einstein was nearing his death in April 1955, his last acts were still fully in character.[2] He remained, to the end, a strong-willed individual obstinately pursuing his Weltbild and Weltanschauung. In his last days, he signed a manifesto with Bertrand Russell and others, intending to bring together the international community of scientists as a unified counterweight to the divisions brought about by the Cold War. At the hospital, shortly before he died of an abdominal aortic aneurysm, he told his stepdaughter Margot, "I have done my thing here," but then had second thoughts and demanded his notepad.[3] At about one o'clock in the morning of April 18, when the aneurysm burst, he suddenly spoke up once more. It is not known what he said because the night nurse did not understand German.

Einstein's arrangements for his passing were emblematic of his enduring struggle for simplicity and against ordinary convention. There was to be no big funeral—only a few family members and friends gathering at the crematorium—no speeches, no religious rites, no flowers, and not even music. As Einstein's ashes were scattered in the wind, Otto Nathan, an old friend and fellow immigrant, recited a few verses of poetry, ending with the lines:

> He gleams like some departing meteor bright,
> Combining, with his own, eternal light.

The grief-stricken Goethe had written these words 150 years earlier, after Friedrich Schiller died.[4] Fittingly, that piece of poetry, Goethe's epilogue to Schiller's "Lied von der Glocke"—joining together the two greatest exponents of Kultur—symbolized Einstein's lifelong theme: synthesis and unity.

NOTES

1. According to radical relativists, Weltanschauungen—identified as arbitrary constructs—in turn, biased statements of fact and thus undermined their validity.

2. The account of Einstein's last days and death follows Holton 1998 and Wickert 1972, 102.

3. He had known of the aneuryism for seven years but refused to have it operated on.

4. To commemorate Schiller's death, his famous "Lied von der Glocke" was performed on stage on August 10, 1805. Goethe added an epilogue. Ten years later, in 1815, the performance was repeated, again with Goethe's epilogue, which was slightly altered. The lines recited by Nathan were from this second version. They are given here in Theodore Martin's translation (1882, 439).

REFERENCES

1882–87. *Brockhaus' Konversations-Lexikon*, 13th ed. Leipzig: F. A. Brockhaus.

1892. *Lehrpläne und Lehraufgaben für die höheren Schulen, nebst Erläuterungen und Ausführungsbestimmungen*. Promulgated by the Prussian "Ministerium der geistlichen, Unterrichts- und Medizinalangelegenheiten." Berlin: Wilhelm Hertz.

1986–96. *Brockhaus Enzyklopädie in vierundzwanzig Bänden*, 19th ed. Mannheim: F. A. Brockhaus.

1922–23. *Brockhaus Handbuch des Wissens in vier Bänden*. Leipzig: F. A. Brockhaus.

1942. *Webster's New International Dictionary*. Springfield, MA: Merriam.

1977–82. *Der Große Brockhaus in zwölf Bänden*, 18th ed. Wiesbaden: F. A. Brockhaus.

1979. *Meyers Enzyklopädisches Lexikon*. Mannheim: Bibliographisches Institut.

Abiko, S. 2003. "On Einstein's Distrust of the Electromagnetic Theory: The Origin of the Light-Velocity Postulate." *Historical Studies in the Physical and Biological Sciences* 33, no. 2: 193–215.

Abir-Am, P. 1993. "From Multidisciplinary Collaboration to Transnational Objectivity: International Space as Constitutive of Molecular Biology. 1930–1970." In *Denationalizing Science: The Context of International Scientific Practice*, ed. E. Crawford, T. Shinn, and S. Sörlin. Dordrecht, Holland: Kluwer, pp. 153–86.

Ackermann, J. 1970. *Heinrich Himmler als Ideologe*. Göttingen: Musterschmidt.

Adams, H. 1918. *The Education of Henry Adams: An Autobiography*. Boston: Houghton Mifflin.

Adler, F. W. 1909. "Die Einheit des physikalischen Weltbildes." *Naturwissenschaftliche Wochenschrift*, N.F., 8, no. 52: 817–22.

Aisenberg, N., and M. Harrington. 1988. *Women of Academe: Outsiders in the Sacred Grove*. Amherst: University of Massachusetts Press.

Ajdukiewicz, K. 1934. "Das Weltbild und die Begriffsapparatur." *Erkenntnis* 4: 259–87.

———. 1935. "Die wissenschaftliche Weltperspektive." *Erkenntnis* 5: 22–30.

Amrine, F., ed. 1996. *Goethe in the History of Science*. 2 vols. New York: Peter Lang.

Amrine, F., F. J. Zucker, and H. Wheeler, eds. 1987. *Goethe and the Sciences: A Reappraisal*. Dordrecht: D. Reidel.

Anderton, K. M. 1993. The Limits of Science: A Social, Political, and Moral Agenda for Epistemology in Nineteenth-Century Germany. PhD thesis, Harvard University.

Arnold, M. 1994. *Culture and Anarchy*, ed. S. Lipman. New Haven, CT: Yale University Press. (Orig. pub. 1869.)

Ash, M. G., and A. Söllner, eds. 1996. *Forced Migration and Scientific Change: Emigré German-Speaking Scientists and Scholars After 1933*. Cambridge: Cambridge University Press.

Assmann, A., and J. Assmann, eds. 1997. *Schleier und Schwelle: Geheimnis und Öffentlichkeit*. Munich: Wilhelm Fink Verlag.

———., eds. 1998. *Schleier und Schwelle: Geheimnis und Offenbarung*. Munich: Wilhelm Fink Verlag.

Assmann, J. 1997. *Moses the Egyptian: The Memory of Egypt in Western Monotheism*. Cambridge, MA: Harvard University Press.

———. 1999. *Das verschleierte Bild zu Sais: Schillers Ballade und ihre griechischen und ägyptischen Hintergründe*. Stuttgart: B. G. Teubner.

Audi, R., ed. 1995. *The Cambridge Dictionary of Philosophy*. Cambridge: Cambridge University Press.

Ault, D. D. 1975. *Visionary Physics: Blake's Response to Newton*. Chicago: University of Chicago Press.

Auroux, S., ed. 1990. *Les notions philosophiques: Dictionnaire*, vol. 2. Paris: Presses universitaires de France.

Austeda, F. 1979. *Lexikon der Philosophie*, 5th ed. Vienna: Hollinek.

———. 1989. *Lexikon der Philosophie*, 6th ed. Vienna: Hollinek.

Austin, J. H. 1978. *Chase, Chance, and Creativity: The Lucky Art of Novelty*. New York: Columbia University Press.

Badash, L. 1972. "The Completeness of Nineteenth-Century Science." *Isis* 63: 48–58.

Barner, W. 1986. "Jüdische Goethe-Verehrung vor 1933." In *Juden in der deutschen Literatur: Ein deutsch-israelisches Symposion*, ed. S. Moses and A. Schöne. Frankfurt: Suhrkamp, pp. 127–51.

Barth, P. 1922. *Die Philosophie der Geschichte als Soziologie*. 2 vols. Leipzig: O. R. Reisland. (Orig. pub. 1897.)

Basalla, G., W. Coleman, and R. H. Kargon, eds. 1970. *Victorian Science: A Self-Portrait from the Presidential Addresses of the British Association for the Advancement of Science*. Garden City, NY: Doubleday.

Bavink, B. 1947. *Das Weltbild der heutigen Naturwissenschaften und seine Beziehungen zu Philosophie und Religion*. Iserlohn: Silva-Verlag.

Bayer, H. 1975. "Zur Soziologie des bürgerlichen Bildungsbegriffs." *Paedagogica Historica* 15: 321–55.

Bayertz, K. 1985. "Spreading the Spirit of Science: Social Determinants of the Popularization of Science in Nineteenth-Century Germany." In *Expository Science: Forms and Functions of Popularization*, ed. T. Shinn and R. Whitley. Dordrecht, Holland: D. Reidel, pp. 209–27.

Behling, L. 1964. "Rembrandts sog. 'Dr. Faustus,' Johann Baptista Portas Magia naturalis und Jacob Böhme." *Oud Holland* 79: 49–77.

Bell, D. 1960. *The End of Ideology: On the Exhaustion of Political Ideas in the Fifties*. Glencoe, IL: Free Press.

Ben-David, J. 1984. *The Scientist's Role in Society: A Comparative Study*. Chicago: University of Chicago Press.

———. 1991. *Scientific Growth: Essays on the Social Organization and Ethos of Science*. Berkeley: University of California Press.

Bergmann, G. 1967. *The Metaphysics of Logical Positivism*. Madison: University of Wisconsin Press.

Berlin, I. 1979. *Concepts and Categories*. New York: Viking Press.

———. 1980. "Einstein and Israel." In *Einstein and Humanism: Selected Papers from the Jerusalem Einstein Centennial Symposium*. New York: Aspen Institute for Humanistic Studies, 7–24.

————. 1992. *The Crooked Timber of Humanity: Chapters in the History of Ideas.* Ed. H. Hardy. New York: Vintage Books.

Bernstein, A. D. 1860. *Naturwissenschaftliche Volkbücher,* 5th ed. Berlin: Dümmler.

Bernstein, J. 1973. *Einstein.* New York: Viking Press.

Bertalanffy, L. v. 1949. *Das biologische Weltbild.* Bern: A. Francke.

Betz, W. 1981. "Zur Geschichte des Wortes 'Weltanschauung.'" In *Kursbuch der Weltanschauungen.* Frankfurt: Ullstein, pp. 18–28.

Beutler, E., ed. 1977. *Johann Wolfgang von Goethe—Die Faustdichtungen.* Munich: Winkler Verlag.

Beyerchen, A. D. 1977. *Scientists under Hitler: Politics and the Physics Community in the Third Reich.* New Haven, CT: Yale University Press.

Billen, A., and M. Skipworth. 1984. *Oxford Type: An Anthology of Isis, the Oxford University Magazine.* London: Robson Books.

Bippart, G. 1848. *Pindar's Leben, Weltanschauung und Kunst.* Jena: Hochhausen.

Blackburn, S. 1994. *The Oxford Dictionary of Philosophy.* Oxford: Oxford University Press.

Blake, W. 1975. *The Marriage of Heaven and Hell.* London: Oxford University Press. (Orig. pub. 1790.)

Blavatsky, H. P. 1972. *Isis Unveiled,* vol. 1: *Science.* Repr., Wheaton, IL: Theosophical Publishing House. (Orig. pub. 1877.)

Bloom, A. 1987. *The Closing of the American Mind: How Higher Education Has Failed Democracy and Impoverished the Souls of Today's Students.* New York: Simon and Schuster.

Boccaccio, G. 2001. *De mulieribus claris,* ed. and trans. V. Brown. Cambridge, MA: Harvard University Press. (Orig. written 1361–62.)

Bohr, N. 1979. "Diskussion mit Einstein über erkenntnistheoretische Probleme in der Atomphysik." In *Albert Einstein als Philosoph und Naturforscher,* ed. P. A. Schilpp. Braunschweig: Vieweg, pp. 115–50.

Bojanowski, M. 1938. "Das Anagramm in Rembrandts 'Faust.'" *Deutsche Vierteljahrsschrift für Literaturwissenschaft und Geistesgeschichte* 16: 527–30.

————. 1956. "Rembrandts 'Faust' Radierung von 1648." *Deutsche Vierteljahrsschrift für Literaturwissenschaft und Geistesgeschichte* 30: 526–32.

Boltzmann, L. 1891–93. *Vorlesungen über Maxwells Theorie der Elektricität und des Lichtes.* 2 vols. Leipzig: Johann Ambrosius Barth.

Bonnet, H. 1952. *Reallexikon der ägyptischen Religionsgeschichte.* Berlin: Walter de Gruyter.

Borkenau, F. 1971. *Der Übergang vom feudalen zum bürgerlichen Weltbild: Studien zur Geschichte der Philosophie der Manufakturperiode.* Darmstadt: Wissenschaftliche Buchgesellschaft. (Orig. pub. 1943.)

Born, I. v. 1784. "Über die Mysterien der Aegyptier." *Journal für Freymaurer* 1: 17–132.

Borscheid, P. 1976. *Naturwissenschaft, Staat und Industrie in Baden.* Stuttgart: Klett.

Bossenbrook, W. J. 1961. *The German Mind.* Detroit: Wayne State University Press.

Botstein, L. 1991. *Judentum und Modernität: Essays zur Rolle der Juden in der deutschen und österreichischen Kultur. 1848–1938.* Wien: Böhlau Verlag.

Bourdieu, P. 1984. *Distinction: A Social Critique of the Judgement of Taste,* trans. R. Nice. Cambridge, MA: Harvard University Press.

Branscomb, L., G. Holton, and G. Sonnert. 2001. "Science for Society—Cutting-Edge Basic Research in the Service of Public Objectives: A Blueprint for an Intellectually Bold and Socially Beneficial Science Policy." *Report on the November 2000 Conference on Basic Research in the Service of Public Objectives.* Cambridge, MA: Belfer Center for Science and International Affairs.

Bratranek, F. T., ed. 1876. *Goethe's Briefwechsel mit den Gebrüdern von Humboldt (1795–1832).* Leipzig: F. A. Brockhaus.

Braudel, F. 1980. *On History.* Chicago: University of Chicago Press.

Bremer, J. G. 1793. *Die symbolische Weisheit der Aegypter aus den verborgensten Denkmälern des Alterthums: Ein Theil der Aegyptischen Maurerey, der zu Rom nicht verbrannt worden,* ed. K. P. Moritz. Berlin: Karl Matzdorff's Buchhandlung.

Brewer, E. C. 1892. *Dictionary of Phrase and Fable. Giving the Derivation, Source, or Origin of Common Phrases, Allusions, and Words That Have a Tale to Tell,* 26th ed. New York: Cassell.

Brian, D. 1996. *Einstein: A Life.* New York: John Wiley.

British Association. 1860. *An Ode, Appropriate to the Meeting at Oxford, June 27th. 1860. By a New Life-Member.* Oxford: T. and G. Shrimpton.

Broda, E. 1979. "Der Einfluß von Ernst Mach und Ludwig Boltzmann auf Albert Einstein." In *Einstein-Centenarium 1979,* ed. H. J. Treder. Berlin: Akademie-Verlag, pp. 227–37.

Broszat, M. 1960. *Der Nationalsozialismus: Weltanschauung, Programm und Wirklichkeit.* Stuttgart: Deutsche Verlags-Anstalt.

Brugger, W., ed. 1976. *Philosophisches Wörterbuch,* 14th ed. Freiburg: Herder.

Bruhns, K., ed. 1872. *Alexander von Humboldt: Eine wissenschaftliche Biographie.* 3 vols. Leipzig: F. A. Brockhaus.

Brush, S. G. 1978. *The Temperature of History: Phases of Science and Culture in the Nineteenth Century.* New York: Burt Franklin.

———. 1980. "The Chimerical Cat: Philosophy of Quantum Mechanics in Historical Perspective." *Social Studies of Science* 10: 393–447.

Büchmann, G. 1926. *Geflügelte Worte: Der Zitatenschatz des deutschen Volkes,* 27th ed. Berlin: Haude & Spenersche Buchhandlung.

Büchner, L. 1856. *Kraft und Stoff. Empirisch-naturphilosophische Studien,* 4th ed. Frankfurt: Meidinger.

———. 1867. *Kraft und Stoff. Empirisch-naturphilosophische Studien,* 9th ed. Leipzig: Theodor Thomas.

———. 1872. *Kraft und Stoff. Empirisch-naturphilosophische Studien,* 12th ed. Leipzig: Theodor Thomas.

Bucky, P. A. 1992. *The Private Albert Einstein.* Kansas City, MO: Andrews and McMeel.

Bucky, T. L. 1990. "Yo-Yo-Test für Einstein." In *Ein Haus für Albert Einstein: Erinnerungen—Briefe—Dokumente,* ed. M. Grüning. Berlin: Verlag der Nation, pp. 444–51.

Burckhardt, J. 1984. "Aus Saltimbanck's politisch-moralischen Schriften." In *Der Maikäfer. Zeitschrift für Nichtphilister,* ed. U. Brandt-Schwarze, A. Kramer, N. Oellers, and H. Rösch-Sondermann, vol. 3. Bonn: Ludwig Röhrscheid Verlag, pp. 203–204. (Orig. pub. 1843, in *Der Maikäfer. Zeitschrift für Nichtphilister* 4, no. 52 [supplement].)

Bürgel, B. H. 1932. *Die Weltanschauung des modernen Menschen: Das All, der Mensch, der Sinn des Lebens.* Berlin: Ullstein.

Bush, V. 1945. *Science, the Endless Frontier.* Washington, DC: US Government Printing Office.

Büttner, N. 1995. "Vom Aussehen eines Negromanten: Die Faustsage in der Bildenden Kunst vor Goethe." In *Faust: Annäherung an einen Mythos,* ed. F. Möbus, F. Schmidt-Möbus, and G. Unverfehrt. Göttingen: Wallstein, pp. 187–207.

Byron, Lord. 1975. "Manfred." In *The Poetical Works of Byron,* Cambridge ed. Boston: Houghton Mifflin, pp. 478–97. (Orig. pub. 1817.)

Cahan, D. 1985. "The Institutional Revolution in German Physics. 1865–1914." *Historical Studies in the Physical Sciences* 15, no. 2: 1–65.

———, ed. 1993. *Letters of Hermann von Helmholtz to His Parents: The Medical Education of a German Scientist. 1837–1846.* Stuttgart: Franz Steiner Verlag.

Calaprice, A., ed. 1996. *The Quotable Einstein.* Princeton, NJ: Princeton University Press.

Calinger, R. 1972. "The German Classical *Weltanschauung* in the Physical Sciences." In *The Influence of Early Enlightenment Thought upon German Classical Science and Letters: Problems for Future Discussion*. New York: Science History Publications, pp. 1–17.

Camus, A. 1958. *The Rebel: An Essay on Man in Revolt*. New York: Vintage Books.

Cannon, S. F. 1978. *Science in Culture: The Early Victorian Period*. New York: Science History Publications.

Carnap, R. 1963. "Intellectual Autobiography." In *The Philosophy of Rudolf Carnap*, ed. P. A. Schilpp. La Salle, IL: Open Court, pp. 3–84.

Carson, C. 1995. "Who Wants a Postmodern Physics?" *Science in Context* 8: 635–55.

Carstensen, H. T., and W. Henningsen. 1988. "Rembrandts sog. Dr. Faustus: Zur Archäologie eines Bildsinns." *Oud Holland* 102: 290–312.

Carus, C. G. 1948. *Goethe: Zu dessen näherem Verständnis und Briefe über Goethes Faust*. Zürich: Rotapfel Verlag.

Cassidy, D. 1995. *Einstein and Our World*. Atlantic Highlands, NJ: Humanities Press.

Cassirer, E. 1932. *Die Philosophie der Aufklärung*. Tübingen: J. C. B. Mohr.

———. 1954. *Philosophie der symbolischen Formen*. 3 vols. Oxford: Bruno Cassirer. (Orig. pub. 1923–29.)

———. 1961. *Freiheit und Form: Studien zur deutschen Geistesgeschichte*. Darmstadt: Wissenschaftliche Buchgesellschaft. (Orig. pub. 1916.)

Castagnetti, G., and H. Goenner. 1997. "Directing a Kaiser-Wilhelm-Institut: Albert Einstein, Organizer of Science?" Paper given at the Boston University Colloquium for Philosophy of Science, March 3.

Centro di Studi Filosofici di Gallarate, ed. 1957. *Enciclopedia Filosofica*. Venice: Istituto per la Collaborazione Culturale.

Chamberlain, H. S. 1917. "Deutsche Weltanschauung." *Deutschlands Erneuerung: Monatsschrift für das deutsche Volk* 1: 6–26.

Chance, J. 1995. *The Mythographic Chaucer: The Fabulation of Sexual Politics*. Minneapolis: University of Minnesota Press.

———, ed. 1999. *The Assembly of Gods*. Kalamazoo, MI: Medieval Institute Publications, Western Michigan University.

Churton, T. 1999. *Gnostics*. New York: Barnes and Noble. (Orig. pub. 1987.)

Clark, R. D., and G. A. Rice. 1982. "Family Constellation and Eminence: The Birth Orders of Novel Prize Winners." *Journal of Psychology* 110: 281–87.

Clark, R. W. 1971. *Einstein: The Life and Times.* New York: World.

Cocalis, S. L. 1978. "The Transformation of *Bildung* from an Image to an Ideal." *Monatshefte für deutschen Unterricht, deutsche Sprache und Literatur* 70: 399–414.

Cohen, N. W. 1984. *Encounter with Emancipation: The German Jews in the United States, 1830–1914.* Philadelphia: Jewish Publication Society of America.

Cohn, J. 1908. *Voraussetzungen und Ziele des Erkennens: Untersuchungen über die Grundfragen der Logik.* Leipzig: Wilhelm Engelmann.

Collins, H. M. 1981. "Stages in the Empirical Programme of Relativism." *Social Studies of Science* 11: 3–10.

Collins, H., and T. Pinch. 1993. *The Golem: What Everyone Should Know About Science.* Cambridge: Cambridge University Press.

Comte, A. 1914. *Entwurf der wissenschaftlichen Arbeiten, welche für eine Reorganisation der Gesellschaft erforderlich sind,* ed. W. Ostwald. Leipzig: Verlag Unesma.

———. 1994. *Rede über den Geist des Positivismus,* trans. and intro. I. Fetscher. Hamburg: Felix Meiner Verlag. (Orig. pub. 1844.)

Conzelmann, H. 1971. "The Mother of Wisdom." In *The Future of Our Religious Past: Essays in Honour of Rudolf Bultmann,* ed. J. M. Robinson. New York: Harper & Row, pp. 230–43.

Cornill, A. 1856. *Arthur Schopenhauer, als Übergangsformation von einer idealistischen in eine realistische Weltanschauung.* Heidelberg: J. C. B. Mohr.

Cornwell, J. 2003. *Hitler's Scientists: Science, War and the Devil's Pact.* New York: Viking.

Cotgrove, S. 1978. "Styles of Thought: Science, Romanticism and Modernization." *British Journal of Sociology* 29: 358–71.

Cranefield, P. F. 1957. "The Organic Physics of 1847 and the Biophysics of Today." *Journal of the History of Medicine and Allied Sciences* 12: 407–23.

———. 1966. "The Philosophical and Cultural Interests of the Biophysics Movement of 1847." *Journal of the History of Medicine and Allied Sciences* 21: 1–7.

Culotta, C. A. 1974. "German Biophysics, Objective Knowledge, and Romanticism." *Historical Studies in the Physical Sciences* 4: 3–38.

Dahlhaus, C. 1990. "Das deutsche Bildungsbürgertum und die Musik." In *Bildungsbürgertum im 19. Jahrhundert, Part II: Bildungsgüter und Bildungswissen,* ed. R. Koselleck. Stuttgart: Klett-Cotta, pp. 220–36.

d'Alembert, J. l. R. 1995. *Preliminary Discourse to the Encyclopedia of Diderot,* trans. R. N. Schwab. Chicago: University of Chicago Press. (Orig. pub. 1751.)

Darrigol, O. 1996. "The Electrodynamic Origins of Relativity Theory." *Historical Studies in the Physical and Biological Sciences* 26, no. 2: 241–312.

Darwin, E. 1804. *The Temple of Nature; or, the Origin of Society: A Poem, with Philosophical Notes.* New York: T. and J. Swords. (Orig. pub. 1803.)

Davie, M. R. 1947. *Refugees in America: Report of the Committee for the Study of Recent Immigration from Europe.* New York: Harper & Brothers.

Degen, P. A. 1989. "Einstein's Weltanschauung and its Spinozistic Elements." In *Science and Religion/Wissenschaft und Religion,* ed. Ä. Bäumer and M. Büttner. Bochum: Universitätsverlag Dr. N. Brockmeyer, pp. 142–50.

Dennert, E. 1907. *Die Weltanschauung des modernen Naturforschers.* Stuttgart: Max Kielmann.

———. 1909. *Das Weltbild im Wandel der Zeit.* Hamburg: Agentur des Rauhen Hauses.

de Pizan, C. 1982. *Le livre de la cité des dames,* trans. E. J. Richards. New York: Persea Books. (Orig. written 1405.)

d'Espagnat, B. 1989. *Reality and the Physicist: Knowledge, Duration and the Quantum World.* Cambridge: Cambridge University Press.

d'Holbach, P. H. T., Baron. 1984. *The System of Nature.* 3 vols. New York: Garland. (Orig. pub. 1770.)

Dilthey, W. 1911. Die Typen der Weltanschauung und ihre Ausbildung in den metaphysischen Systemen. In *Weltanschauung* by Wilhelm Dilthey et al. Berlin: Reichl, pp. 1–51.

———. 1960. *Weltanschauungslehre: Abhandlungen zur Philosophie der Philosophie.* In *Gesammelte Schriften,* vol. 8. Stuttgart: B. G. Teubner.

Dingler, H. 1951. *Das physikalische Weltbild.* Meisenheim/Glan: Westkulturverlag A. Hain.

Dohm, C. W. 1781. *Über die bürgerliche Verbesserung der Juden.* Berlin: Nicolai.

Dornseiff, F. 1945–46. "Weltanschauung: Kurzgefaßte Wortgeschichte." *Die Wandlung* 1: 1086–88.

Driesch, H. 1930. *Relativitätstheorie und Weltanschauung.* Leipzig: Quelle & Meyer.

Du Bois-Reymond, E. 1873. *Über die Grenzen des Naturerkennens.* Leipzig: Veit.

———. 1912. *Reden.* 2 vols. Leipzig: Veit.

———, ed. 1982. *Two Great Scientists of the Nineteenth Century: Correspondence of Emil Du Bois-Reymond and Carl Ludwig.* Baltimore: Johns Hopkins University Press.

Duggon, S. P. H., and B. Drury. 1948. *The Story of the Emergency Committee in Aid of Displaced Scholars*. New York: Macmillan.

Duhem, P. 1962. *The Aim and Structure of Physical Theory*, trans. P. P. Wiener. New York: Atheneum. (Orig. pub. 1906.)

———. 1991. *German Science*, trans. J. Lyon. La Salle, IL: Open Court. (Orig. pub. 1915.)

Dühring, E. 1875. *Cursus der Philosophie als streng wissenschaftlicher Weltanschauung und Lebensgestaltung*. Leipzig: Erich Koschny.

Dukas, H., and B. Hoffmann, eds. 1979. *Albert Einstein: The Human Side— New Glimpses from His Archives*. Princeton, NJ: Princeton University Press.

Dupré, F. 1926. *Weltanschauung und Menschenzüchtung*. Berlin-Lichterfelde: Selbstverlag des Verfassers.

Durozoi, G., and A. Roussel. 1987. *Dictionnaire de philosophie*. Paris: Nathan.

Eamon, W. 1994. *Science and the Secrets of Nature: Books of Secrets in Medieval and Early Modern Culture*. Princeton, NJ: Princeton University Press.

Eberle, F., and T. Stammen, eds. 1989. *Die Französische Revolution in Deutschland: Zeitgenössische Texte deutscher Autoren*. Stuttgart: Phillip Reclam.

Ecker, A. 1883. *Lorenz Oken: A Biographical Sketch*. London: Kegan Paul, Trench.

Edwards, P., ed. 1967. *The Encyclopedia of Philosophy*. New York: Macmillan.

Einstein, A. 1917. "Helmholtz, H. v., Zwei Vorträge über Goethe. Review." *Naturwissenschaften* 5: 675.

———. 1924. "Elsbachs Buch: Kant und Einstein." *Deutsche Literaturzeitung*, N.F., 1: 1685–92.

———. 1929. *Gelegentliches*. Berlin: Soncino-Gesellschaft der Freunde des jüdischen Buches.

———. 1949. "Remarks Concerning the Essays Brought Together in This Cooperative Volume." In *Albert Einstein: Philosopher-Scientist*, ed. P. A. Schilpp. Evanston, IL: Library of Living Philosophers, pp. 665–88.

———. 1954. *Ideas and Opinions*. New York: Dell.

———. 1955a. *Mein Weltbild*. Frankfurt: Ullstein. (Orig. pub. 1934.)

———. 1955b. "Erinnerungen—Souvenirs." *Schweizerische Hochschulzeitung* 28: 145–53.

———. 1956. "Autobiographische Skizze." In *Helle Zeit—Dunkle Zeit: In Memoriam Albert Einstein*, ed. C. Seelig. Zürich: Europa Verlag, pp. 9–17.

———. 1979a. "Autobiographisches." In *Albert Einstein als Philosoph und Naturforscher*, ed. P. A. Schilpp. Braunschweig: Vieweg, pp. 1–35. (Orig. pub. 1949.)

———. 1979b. "Bemerkungen zu den in diesem Bande vereinigten Arbeiten." In *Albert Einstein als Philosoph und Naturforscher*, ed. P. A. Schilpp. Braunschweig: Vieweg, pp. 493–511. (Orig. pub. 1949.)

———. 1979c. *Aus meinen späten Jahren*. Stuttgart: Deutsche Verlags-Anstalt.

———. 1987. *Letters to Solovine*. Introduction by M. Solovine. New York: Philosophical Library.

Einstein, A., H. Born, and M. Born. 1969. *Briefwechsel. 1916–1955*. Munich: Nymphenburger Verlagshandlung.

Einstein, A., and M. Besso. 1972. *Correspondance, 1903–1955*, trans. and intro. P. Speziali. Paris: Hermann.

Einstein, A., and L. Infeld. 1938. *The Evolution of Physics: The Growth of Ideas from Early Concepts to Relativity and Quanta*. New York: Simon and Schuster.

Einstein, A. J. 1991a. "Report on His Family." In *Zeugnisse zur Geschichte der Juden in Ulm: Erinnerungen und Dokumente*, ed. Stadtarchiv Ulm. Ulm: Stadtarchiv Ulm, 168–73.

Einstein, E. R. 1991b. *Hans Albert Einstein: Reminiscences of His Life and Our Life Together*. Iowa City: University of Iowa.

Eisler, R. 1930a. *Wörterbuch der philosophischen Begriffe*, vol. 3. Berlin: E. S. Mittler and Sohn.

———, ed. 1930b. *Kant-Lexikon*. Berlin: E. S. Mittler.

Elias, N. 1939. *Über den Prozess der Zivilisation*. 2 vols. Basel: Verlag Haus zum Falken.

Eliot, T. S. 1949. *Notes towards the Definition of Culture*. New York: Harcourt, Brace.

Engelhardt, D. v. 1990. "Der Bildungsbegriff in der Naturwissenschaft des 19. Jahrhunderts." In *Bildungsbürgertum im 19. Jahrhundert, Part II: Bildungsgüter und Bildungswissen*, ed. R. Koselleck. Stuttgart: Klett-Cotta, pp. 106–16.

Engelhardt, U. 1986. *"Bildungsbürgertum": Begriffs- und Dogmengeschichte eines Etiketts*. Stuttgart: Klett-Cotta.

Engels, F. 1970. *Herrn Eugen Dührings Umwälzung der Wissenschaft*. Berlin: Dietz. (Orig. pub. 1878.)

Epstein, P. 1998. *Kabbalah: The Way of the Jewish Mystic*. New York: Barnes and Noble. (Orig. pub. 1978.)

Erikson, E. H. 1970. "Autobiographic Notes on the Identity Crisis." *Daedalus* 99: 730–59.

———. 1982. "Psychoanalytic Reflections on Einstein's Centenary." In *Albert Einstein: Historical and Cultural Perspectives*, ed. G. Holton and Y. Elkana. Princeton, NJ: Princeton University Press, pp. 151–73.

Falkenberg, H. 1911. "Die Listen der besten Bücher: Ein bibliographischer Versuch." *Zeitschrift für Bücherfreunde*, N.F., 3, no. 1: 45–47.

Farrar, W. V. 1975. "Science and the German University System." In *The Emergence of Science in Western Europe*, ed. M. Crosland. London: Macmillan, pp. 179–92.

Fauth, P., ed. 1913. *Hörbigers Glacial-Kosmogonie: Eine neue Entwickelungsgeschichte des Weltalls und des Sonnensystems*. Kaiserslautern: Hermann Kaysers Verlag.

Feigl, H., ed. 1931. *Jahrbuch deutscher Bibliophilen und Literaturfreunde*, vols. 16 and 17. Zürich: Amalthea-Verlag.

Fermi, L. 1971. *Illustrious Immigrants: The Intellectual Migration from Europe, 1930–41*. Chicago: University of Chicago Press.

Feuer, L. S. 1963. *The Scientific Intellectual: The Psychological and Sociological Origins of Modern Science*. New York: Basic Books.

———. 1974. *Einstein and the Generations of Science*. New York: Basic Books.

Feuerbach, L. 1909. *Das Wesen des Christentums*. Leipzig: A. Kröner. (Orig. pub. 1841.)

Fichte, J. G. 1812. *Über die einzig mögliche Störung der akademischen Freiheit. Eine Rede beim Antritte seines Rektorats an der Universität zu Berlin den 19ten Oktober 1811 gehalten*. Berlin: L. W. Wittich.

———. 1845. "Versuch einer Kritik aller Offenbarung." In *Sämmtliche Werke*, ed. I. H. Fichte, vol. 5: *2. Abt. B. Zur Religionsphilosophie, 3. Bd.* Berlin: Veit. (Orig. pub. 1792.)

———. 1970. *Science of Knowledge, with the First and Second Introductions*, trans. P. Heath and J. Lachs. New York: Meredith Corporation. (Orig. pub. 1794.)

———. 1979. *Grundlage des Naturrechts nach Prinzipien der Wissenschaftslehre*. Hamburg: Felix Meiner. (Orig. pub. 1796.)

Fisch, J. 1992. "Zivilisation, Kultur." In *Geschichtliche Grundbegriffe: Historisches Lexikon zur politisch-sozialen Sprache in Deutschland*, ed. O. Brunner, W. Conze, and R. Koselleck, vol. 7. Stuttgart: Ernst Klett, pp. 679–774.

Flamm, D. 1983. "Ludwig Boltzmann and His Influence on Science." *Studies in the History and Philosophy of Science* 14: 255–78.

Fleck, L. 1980. *Entstehung und Entwicklung einer wissenschaftlichen Tatsache. Einführung in die Lehre vom Denkstil und Denkkollektiv*. Frankfurt: Suhrkamp. (Orig. pub. 1935.)

Fleming, D., and B. Bailyn, eds. 1969. *The Intellectual Migration: Europe and America. 1930–1960*. Cambridge, MA: Belknap Press.

Flückinger, M. 1974. *Albert Einstein in Bern: Das Ringen um ein neues Weltbild. Eine dokumentarische Darstellung über den Aufstieg eines Genies.* Bern: Paul Haupt.

Fludd, R. 1617–21. *Utriusque cosmi maioris scilicet et minoris metaphysica, physica atque technica historia.* Oppenheim: Hieronymus Galler.

Fölsing, A. 1993. *Albert Einstein: Eine Biographie.* Frankfurt: Suhrkamp.

Forman, P. 1971. "Weimar Culture, Causality, and Quantum Theory, 1918–1927: Adaption by German Physicists and Mathematicians to a Hostile Intellectual Environment." *Historical Studies in the Physical Sciences* 3: 1–115.

———. 1981. "Einstein and Research." Introduction to *the Joys of Research*, ed. W. Shropshire Jr. Washington, DC: Smithsonian Institution Press, pp. 13–24.

Foulquié, P., and R. Saint-Jean. 1969. *Dictionnaire de la langue philosophique.* Paris: Presses universitaires de France.

Frank, P. 1928. "Über die 'Anschaulichkeit' physikalischer Theorien." *Die Naturwissenschaften* 16: 121–28.

———. 1932. *Das Kausalgesetz und seine Grenzen.* Vienna: Julius Springer.

———. 1947. *Einstein: His Life and Times*, trans. G. Rosen, ed. S. Kusaka. New York: A. A. Knopf.

———. 1949a. *Einstein: Sein Leben und seine Zeit.* Munich: Paul List.

———. 1949b. *Modern Science and its Philosophy.* Cambridge, MA: Harvard University Press.

———. 1950. *Relativity: A Richer Truth.* Foreword by A. Einstein. Boston: Beacon Press.

———. 1979. "Anecdotes." In *Einstein: a Centenary Volume*, ed. A. P. French. London: Heinemann.

Franquiz, J. A. 1964. "Albert Einstein's Philosophy of Religion." *Journal for the Scientific Study of Religion* 4: 64–70.

Frantzius, F. v. 1916. *Deutschland, der Träger der Welt-Kultur: Wissenswerte Tatsachen für jeden Deutschen und Deutsch-Amerikaner.* Chicago: Frantzius.

Freud, S. 1960. *Briefe 1873–1939*, select. and ed. E. L. Freud. Frankfurt: S. Fischer Verlag.

Friedlaender, F. 1932. *Heine und Goethe.* Berlin: Walter de Gruyter.

Frobenius, L. 1898. *Die Weltanschauung der Naturvölker.* Weimar: Emil Felber.

Frühwald, W. 1990. "Büchmann und die Folgen: Zur sozialen Funktion des Bildungszitates in der deutschen Literatur." In *Bildungsbürgertum im 19. Jahrhundert, Part II: Bildungsgüter und Bildungswissen,* ed. R. Koselleck. Stuttgart: Klett-Cotta, pp. 197–219.

————. 1991. "'Von der Poesie im Recht.' Über die Brüder Grimm und die Rechtsauffassung der deutschen Romantik." In *Die deutsche literarische Romantik und die Wissenschaften,* ed. N. Saul. Munich: Iudicium Verlag, pp. 282–305.

Fuchs, K., and H. Raab. 1972. *Dtv-Wörterbuch zur Geschichte.* Munich: Deutscher Taschenbuch-Verlag.

Gabor, A. 1995. *Einstein's Wife: Work and Marriage in the Lives of Five Great Twentieth-Century Women.* New York: Viking.

Galilei, G. 1967. *Dialogue Concerning the Two Chief World Systems.* Foreword by A. Einstein. Berkeley: University of California Press. (Orig. written 1630.)

Galison, P. L. 1987. *How Experiments End.* Chicago: University of Chicago Press.

————. 1990. "Aufbau/Bauhaus: Logical Positivism and Architectural Modernism." *Critical Inquiry* 16: 709–52.

————. 1998. "The Americanization of Unity." *Daedalus* 127, no. 1 (Winter): 45–71.

————. 2003. *Einstein's Clocks, Poincaré's Maps: Empires of Time.* New York: W. W. Norton.

Galton, F. 1869. *Hereditary Genius: An Inquiry into its Laws and Consequences.* London: Macmillan.

————. 1875. *English Men of Science: Their Nature and Nurture.* New York: Appleton.

Gans, H. J. 1999. *Popular Culture and High Culture: An Analysis and Evaluation of Taste.* New York: Basic Books. (Orig. pub. 1974.)

Gardner, H. 1993. *Creating Minds: An Anatomy of Creativity Seen through the Lives of Freud, Einstein, Picasso, Stravinsky, Eliot, Graham, and Gandhi.* New York: Basic Books.

Gay, P. 1968. *Weimar Culture: The Outsider as Insider.* New York: Harper & Row.

Gebhard, B. 1962. *Handbuch der deutschen Geschichte,* 8th ed. Stuttgart: Union Verlag.

Gent, W. 1931. *Weltanschauung: Eine systematische und problemgeschichtliche Untersuchung erläutert am Beispiel der Weltanschauung Lessings.* Darmstadt: L. C. Wittich.

German, R., P. Filzer, and J. Einstein. 1987. *Bad Buchau und der Federsee: Im Herzen Oberschwabens.* Bad Buchau: Federsee-Verlag.

Giesen, B. 1998. "Cosmopolitans, Patriots, Jacobins, and Romantics." *Daedalus* (Summer): 221–50.

Gilbert, F. 1980. "Einstein's Europe." In *Some Strangeness in the Proportion: A Centennial Symposium to Celebrate the Achievements of Albert Einstein,* ed. H. Woolf. Reading, MA: Addison-Wesley, pp. 13–27.

Gillispie, C. C., editor-in-chief. 1970–80. *Dictionary of Scientific Biography.* New York: Scribner.

Gilson, E. 1950. *Reason and Revelation in the Middle Ages.* New York: Charles Scribner's Sons. (Orig. pub. 1938.)

———. 1955. *History of Christian Philosophy in the Middle Ages.* New York: Random House.

Glanz, R. 1970. *Studies in Judaica Americana.* New York: Ktav Publishing House.

Glass, H. B. 1974. "Science: Endless Horizons or Golden Age?" In *The Maturing of American Science: A Portrait of Science in Public Life Drawn from the Presidential Addresses of the American Association for the Advancement of Science 1920–1970,* ed. R. H. Kargon. Washington, DC: American Association for the Advancement of Science, pp. 237–50.

Glover, J. A., R. R. Ronning, and C. R. Reynolds, eds. 1989. *Handbook of Creativity.* New York: Plenum.

Goenner, H., and G. Castagnetti. 1996. Albert Einstein as a Pacifist and Democrat during the First World War. Preprint 35. Berlin: Max-Planck-Institut für Wissenschaftsgeschichte.

Goethe, J. W. v. 1962. *Goethes Werke.* Hamburger Ausgabe, 4th ed. Hamburg: Christian Wegner Verlag.

———. 1984. *Faust I and II,* ed. and trans. S. Atkins. Cambridge, MA: Suhrkamp/Insel Publishers Boston.

Goffman, E. 1959. *The Presentation of Self in Everyday Life.* Garden City, NY: Doubleday.

Gogarten, F. 1937. *Weltanschauung und Glaube.* Berlin: Furche-Verlag.

Gombert, A. 1901. "Über Richard M. Meyers Vierhundert Schlagworte." *Zeitschrift für deutsche Wortforschung* 2: 256–76.

———. 1902. "Robert Arnold über Richard M. Meyers Vierhundert Schlagworte." *Zeitschrift für deutsche Wortforschung* 3: 144–58.

———. 1906. "Kleine Bemerkungen zur Wortgeschichte." *Zeitschrift für deutsche Wortforschung* 8: 121–40.

Gomperz, H. 1905. *Weltanschauungslehre: Ein Versuch, die Hauptprobleme der allgemeinen theoretischen Philosophie geschichtlich zu entwickeln und sachlich zu bearbeiten*, vol. 1. Jena: Eugen Diederichs.

———. 1908. *Weltanschauungslehre: Ein Versuch, die Hauptprobleme der allgemeinen theoretischen Philosophie geschichtlich zu entwickeln und sachlich zu bearbeiten*, vol. 2. Jena: Eugen Diederichs, part 1.

Goodman, N. 1978. *Ways of Worldmaking*. Hassocks, UK: Harvester Press.

Görres, J. 1807. *Die Teutschen Volksbücher: Nähere Würdigung der schönen Historien-, Wetter- und Arzneybüchlein*. Heidelberg: Mohr und Zimmer.

———. 1926. "Wachstum der Historie." In *Gesammelte Schriften*, ed. W. Schellberg, vol. 3. Köln: Gilde-Verlag, pp. 363–440. (Orig. written 1808.)

Götze, A. 1924. "Weltanschauung." *Euphorion* 25: 42–51.

Gower, B. 1973. "Speculation in Physics: The History and Practice of *Naturphilosophie*." *Studies in History and Philosophy of Science* 3: 301–56.

Graupe, H. M. 1977. *Die Entstehung des modernen Judentums: Geistesgeschichte der deutschen Juden, 1650–1942*. Hamburg: Helmut Buske Verlag.

Gregory, F. 1977. *Scientific Materialism in Nineteenth Century Germany*. Dordrecht, Holland: D. Reidel.

———. 1989. "Kant, Schelling, and the Administration of Science in the Romantic Era." *Osiris*, 2nd ser., 5: 17–35.

Grimm, J., and W. Grimm. 1955. *Deutsches Wörterbuch*, vol. 14. Leipzig: S. Hirzel.

Groethuysen, B. 1927–30. *Die Entstehung der bürgerlichen Welt- und Lebensanschauung in Frankreich*. 2 vols. Halle/Saale: Max Niemeyer.

Groß, W. 1936. *Rasse, Weltanschauung, Wissenschaft*. Berlin: Junker und Dünnhaupt.

Grossmann, M. 1929. "Fachbildung, Geisteskultur und Phantasie." In *Festschrift Prof. Dr. A. Stodola zum 70. Geburtstag*, ed. E. Honegger. Zürich: Orell Füssli Verlag, pp. 187–90.

Grundmann, S. 1965. "Die Auslandsreisen Albert Einsteins und die Außenpolitik der deutschen Monopolbourgeoisie nach dem ersten Weltkrieg." *NTM—Schriftenreihe für Geschichte der Naturwissenschaften, Technik und Medizin* 2, no. 6: 1–9.

Gruppe, O. 1921. *Geschichte der klassischen Mythologie und Religionsgeschichte während des Mittelalters im Abendland und während der Neuzeit*. Leipzig: B. G. Teubner.

Haas, A. E. 1920. *Das Naturbild der modernen Physik*. Berlin: Walter de Gruyter.

Hadot, P. 1982. "Zur Idee der Naturgeheimnisse: Beim Betrachten des Widmungsblattes in den Humboldtschen 'Ideen zu einer Geographie der Pflanzen.'" In *Abhandlungen der geistes- und sozialwissenschaftlichen Klasse der Akademie der Wissenschaften und der Literatur, Mainz*, no. 8. Wiesbaden: Steiner.

Haeckel, E. 1903. *Die Welträthsel: Gemeinverständliche Studien über monistische Philosophie*. Stuttgart: Alfred Kröner. (Orig. pub. 1899.)

Hager, N., and U. Röseberg. 1977. "Philosophisch-weltanschauliche Aspekte des Weltbildes der klassischen Physik." *Deutsche Zeitschrift für Philosophie* 25: 577–86.

Hall, P. 2001. *Cities in Civilization*. New York: Fromm International. (Orig. pub. 1998.)

Hallier, E. 1875. *Die Weltanschauung des Naturforschers*. Jena: H. Dufft.

Hamlin, C. 1995. "Faust und der Erdgeist: Goethes radikale Transformation der Teufelsbeschwörung." Vortrag beim Graduiertenkolleg "Religion und Normativität," Universität Heidelberg, June 12. (Rev. April 1998.)

Hansen, E. F. 1992. *Wissenschaftswahrnehmung und -umsetzung im Kontext der deutschen Frühromantik: Zeitgenössische Naturwissenschaft und Philosophie im Werk Friedrich von Hardenbergs (Novalis)*. Frankfurt: Peter Lang.

Hansen, W. 1938. *Die Entwicklung des kindlichen Weltbildes*. Munich: Kösel-Pustet.

Hardtwig, W. 1989. "Eliteanspruch und Geheimnis in den Geheimgesellschaften des 18. Jahrhunderts." In *Aufklärung und Geheimgesellschaften: Zur politischen Funktion und Sozialstruktur der Freimaurerlogen im 18. Jahrhundert*, ed. H. Reinhalter. Munich: R. Oldenbourg, pp. 63–86.

Harless, C. F. 1830. *Die Verdienste der Frauen um Naturwissenschaft und Heilkunde*. Göttingen: Vandenhoeck-Ruprecht.

Harnack, A. v. 1930. "Die Kaiser Wilhelm-Gesellschaft zur Förderung der Wissenschaften." In *Das akademische Deutschland*, ed. M. Doeberl, O. Scheel, W. Schlink, H. Sperl, E. Spranger, H. Bitter, and P. Frank, vol. 3. Berlin: C. A. Weller, pp. 609–18.

Harrington, A. 1996. *Reenchanted Science: Holism in German Culture from Wilhelm II to Hitler*. Princeton, NJ: Princeton University Press.

Hartman, G. H. 1997. *The Fateful Question of Culture*. New York: Columbia University Press.

Hartmann, E. v. 1902. *Die Weltanschauung der modernen Physik*. Leipzig: Hermann Haacke.

Harwood, J. 1993. *Styles of Scientific Thought: The German Genetics Community, 1900–1933*. Chicago: University of Chicago Press.

Haym, R. 1857. *Hegel und seine Zeit*. Berlin: Rudolph Gaertner.

Hazard, P. 1963. *The European Mind, 1680–1715*, trans. J. L. May. Cleveland, OH: World. (Orig. pub. 1935.)

Hebbel, F. 1904. "Abfertigung eines aesthetischen Kannegiessers." In *Sämtliche Werke*, ed. R. M. Werner, Abt. I, vol. 11. Berlin: B. Behr, pp. 387–409. (Orig. pub. 1851.)

Hebler, C. 1869. "Die Lehre des Copernicus und die moderne Weltanschauung." In *Philosophische Aufsätze* by C. Hebler. Leipzig: Fues's Verlag, pp. 1–34.

Hegel, G. W. F. 1843. *Die Wissenschaft der Logik*. In *Werke*, vol. 6. Berlin: Duncker and Humblot. (Orig. pub. 1812.)

———. 1928. *Grundlinien der Philosophie des Rechts oder Naturrecht und Staatswissenschaft im Grundrisse*, ed. E. Gans. in *Sämtliche Werke*, ed. H. Glockner, vol. 7. Stuttgart: Fr. Frommanns Verlag. (Orig. pub. 1821.)

———. 1952. *Phänomenologie des Geistes*, ed. J. Hoffmeister. Hamburg: Felix Meiner. (Orig. pub. 1807.)

———. 1975. *Aesthetics: Lectures on Fine Art*, trans. T. M. Knox. 2 vols. Oxford: Clarendon Press. (Orig. pub. 1835–38.)

———. 1985. *Ästhetik*, ed. F. Bassenge. 2 vols. Berlin: Verlag das europäische buch. (Orig. pub. 1835–38.)

Heidegger, M. 1957. "Die Zeit des Weltbildes." In *Holzwege* by M. Heidegger. Frankfurt: Vittorio Klostermann, pp. 69–104. (Orig. written 1938.)

———. 2002. "The Age of the World Picture." In *Off the Beaten Track*, by M. Heidegger, ed. and trans. J. Young and K. Haynes. Cambridge: Cambridge University Press, pp. 57–85. (Orig. written 1938.)

Heilbron, J. L. 1986. *The Dilemmas of an Upright Man: Max Planck as Spokesman for German Science*. Berkeley: University of California Press.

Heilbut, A. 1983. *Exiled in Paradise: German Refugee Artists and Intellectuals in America, from the 1930s to the Present*. New York: Viking Press.

Heim, K. 1951. *Die Wandlung im naturwissenschaftlichen Weltbild*. Hamburg: Furche-Verlag.

Heisenberg, W. 1955. *Das Naturbild der heutigen Physik*. Hamburg: Rowohlt.

———. 1958. "The Representation of Nature in Contemporary Physics." *Daedalus* (Summer): 95–108.

Helmholtz, H. v. 1884. *Vorträge und Reden*. 2 vols. Braunschweig: Friedrich Vieweg und Sohn.

————. 1995. *Science and Culture: Popular and Philosophical Essays*, ed. and intro. D. Cahan. Chicago: University of Chicago Press.

Hendry, J. 1980. "Weimar Culture and Quantum Causality." *History of Science* 18: 155–80.

Hennemann, G. 1967–68. "Der dänische Physiker Hans Christian Oersted und die Naturphilosophie der Romantik." *Philosophia Naturalis* 10: 112–22.

Hentschel, K. 1990. *Interpretationen und Fehlinterpretationen der speziellen und der allgemeinen Relativitätstheorie durch Zeitgenossen Albert Einsteins.* Basel: Birkhäuser.

Herder, J. G. 1784–91. *Ideen zur Philosophie der Geschichte der Menschheit.* 4 vols. Riga: Johann Friedrich Hardknoch.

————. 1967. "Vom Erkennen und Empfinden, den zwo Hauptkräften der menschlichen Seele." In *Sämtliche Werke*, ed. B. Suphan, vol. 8. Hildesheim: Georg Olms Verlagsbuchhandlung, pp. 263–333. (Orig. written 1775.)

————. 1968. "Versuch einer Geschichte der lyrischen Dichtkunst." In *Sämtliche Werke*, ed. B. Suphan, vol. 32. Hildesheim: Georg Olms Verlagsbuchhandlung, pp. 85–140. (Orig. written 1764.)

Herf, J. 1984. *Reactionary Modernism: Technology, Culture and Politics in Weimar and the Third Reich.* Cambridge: Cambridge University Press.

Hermann, A. 1994. *Einstein: Der Weltweise und sein Jahrhundert.* Munich: Piper.

Hermanns, W. 1983. *Einstein and the Poet: In Search of the Cosmic Man.* Brookline Village, MA: Branden Press.

Herneck, F. 1963. *Albert Einstein: Ein Leben für Wahrheit, Menschlichkeit und Frieden.* Berlin: Buchverlag Der Morgen.

————. 1976. *Einstein und sein Weltbild.* Berlin: Buchverlag Der Morgen.

————. 1978. *Einstein privat: Herta W. erinnert sich an die Jahre 1927 bis 1933.* Berlin: Buchverlag Der Morgen.

Herrmann, C. 1926. *Die Weltanschauung Gerhart Hauptmanns in seinen Werken.* Berlin: Gebrüder Paetel.

Herrmann, U. 1990. "Über 'Bildung' im Gymnasium des wilhelminischen Kaiserreichs." In *Bildungsbürgertum im 19. Jahrhundert, Part II: Bildungsgüter und Bildungswissen*, ed. R. Koselleck. Stuttgart: Klett-Cotta, pp. 346–68.

Hessing, S., ed. 1933. *Spinoza-Festschrift.* Heidelberg: Carl Winter.

Highfield, R., and P. Carter. 1993. *The Private Lives of Albert Einstein.* London: Faber and Faber.

Hildebrand, R. 1910. *Gedanken über Gott, die Welt und das Ich: Ein Vermächtnis.* Jena: Diederichs.

Hiroshige, T. 1976. "The Ether Problem, the Mechanistic Worldview, and the Origin of the Theory of Relativity." *Historical Studies in the Physical Sciences* 7: 3–82.

Hoffmann, B., with H. Dukas. 1972. *Albert Einstein: Creator and Rebel.* New York: Viking.

Hoffmeister, J., ed. 1955. *Wörterbuch der philosophischen Begriffe,* 2nd ed. Hamburg: Felix Meiner.

Hofmann, P. 1914. *Die antithetische Struktur des Bewusstseins: Grundlegung einer Theorie der Weltanschauungsformen.* Berlin: Georg Reimer.

Hollinger, D. A. 1996. *Science, Jews, and Secular Culture: Studies in Mid-Twentieth-Century American Intellectual History.* Princeton, NJ: Princeton University Press.

———. 2002. "Why Are Jews Preeminent in Science and Scholarship? The Veblen Thesis Reconsidered." *Aleph* 2: 145–63.

Hollitscher, W. 1985. *Naturbild und Weltanschauung.* Vienna: Globus.

Holton, G. 1973. *Thematic Origins of Scientific Thought: Kepler to Einstein.* Cambridge, MA: Harvard University Press.

———. 1986. *The Advancement of Science, and its Burdens. The Jefferson Lecture and Other Essays.* Cambridge: Cambridge University Press.

———. 1993a. *Science and Anti-Science.* Cambridge, MA: Harvard University Press.

———. 1993b. "From the Vienna Circle to Harvard Square: The Americanization of a European World Conception." In *Scientific Philosophy: Origins and Developments,* ed. F. Stadler. Dordrecht, Holland: Kluwer, pp. 47–73.

———. 1995. "Einstein and Books." In *No Truth Except in the Details: Essays in Honor of Martin J. Klein,* ed. A. J. Kox and D. M. Siegel. Dordrecht, Holland: Kluwer, pp. 273–79.

———. 1996. *Einstein, History, and Other Passions: The Rebellion against Science at the End of the Twentieth Century.* Reading, MA: Addison-Wesley.

———. 1998. "Einstein and the Cultural Roots of Modern Science." *Daedalus* 127, no. 1 (Winter): 1–44.

Holton, G., H. Chang, and E. Jurkowitz. 1998. "How a Scientific Discovery Is Made: A Case History." *American Scientist* 84, no. 10: 364–75.

Holton, G., and G. Sonnert. 1999. "A Vision of Jeffersonian Science." *Issues in Science and Technology* 16, no. 1: 61–65.

Honderich, T., ed. 1995. *The Oxford Companion to Philosophy.* Oxford: Oxford University Press.

Hoppe, B. 1979. *Aus der Frühzeit der chemischen Konstitutionsforschung: Die Tropanalkaloide Atropin und Cocain in Wissenschaft und Wirtschaft.* Munich: R. Oldenbourg.

Horgan, J. 1996. *The End of Science: Facing the Limits of Knowledge in the Twilight of the Scientific Age.* Reading, MA: Addison-Wesley.

Hörisch, J. 1998. "Vom Geheimnis zum Rätsel. Die offenbar geheimen und profan erleuchteten Namen Walter Benjamins." In *Schleier und Schwelle: Geheimnis und Offenbarung*, ed. A. Assmann and J. Assmann. Munich: Wilhelm Fink Verlag, pp. 161–78.

Horn, G. 1994. "'Horen'—'Propyläen'—'Athenäum': Zeitschriften im Widerstreit." In *Evolution des Geistes: Jena um 1800: Natur und Kunst, Philosophie und Wissenschaft im Spannungsfeld der Geschichte*, ed. F. Strack. Stuttgart: Klett-Cotta, pp. 306–22.

Hornung, E. 2001. *The Secret Lore of Egypt: Its Impact on the West.* Ithaca, NY: Cornell University Press.

Hörz, H., and S. Wollgast. 1986. "Hermann von Helmholtz und Emil du Bois-Reymond: Wissenschaftsgeschichtliche Einordnung in die naturwissenschaftlichen und philosophischen Bewegungen ihrer Zeit." In *Dokumente einer Freundschaft: Briefwechsel zwischen Hermann von Helmholtz und Emil du Bois-Reymond. 1846–1894*, ed. C. Kirsten. Berlin: Akademie-Verlag, pp. 11–64.

Howard, D. 1984. "Realism and Conventionalism in Einstein's Philosophy of Science: The Einstein-Schlick Correspondence." *Philosophia Naturalis* 21: 616–29.

———. 1990. "Einstein and Duhem." *Synthese* 83: 363–84.

———. 1994. "Einstein, Kant, and the Origins of Logical Empiricism." In *Logic, Language, and the Structure of Scientific Theories*, ed. W. Salmon and G. Wolters. Pittsburgh: University of Pittsburgh Press, pp. 45–105.

———. 1997. "A Peek Behind the Veil of Maya: Einstein, Schopenhauer, and the Historical Background of the Conception of Space as a Ground for the Individuation of Physical Systems." In *The Cosmos of Science: Essays of Exploration*, ed. J. Earman and J. D. Norton. Pittsburgh, PA: University of Pittsburgh Press, pp. 87–150.

Hübner, J., ed. 1987. *Der Dialog zwischen Theologie und Naturwissenschaft: Ein bibliographischer Bericht.* Munich: Kaiser.

Huch, R. 1920. *Die Romantik.* 2 vols. Leipzig: H. Haessel.

Hughes, H. S. 1975. *The Sea Change: The Migration of Social Thought, 1930–1965.* New York: Harper & Row.

Humboldt, A. v. 1845. *Kosmos. Entwurf einer physischen Weltbeschreibung*, vol. 1. Stuttgart and Tübingen: J. G. Cotta'scher Verlag.

———. 1847. *Kosmos. Entwurf einer physischen Weltbeschreibung,* vol. 2. Stuttgart and Tübingen: J. G. Cotta'scher Verlag.

———. 1978. *Kosmos,* ed. H. Beck. Stuttgart: Brockhaus.

Humboldt, W. v. 1836–39. *Über die Kawi-Sprache auf der Insel Java, nebst einer Einleitung über die Verschiedenheit des menschlichen Sprachbaues und ihren Einfluss auf die geistige Entwickelung des Menschengeschlechts.* 3 vols. Berlin: Druckerei der Königlichen Akademie der Wissenschaften.

Huntington, S. P. 1996. *The Clash of Civilizations and the Remaking of World Order.* New York: Simon & Schuster.

Husserl, E. 1911. "Philosophie als strenge Wissenschaft." *Logos* 1: 289–341.

———. 1970. "Philosophy and the Crisis of European Humanity." In E. Husserl, *The Crisis of European Sciences and Transcendental Phenomenology: An Introduction to Phenomenological Philosophy,* trans. D. Carr. Evanston, IL: Northwestern University Press, pp. 269–99. (Orig. pub. 1935.)

Illy, J. 1979. "Albert Einstein in Prague." *Isis* 70: 76–84.

Infeld, L. 1950. *Albert Einstein: His Work and its Influence on Our World.* New York: Scribner.

Jackman, J. C., and C. M. Borden. 1983. *The Muses Flee Hitler: Cultural Transfer and Adaption, 1930–1945.* Washington, DC: Smithsonian Institution Press.

Jackson, M. W. 2003. "Harmonious Investigators of Nature: Music and the Persona of the German *Naturforscher* in the Nineteenth Century." *Science in Context* 16, nos. 1–2: 121–45.

James, W. 2003. *The Varieties of Religious Experience: A Study in Human Nature.* New York: Signet Classic. (Orig. pub. 1902.)

Jammer, M. 1966. *The Conceptual Development of Quantum Mechanics.* New York: McGraw-Hill.

———. 1995. *Einstein und die Religion.* Konstanz: Universitätsverlag.

———. 1999. *Einstein and Religion: Physics and Theology.* Princeton, NJ: Princeton University Press.

Jaspers, K. 1919. *Psychologie der Weltanschauungen.* Berlin: Julius Springer.

Jeggle, U. 1969. *Judendörfer in Württemberg.* Tübingen: Tübinger Vereinigung für Volkskunde.

Jeremias, A. 1929. *Die Weltanschauung der Sumerer.* Leipzig: J. C. Hinrichs'sche Buchhandlung.

Jerome, F. 2002. *The Einstein File: J. Edgar Hoover's Secret War against the World's Most Famous Scientist.* New York: St. Martin's Press.

Joachimi, M. 1905. *Die Weltanschauung der deutschen Romantik.* Jena: Eugen Diederichs.

Joel, K. 1928. *Wandlungen der Weltanschauung: Eine Philosophiegeschichte als Geschichtsphilosophie.* Tübingen: J. C. B. Mohr.

Johnston, William M. 1972. *The Austrian Mind: An Intellectual and Social History, 1848–1938.* Berkeley: University of California Press.

Jonas, H. 1934. *Gnosis und spätantiker Geist.* Göttingen: Vandenhoeck & Ruprecht, 1964.

——. 1958. *The Gnostic Religion: The Message of the Alien God and the Beginnings of Christianity.* Boston, MA: Beacon Press.

Jones, W. T. 1973. *The Romantic Syndrome: Toward a New Method in Cultural Anthropology and History of Ideas,* 2nd ed. The Hague: Marinus Nijhoff.

Jungnickel, C., and R. McCormmach. 1986. *Intellectual Mastery of Nature: Theoretical Physics from Ohm to Einstein.* 2 vols. Chicago: University of Chicago Press.

Kagan, J. 2002. *Surprise, Uncertainty, and Mental Structures.* Cambridge, MA: Harvard University Press.

Kahn, L. 1985. "Heine's Jewish Writer Friends: Dilemmas of a Generation, 1817–33." In *The Jewish Response to German Culture,* ed. J. Reinharz and W. Schatzberg. Hanover, NH: University Press of New England, pp. 120–36.

Kahn, L., and D. D. Hook. 1992. "The Impact of Heine on Nineteenth-Century German-Jewish Writers." In *The Jewish Reception of Heinrich Heine,* ed. M. H. Gelber. Tübingen: Max Niemeyer Verlag, pp. 53–65.

Kainz, F. 1943. "Klassik und Romantik." In *Deutsche Wortgeschichte. Festschrift für Alfred Götze,* ed. F. Maurer and F. Stroh, vol. 2. Berlin: Walter de Gruyter, pp. 191–318.

Kanitscheider, B. 1988. *Das Weltbild Albert Einsteins.* Munich: C. H. Beck.

Kant, I. 1922. *Kritik der Urteilskraft.* in *Immanuel Kants Werke,* ed. E. Cassirer, vol. 5. Berlin: Bruno Cassirer, pp. 233–568. (Orig. pub. 1790.)

——. 1925. *Kritik der reinen Vernunft.* Leipzig: Alfred Kröner Verlag. (Orig. pub. 1781.)

——. 1975a. *De mundi sensibilis atque intelligibilis forma et principiis.* In *Werke in zehn Bänden,* ed. W. Weischedel, vol. 5. Darmstadt: Wissenschaftliche Buchgesellschaft, pp. 7–107. (Orig. pub. 1770.)

——. 1975b. "Idee zu einer allgemeinen Geschichte in weltbürgerlicher Absicht." In *Werke in zehn Bänden,* ed. W. Weischedel, vol. 9. Darmstadt: Wissenschaftliche Buchgesellschaft, pp. 31–50. (Orig. pub. 1784.)

Kapitza, P. 1968. *Die frühromantische Theorie der Mischung: Über den Zusammenhang von romantischer Dichtungstheorie und zeitgenössischer Chemie.* Munich: Max Hueber Verlag.

Kassner, R. 1951. *Physiognomik*. Wiesbaden: Insel, 1932.

———. 1978. *Das physiognomische Weltbild*. In *Sämtliche Werke*, vol. 4. Pfullingen: Neske, pp. 301–528. (Orig. pub. 1930.)

Kater, M. H. 1974. *Das "Ahnenerbe" der SS 1935–1945: Ein Beitrag zur Kulturpolitik des Dritten Reiches*. Stuttgart: Deutsche Verlags-Anstalt.

Katz, H. 1892–93. *Die Weltanschauung Friedrich Nietzsches*. Dresden and Leipzig: C. Pierson.

Katz, J. 1985. "German Culture and the Jews." In *The Jewish Response to German Culture*, ed. J. Reinharz and W. Schatzberg. Hanover, NH: University Press of New England, pp. 85–99.

Kayser, R. 1946. *Spinoza: Portrait of a Spiritual Hero*. New York: Philosophical Library.

Kaznelson, S., ed. 1962. *Juden im deutschen Kulturbereich: Ein Sammelwerk*. Berlin: Jüdischer Verlag.

Kee, H. C. 1980. "Myth and Miracle: Isis, Wisdom, and the Logos of John." In *Myth, Symbol, and Reality*, ed. A. M. Olson. Notre Dame, IN: University of Notre Dame Press, pp. 145–64.

Keller, E. F. 1985. *Reflections on Gender and Science*. New Haven, CT: Yale University Press.

Kent, D. 1953. *The Refugee Intellectual*. New York: Harper & Brothers.

Kern, B. 1911. *Weltanschauungen und Welterkenntnis*. Berlin: August Hirschwald.

Kessler, H. G. 1982. *Tagebücher. 1918–1937*. Frankfurt: Insel.

Kirchweger, A. J. 1723. *Aurea Cateni Homeri, oder, Eine Beschreibung von dem Ursprung der Natur und natürlichen Dingen*. Frankfurt: J. G. Böhme.

Kirsten, C., and H.-J. Treder, eds. 1979. *Albert Einstein in Berlin, 1913–1933*. 2 vols. Berlin: Akademie-Verlag.

Klages, L. 1936. *Grundlegung der Wissenschaft vom Ausdruck*. Leipzig: Johann Ambrosius Bart.

Klaus, G., and M. Buhr, eds. 1975. *Philosophisches Wörterbuch*, 11th ed. Berlin: Verlag das europäische buch.

Klein, F. F. 1931. *Lessings Weltanschauung*. Vienna: C. Gerold's Sohn.

Klemm, G. F. 1843–52. *Allgemeine Cultur-Geschichte der Menschheit. Nach den beßten Quellen bearbeitet und mit xylographischen Abbildungen der verschiedenen Nationalphysiognomien, Geräthe, Waffen, Trachten, Kunstproducte u.s.w. versehen*. 10 vols. Leipzig: B. G. Teubner.

———. 1854–55. *Allgemeine Culturwissenschaft: Die materiellen Grundlagen menschlicher Cultur*. 2 vols. Leipzig: Romberg.

Kloppenberg, J. S. 1982. "Isis and Sophia in the Book of Wisdom." *Harvard Theological Review* 75: 57–84.

Kluckhohn, P., ed. 1932. *Weltanschauung der Frühromantik*. Leipzig: P. Reclam.

Knight, C., and R. Lomas. 1998. *The Hiram Key: Pharaohs, Freemasons and the Discovery of the Secret Scrolls of Jesus*. New York: Barnes and Noble. (Orig. pub. 1996.)

Knight, D. M. 1966–67. "The Scientist as Sage." *Studies in Romanticism* 6: 65–88.

———. 1975. "German Science in the Romantic Period." In *The Emergence of Science in Western Europe*, ed. M. Crosland. London: Macmillan, pp. 161–78.

———. 1990. "Romanticism and the Sciences." In *Romanticism and the Sciences*, ed. A. Cunningham and N. Jardine. Cambridge: Cambridge University Press, pp. 13–24.

Koch, W. 1933. *Stefan George: Weltbild, Naturbild, Menschenbild*. Halle/Saale: Max Niemeyer.

Kocka, J. 1975. *Unternehmer in der deutschen Industrialisierung*. Göttingen: Vandenhoek und Ruprecht.

Kohlstedt, S. G. 1999. "Creating a Forum for Science: AAAS in the Nineteenth Century." In *The Establishment of Science in America: 150 Years of the American Association for the Advancement of Science*, ed. S. G. Kohlstedt, M. M. Sokal, and B. V. Lewenstein. New Brunswick, NJ: Rutgers University Press, pp. 7–49.

Kohut, A. 1905. *Justus von Liebig: Sein Leben und Wirken*. Gießen: Emil Roth.

Kondo, H. 1969. *Albert Einstein and the Theory of Relativity*. New York: Franklin Watts.

Koselleck, R. 1990. "Einleitung—Zur anthropologischen und semantischen Struktur der Bildung." In *Bildungsbürgertum im 19. Jahrhundert, Part II: Bildungsgüter und Bildungswissen*, ed. R. Koselleck. Stuttgart: Klett-Cotta, pp. 11–46.

Kosing, A. 1985. *Wörterbuch der Philosophie*. Berlin: Verlag Das Europäische Buch.

Kracauer, S. 1969. *History: The Last Things before the Last*. New York: Oxford University Press.

Kraft, P., and P. Kroes. 1984. "Adaption of Scientific Knowledge to an Intellectual Environment. Paul Forman's 'Weimar Culture, Causality, and Quantum Theory. 1918–1927': Analysis and Criticism." *Centaurus* 27: 76–99.

Kragh, H. 1999. *Quantum Generations: A History of Physics in the Twentieth Century*. Princeton, NJ: Princeton University Press.

Kretschmer, E. 1958. *Geniale Menschen*, 5th ed. Berlin: Springer. (Orig. pub. 1929.)

———. 1961. *Körperbau und Charakter: Untersuchungen zum Konstitutionsproblem und zur Lehre von den Temperamenten*, 23rd/24th ed. Berlin: Springer. (Orig. pub. 1931.)

Krieck, E. 1934. *Wissenschaft, Weltanschauung, Hochschulreform*. Leipzig: Armanen-Verlag.

———. 1942. *Natur und Naturwissenschaft*. Leipzig: Quelle & Meyer.

Kroeber, A. L. 1944. *Configurations of Culture Growth*. Berkeley: University of California Press.

Kroeber, A. L., and C. Kluckhohn. 1963. *Culture: A Critical Review of Concepts and Definitions*. With the assistance of W. Untereiner and appendixes by A. G. Meyer. New York: Vintage Books. (Orig. pub. 1952.)

Krohn, C.-D. 1993. *Intellectuals in Exile: Refugee Scholars and the New School for Social Research*. Amherst: University of Massachusetts Press.

Kuhlenbeck, L. 1899. *Bruno, der Märtyrer der neuen Weltanschauung: Sein Leben, seine Lehre und sein Tod auf dem Scheiterhaufen*. Leipzig: H. W. Theodor Dieter.

Kuper, A. 1999. *Culture: The Anthropologists' Account*. Cambridge, MA: Harvard University Press.

Kurz, H. 1873. *Die beiden Tubus*. In *Deutscher Novellenschatz*, ed. P. Heyse and H. Kurz, vol. 18. Munich: Rudolph Oldenbourg, pp. 149–277. (Orig. pub. 1859.)

Kuznetsov, B. 1965. *Einstein*. Moscow: Progress Publishers.

———. 1972. *Einstein and Dostoyevsky: A Study of the Relation of Modern Physics to the Main Ethical and Aesthetic Problems of the Nineteenth Century*. London: Hutchinson.

———. 1979a. *Einstein: Leben-Tod-Unsterblichkeit*. Berlin: Akademie-Verlag.

———. 1979b. "Einstein, Science and Culture." In *Einstein: A Centenary Volume*, ed. A. P. French. London: Heinemann, pp. 167–83.

Ladenburg, A. 1911a. "Über den Einfluß der Naturwissenschaften auf die Weltanschauung." In *Naturwissenschaftliche Vorträge in gemeinverständlicher Darstellung*. Leipzig: Akademische Verlagsgesellschaft, pp. 299–52.

———. 1911b. "Epilog zur Kasseler Rede." In *Naturwissenschaftliche Vorträge in gemeinverständlicher Darstellung*. Leipzig: Akademische Verlagsgesellschaft, pp. 253–64.

Ladenburger, G. 1987. "Die Geschichte von Stadt und Stift Buchau." In *Bad Buchau und der Federsee: Im Herzen Oberschwabens*. Bad Buchau: Federsee-Verlag, pp. 88–97.

Lanczos, C. 1965. *Albert Einstein and the Cosmic World Order*. New York: John Wiley.

———. 1974. *The Einstein Decade (1905–1915)*. New York: Academic Press.

Latour, B., and S. Woolgar. 1979. *Laboratory Life: The Social Construction of Scientific Facts*. Beverly Hills, CA: Sage.

Laue, M. v. 1961. "Das physikalische Weltbild." In *Gesammelte Schriften und Vorträge* by M. v. Laue, vol. 3. Braunschweig: F. Vieweg and Sohn, pp. 25–47.

Lecher, E. 1912. *Physikalische Weltbilder*. Leipzig: Theodor Thomas Verlag.

Legrand, G. 1972. *Dictionnaire de philosophie*. Paris: Bordas.

Leisegang, H. 1928. *Denkformen*. Berlin: Walter de Gruyter.

———. 1931. *Lessings Weltanschauung*. Leipzig: Felix Meiner.

Lenard, P. 1936. *Deutsche Physik in vier Bänden*, vol. 1. Munich: J. F. Lehmanns Verlag.

Lenin, V. I. 1927. *Materialism and Empirio-Criticism: Critical Notes Concerning a Reactionary Philosophy*. New York: International Publishers.

Lenning, C. 1822–28. *Encyclopädie der Freimauerei*. 3 vols. Leipzig: F. A. Brockhaus.

———. 1900–1901. *Allgemeines Handbuch der Freimauerei*, 3rd rev. ed. 2 vols. Leipzig: Max Hesse's Verlag.

Lenoir, T. 1980. "Kant, Blumenbach, and Vital Materialism in German Biology." *Isis* 71: 77–108.

Lessing, G. E. 1981. *Freimäurergespräche und anderes: Ausgewählte Schriften*, ed. C. Träger. Munich: C. H. Beck. (Orig. written 1776–78.)

Levenson, T. 2003. *Einstein in Berlin*. New York: Bantam Books.

Levin, H. 1965. "Semantics of Culture." *Daedalus* 94, no. 1: 1–13.

Linden, W. 1940. *Alexander von Humboldt: Weltbild der Naturwissenschaft*. Hamburg: Hoffmann und Campe.

Lindroth, S. 1992. "Berzelius and His Time." In *Enlightenment Science in the Romantic Era: The Chemistry of Berzelius and its Cultural Setting*, ed. E. M. Melhado and T. Frängsmyr. Cambridge: Cambridge University Press, pp. 9–34.

Lindsay, A. D. 1915. "German Philosophy." In *German Culture: The Contribution of the Germans to Knowledge, Literature, Art, and Life*, ed. W. P. Paterson. New York: Charles Scribner's Sons, pp. 35–63.

Linse, U. 1976. "Die Jugendkulturbewegung." In *Das wilhelminische Bildungsbürgertum: Zur Sozialgeschichte seiner Ideen*, ed. K. Vondung. Göttingen: Vandenhoeck & Ruprecht, pp. 119–37.

Linser, H. 1948. *Chemismus des Lebens: Das biologische Weltbild der Gegenwart*. Vienna: Universum.

Linton, R. 1955. *The Tree of Culture*. New York: Alfred A. Knopf.

Lippert, J. 1886–87. *Kulturgeschichte der Menschheit in ihrem organischen Aufbau*. 2 vols. Stuttgart: F. Enke.

Lippert, P. 1927. *Die Weltanschauung des Katholizismus*, 2nd ed. Leipzig: Emmanuel Reinicke.

Lipps, G. F. 1911. *Weltanschauung und Bildungsideal: Untersuchungen zur Begründung der Unterrichtslehre*. Leipzig: B. G. Teubner.

Litt, T. 1928. *Wissenschaft, Bildung, Weltanschauung*. Leipzig: B. G. Teubner.

———. 1930. *Die Philosophie der Gegenwart und ihr Einfluß auf das Bildungsideal*. Leipzig: B. G. Teubner.

Loewe, H. 1917. *Die Entwicklung des Schulkampfs in Bayern bis zum vollständigen Sieg des Neuhumanismus*. Berlin: Weidmannsche Buchhandlung.

Lombroso, C. 1891. *The Man of Genius*. London: W. Scott.

Luckmann, B. 1981. "Exil oder Emigration: Aspekte der Amerikanisierung an der 'New School for Social Research' in New York." In *Leben im Exil: Probleme der Integration deutscher Flüchtlinge im Ausland. 1933–1945*, ed. W. Frühwald and W. Schieder. Hamburg: Hoffmann und Campe, pp. 227–34.

Lukacs, G. 1954. *Die Zerstörung der Vernunft*. Berlin: Aufbau-Verlag.

Lüttge, E. 1900. *Die Bildungsideale der Gegenwart in ihrer Bedeutung für Erziehung und Unterricht: Ein Beitrag zur Würdigung sozialpädagogischer Reformbestrebungen*. Leipzig: Ernst Wunderlich.

Mach, E. 1889. *Die Mechanik in ihrer Entwickelung, historisch-kritisch dargestellt*, 2nd ed. Leipzig: F. A. Brockhaus.

———. 1910. "Die Leitgedanken meiner naturwissenschaftlichen Erkenntnislehre und ihre Aufnahme durch die Zeitgenossen." *Physikalische Zeitschrift* 11: 599–606.

———. 1923. *Die Principien der Wärmelehre, historisch-kritisch entwickelt*, 4th ed. Leipzig: Johann Ambrosius Barth.

MacIver, R. M. 1931. *Society: Its Structure and Changes*. New York: R. Long and R. R. Smith.

MacLeod, R. 1982. "The 'Bankruptcy of Science' Debate: The Creed of Science and its Critics, 1885–1990." *Science, Technology, and Human Values* 7 (Fall): 2–15.

Maddox, J. R. 1998. *What Remains to Be Discovered: Mapping the Secrets of the Universe, the Origins of Life, and the Future of the Human Race*. London: Macmillan.

Makkreel, R. A. 1975. *Dilthey: Philosopher of the Human Studies*. Princeton, NJ: Princeton University Press.

Mandelkow, K. R. 1990. "Die bürgerliche Bildung in der Rezeptions-geschichte der deutschen Klassik." In *Bildungsbürgertum im 19. Jahrhundert, Part II: Bildungsgüter und Bildungswissen*, ed. R. Koselleck. Stuttgart: Klett-Cotta, 181–96.

Mann, K. 1974. *Meine ungeschriebenen Memoiren*, ed. E. Plessen and M. Mann. Frankfurt: S. Fischer.

Mann, T. 1920. *Betrachtungen eines Unpolitischen*. Berlin: S. Fischer.

Mannheim, K. 1936. *Ideology and Utopia: An Introduction to the Sociology of Knowledge*. New York: Harcourt Brace Jovanovich.

Marianoff, D. 1944. *Einstein: An Intimate Study of a Great Man*, with P. Wayne. Garden City, NY: Doubleday, Doran.

Martin, T., trans. 1882. "Epilogue to Schiller's Song of the Bell." In *The Poems of Goethe—Household Edition*. Chicago: Belford, Clarke, pp. 436–39.

Marx, K. 1957. *Kritik des Hegelschen Staatsrechts*, ed. Institut für Marxismus-Leninismus beim ZK der SED. Berlin: Dietz Verlag, pp. 203–333. (Orig. written 1843.)

Mason, S. F. 1962. *A History of the Sciences*. New York: Collier Books.

Mauthner, F. 1911. *Wörterbuch der Philosophie*, vol. 2. Munich: Georg Müller.

Mayr, E. 1965. "Comments." *Boston Studies in the Philosophy of Science* 2: 151–55.

Mazumdar, P. M. H. 1995. *Species and Specificity: An Interpretation of the History of Immunology*. Cambridge: Cambridge University Press.

McCormmach, R. 1970a. "H. A. Lorentz and the Electromagnetic View of Nature." *Isis* 61: 459–97.

———. 1970b. "Einstein, Lorentz, and the Electron Theory." *Historical Studies in the Physical Sciences* 2: 41–87.

———. 1976. "On Academic Scientists in Wilhelmian Germany." In *Science and Its Public: The Changing Relationship*, ed. G. Holton and W. A. Blanpied. Dordrecht, Holland: Reidel, pp. 157–71.

Meier, H. G. 1967. "Weltanschauung": Studien zu einer Geschichte und Theorie des Begriffs. PhD dissertation, Münster.

Meisner, E. 1927. *Weltanschauung eines Technikers*. Berlin: Carl Heymann.

Mendelsohn, E. 1964a. "The Emergence of Science as a Profession in Nineteenth-Century Europe." In *The Management of Scientists*, ed. K. Hill. Boston, MA: Beacon Press, pp. 3–48.

———. 1964b. "The Biological Sciences in the Nineteenth Century: Some Problems and Sources." *History of Science* 3: 39–59.

———. 1965. "Physical Models and Physiological Concepts: Explanation in Nineteenth-Century Biology." *Boston Studies in the Philosophy of Science* 2: 127–50.

Merchant, C. 1980. *The Death of Nature: Women, Ecology, and the Scientific Revolution*. San Francisco, CA: Harper & Row.

———. 1982. "Isis' Consciousness Raised." *Isis* 73: 398–409.

Merkelbach, R. 1995. *Isis regina—Zeus Sarapis: Die griechisch-ägyptische Religion nach den Quellen dargestellt*. Stuttgart: B. G. Teubner.

Merton, R. K. 1936. "Civilization and Culture." *Sociology and Social Research* 21: 103–13.

———. 1970. *Science, Technology and Society in Seventeenth Century England*. New York: Howard Fertig. (Orig. pub. 1938.)

———. 1942. "A Note on Science and Democracy." *Journal of Legal and Political Sociology* 1: 115–26.

———. 1968. *Social Theory and Social Structure*. New York: Free Press. (Orig. pub. 1949.)

———. 1965. *On the Shoulders of Giants: A Shandean Postscript*. New York: Free Press.

———. 1973. *The Sociology of Science: Theoretical and Empirical Investigations*, ed. N. W. Storer. Chicago: University of Chicago Press.

———. 1996. *On Social Structure and Science*, ed. P. Sztompka. Chicago: University of Chicago Press.

Merz, J. T. 1965. *A History of European Thought in the Nineteenth Century*. 4 vols. New York: Dover. (Orig. pub. 1904–12.)

Meyenn, K. v. 1994. "Ist die Quantentheorie milieubedingt?" In *Quantenmechanik und Weimarer Republik*, ed. K. v. Meyenn. Braunschweig: Vieweg, pp. 3–58.

Meyer, M. A. 1967. *The Origins of the Modern Jew: Jewish Identity and European Culture in Germany, 1749–1824*. Detroit: Wayne State University Press.

Michel, E. 1920. *Weltanschauung und Naturdeutung: Vorlesungen über Goethes Naturanschauung*. Jena: Eugen Diederichs.

Michelmore, P. 1962. *Einstein: Profile of the Man*. New York: Dodd, Mead.

Miller, A. I. 1996. *Insights of Genius: Imagery and Creativity in Science and Art*. New York: Copernicus.

Millikan, R. A. 1949. "Albert Einstein on His Seventieth Birthday." *Reviews of Modern Physics* 21: 343–44.

Milton, J. 1993. *Paradise Lost*, ed. R. Flannagan. New York: Macmillan. (Orig. pub. 1667.)

Mises, R. v. 1930. "Über das naturwissenschaftliche Weltbild der Gegenwart." *Die Naturwissenschaften* 18: 885–93.

———. 1939. *Kleines Lehrbuch des Positivismus: Einführung in die empiristische Wissenschaftsauffassung*. The Hague: W. P. van Stockum & Zoon.

————. 1951. *Positivism: A Study in Human Understanding.* Cambridge, MA: Harvard University Press.

Mitroff, I. I. 1974. "Norms and Counter-Norms in a Select Group of Apollo Moon Scientists: A Case Study of the Ambivalence of Scientists." *American Sociological Review* 39: 579–95.

Mitzka, W., ed. 1957. *Trübners Deutsches Wörterbuch.* Berlin: Walter de Gruyter.

Mohn, J. 1970. *Der Leidensweg unter dem Hakenkreuz.* Bad Buchau: Stadt Bad Buchau.

Möller, H. 1984. *Exodus der Kultur: Schriftsteller, Wissenschaftler und Künstler in der Emigration nach 1933.* Munich: C. H. Beck.

Montfaucon, Bernard de. 1743. "Sur les Anciennes Divinités de l'Égypte." *Histoire de l'Academie Royale des Inscriptions et Belles Lettres* 14: 7–14.

Moore, S. F. 1994a. *Anthropology and Africa: Changing Perspectives on a Changing Scene.* Charlottesville: University Press of Virginia.

————. 1994b. "The Ethnography of the Present and the Analysis of the Process." In *Assessing Cultural Anthropology*, ed. R. Borofsky. New York: McGraw-Hill, pp. 362–74.

Moore, W. 1989. *Schrödinger: Life and Thought.* Cambridge: Cambridge University Press.

Morfaux, L.-M. 1980. *Vocabulaire de la philosophie et des sciences humaines.* Paris: Armand Colin.

Morrell, J., and A. Thackray. 1981. *Gentlemen of Science: Early Years of the British Association for the Advancement of Science.* Oxford: Clarendon Press.

————, eds. 1984. *Gentlemen of Science: Early Correspondence of the British Association for the Advancement of Science.* London: Royal Historical Society.

Morrison, R. D. 1979. "Albert Einstein: The Methodological Unity Underlying Science and Religion." *Zygon* 14: 255–78.

Mosse, G. L. 1964. *The Crisis of German Ideology: Intellectual Origins of the Third Reich.* New York: Grosset & Dunlap.

————. 1985a. *German Jews beyond Judaism.* Bloomington: Indiana University Press.

————. 1985b. "Jewish Emancipation: Between *Bildung* and Respectability." In *The Jewish Response to German Culture*, ed. J. Reinharz and W. Schatzberg. Hanover, NH: University Press of New England, pp. 1–16.

————. 1990. "Das deutsch-jüdische Bildungsbürgertum." In *Bildungsbürgertum im 19. Jahrhundert, Part II: Bildungsgüter und Bildungswissen*, ed. R. Koselleck. Stuttgart: Klett-Cotta, pp. 168–80.

Moszkowski, A. 1922. *Einstein: Einblicke in seine Gedankenwelt.* Berlin: F. Fontane.

Mourral, I., and L. Millet. 1993. *Petite encyclopédie philosophique*. Paris: Editions Universitaires.

Muir, H. 2003. "Einstein and Newton Showed Signs of Autism." *NewScientist.com*, http://www.newscientist.com/. [Accessed April 30, 2003.]

Mulkay, M. 1976. "Norms and Ideology in Science." *Social Science Information* 15: 637–56.

Müller, K. O. 1912. *Die oberschwäbischen Reichsstädte: Ihre Entstehung und ältere Verfassung*. Stuttgart: W. Kohlhammer.

Müller-Freienfels, R. 1923. *Persönlichkeit und Weltanschauung*. Leipzig: B. G. Teubner. (Orig. pub. 1919.)

Münz, L., ed. 1952. *Rembrandt's Etchings: Reproductions of the Whole Original Etched Work*. London: Phaidon Press.

Nagel, B. 1991. *Die Welteislehre: Ihre Geschichte und ihre Rolle im "Dritten Reich."* Stuttgart: Verlag für Geschichte der Naturwissenschaften und der Technik.

Nathan, O., and H. Norden, eds. 1960. *Einstein on Peace*. New York: Schocken.

Neuberg, A. 1951. *Das Weltbild der Physik: In seinen Grundzügen und Hauptergebnissen dargestellt*. 5th ed. Göttingen: Vandenhoeck & Ruprecht.

Neugebauer-Wölk, M., ed. 1999. *Aufklärung und Esoterik*. Hamburg: Felix Meiner.

Neukirch, J. H. 1853. *Dichterkanon: Ein Versuch, die vollendetsten Werke der Dichtkunst aller Zeiten und Nationen auszuzeichnen*. Kiew: University Press.

Neurath, O. 1973. *Empiricism and Sociology*, ed. M. Neurath and B. S. Cohen. Dordrecht, Holland: D. Reidel.

Newton, I. 1934. *Mathematical Principles*, originally trans. A. Motte; rev. trans. F. Cajori. Berkeley: University of California Press. (Orig. pub. 1687.)

Nicolson, M. 1987. "Alexander von Humboldt, Humboltian Science and the Origins of the Study of Vegetation." *History of Science* 25: 167–94.

Nietzsche, F. W. 1987. *Die Geburt der Tragödie aus dem Geiste der Musik*. Frankfurt am Main: Insel Verlag. (Orig. pub. 1872.)

———. 1984. *Vom Nutzen und Nachteil der Historie für das Leben*. Zürich: Diogenes. (Orig. pub. 1874.)

Niewyk, D. L. 1980. *The Jews in Weimar Germany*. Baton Rouge: Louisiana State University Press.

Noble, D. E. 1992. *A World without Women: The Christian Clerical Culture of Western Science*. New York: Alfred A. Knopf.

Noiré, L. 1874. *Die Welt als Entwicklung des Geistes: Bausteine zu einer monistischen Weltanschauung.* Leipzig: Veit mp.

Nordenbo, S. E. 2002. "*Bildung* and the Thinking of *Bildung.*" *Journal of Philosophy of Education* 36: 341–52.

Nostradamus. 1555. *Les propheties de M. Michel Nostradamus. Reueuës & corrigees sur la copie imprimee à Lyon par Benoist Rigaud en l'an 1568.* Troyes: Pierre du Ruau, n.d.; facsimile ed. Buenos Aires: Impr. de la Biblioteca nacional, 1943.

Novalis. 1907. *Fragmente vermischten Inhalts.* In *Schriften,* ed. J. Minor, vol. 2. Jena: E. Diederichs.

———. 1960a. *Heinrich von Ofterdingen.* In *Schriften,* ed. P. Kluckhohn and R. Samuel, vol. 1. Stuttgart: W. Kohlhammer, pp. 181–369. (Orig. written 1799–1800.)

———. 1960b. *Die Lehrlinge zu Sais.* In *Schriften,* ed. P. Kluckhohn and R. Samuel, vol. 1. Stuttgart: W. Kohlhammer, pp. 69–111. (Orig. written 1798.)

———. 1960c. "Dialogen." In *Schriften,* ed. P. Kluckhohn and R. Samuel, vol. 2. Stuttgart: W. Kohlhammer, 661–71. (Orig. written 1798–99.)

———. 1960d. "Blüthenstaub." In *Schriften,* ed. P. Kluckhohn and R. Samuel, vol. 2. Stuttgart: W. Kohlhammer, pp. 413–70. (Orig. written 1798.)

———. 1960e. "Die Christenheit oder Europa: Ein Fragment." In *Schriften,* ed. P. Kluckhohn and R. Samuel, vol. 3. Stuttgart: W. Kohlhammer, pp. 507–24. (Orig. written 1799.)

———. 1976. *Schriften,* ed. P. Kluckhohn and R. Samuel, vol. 4. Stuttgart: W. Kohlhammer.

Nowotny, H., P. Scott, and M. Gibbons. 2001. *Re-Thinking Science: Knowledge and the Public in an Age of Uncertainty.* Cambridge, UK: Polity.

Nye, M. J. 1996. *Before Big Science: The Pursuit of Modern Chemistry and Physics, 1800–1940.* New York: Twayne.

Oevermann, U. 2003. "Wissenschaft als Beruf." In *Studienstiftung des deutschen Volkes—Jahresbericht 2002.* Bonn: Studienstiftung des deutschen Volkes, pp. 20–38.

Oken, L. 1809. *Lehrbuch der Naturphilosophie,* vol. 1. Jena: F. Frommann.

Ostwald, W. 1909. *Energetische Grundlagen der Kulturwissenschaft.* Leipzig: Dr. Werner Klinkhardt.

———. 1910. *Grosse Männer.* Leipzig: Akademische Verlagsanstalt.

Ozment, S. 1980. *The Age of Reform, 1250–1550: An Intellectual and Religious History of Late Medieval and Reformation Europe.* New Haven, CT: Yale University Press.

Pagel, W. 1982. *Paracelsus: An Introduction to Philosophical Medicine in the Era of the Renaissance*, 2nd ed. Basel: Karger.

Pais, A. 1982. *'Subtle is the Lord . . .': The Science and the Life of Albert Einstein*. Oxford: Oxford University Press.

Park, R. E. 1928. "Human Migration and the Marginal Man." *American Journal of Sociology* 33: 881–93.

Park, R. E., and E. W. Burgess. 1921. *An Introduction to the Science of Sociology*. Chicago: University of Chicago Press.

Paterson, W. P., ed. 1915. *German Culture: The Contribution of the Germans to Knowledge, Literature, Art, and Life*. New York: Charles Scribner's Sons.

Paty, M. 1986. "Einstein and Spinoza." In *Spinoza and the Sciences*, ed. M. Grene and D. Nails. Dordrecht, Holland: D. Reidel, pp. 267–302.

Paul, J. 1963. *Werke*, vol. 5. Munich: Hanser Verlag.

Paul, R. 1984. "German Academic Science and the Mandarin Ethos. 1850–1880." *British Journal for the History of Science* 17: 1–29.

Paulsen, F. 1902. *Die deutschen Universitäten und das Universitätsstudium*. Berlin: A. Asher.

———. 1906. *Das deutsche Bildungswesen in seiner geschichtlichen Entwickelung*. Leipzig: B. G. Teubner.

Pepper, S. C. 1942. *World Hypotheses: A Study in Evidence*. Berkeley: University of California Press.

———. 1945. *The Basis of Criticism in the Arts*. Cambridge, MA: Harvard University Press.

———. 1967. *Concept and Quality: A World Hypothesis*. La Salle, IL: Open Court.

Pesic, P. 1996. "Einstein and Spinoza: Determinism and Identicality Reconsidered." *Studia Spinozana* 12: 195–203.

———. 2000. *Labyrinth: A Search for the Hidden Meaning of Science*. Cambridge, MA: MIT Press.

———. 2002. *Seeing Double: Shared Identities in Physics, Philosophy, and Literature*. Cambridge, MA: MIT Press.

Petersen, J. 1944. *Die Wissenschaft von der Dichtung: System und Methodenlehre der Literaturwissenschaft*. Berlin: Junker und Dünnhaupt.

Pfannenstiel, M. 1958. *Kleines Quellenbuch zur Geschichte der Gesellschaft Deutscher Naturforscher und Ärzte: Gedächtnisschrift für die hundertste Tagung der Gesellschaft*. Berlin: Springer.

Pfister, O. 1928. *Psychoanalyse und Weltanschauung*. Leipzig: Internationaler Psychoanalytischer Verlag.

Pick, L. 1912. *Die Weltanschauung des Judentums*. Berlin: Boas.

Picker, H. 1965. *Hitlers Tischgespräche im Führerhauptquartier. 1941–1942,* ed. P. E. Schramm. Stuttgart: Seewald Verlag.

Planck, M. 1910. "Zur Machschen Theorie der physikalischen Erkenntnis." *Physikalische Zeitschrift* 11: 1186–90.

———. 1932. "The Scientist's Picture of the Physical Universe." In *Where Is Science Going?* by M. Planck. New York: Norton, pp. 84–106.

———. 1970a. "Die Physik im Kampf um die Weltanschauung." In *Vorträge und Erinnerungen* by M. Planck. Darmstadt: Wissenschaftliche Buchgesellschaft, pp. 285–300.

———. 1970b. "Das Weltbild der neuen Physik." In *Vorträge und Erinnerungen* by M. Planck. Darmstadt: Wissenschaftliche Buchgesellschaft, pp. 206–27.

———. 1970c. "Die Stellung der neueren Physik zur mechanischen Naturanschauung." In *Vorträge und Erinnerungen* by M. Planck. Darmstadt: Wissenschaftliche Buchgesellschaft, pp. 52–68.

———. 1970d. "Die Einheit des physikalischen Weltbildes." In *Vorträge und Erinnerungen* by M. Planck. Darmstadt: Wissenschaftliche Buchgesellschaft, pp. 28–51.

Poggi, S., and M. Bossi, eds. 1994. *Romanticism in Science: Science in Europe, 1790–1840.* Boston Studies in the Philosophy of Science, vol. 152. Dordrecht, Holland: Kluwer.

Poincaré, H. 1958. *The Value of Science.* New York: Dover. (Orig. pub. 1904.)

Portig, G. 1904. *Die Grundzüge der monistischen und dualistischen Weltanschauung unter Berücksichtigung des neuesten Standes der Naturwissenschaft.* Stuttgart: Max Kielmann.

Prager, H. 1925. *Die Weltanschauung Dostojewskis.* Hildesheim: F. Borgmeyer.

Precht, H. 1960. *Das wissenschaftliche Weltbild und seine Grenzen.* Munich: Ernst Reinhardt.

Prechtl, P., and F.-P. Burkard, eds. 1996. *Metzler Philosophie Lexikon: Begriffe und Definitionen.* Stuttgart: Metzler.

Proksch, E. J. 1968. *Die Weltanschauung des Christen.* Stein am Rhein: Christiana Verlag.

Proskauer, H. O., ed. 1982. *Goethes Faust: Erster Teil.* Basel: Zbinden Verlag.

Pross, W. 1991. "Lorenz Oken—Naturforschung zwischen Naturphilosophie und Naturwissenschaft." In *Die deutsche literarische Romantik und die Wissenschaften,* ed. N. Saul. Munich: Iudicium Verlag, pp. 44–71.

Pyenson, L. 1985. *The Young Einstein: The Advent of Relativity.* Bristol: Adam Hilger.

Querner, H., and H. Schipperges. 1972. *Wege der Naturforschung 1822–1972 im Spiegel der Versammlungen Deutscher Naturforscher und Ärzte.* Berlin: Springer.

Ranke, F. 1953. *Gott, Welt und Humanität in der deutschen Dichtung des Mittelalters*. Basel: Benno Schwabe.

Ranke, L. v. 1971. *Über die Epochen der neueren Geschichte*. Munich: R. Oldenbourg. (Original lecture given 1854.)

Ratke, H. 1929. *Systematisches Handlexikon zu Kants Kritik der reinen Vernunft*. Leipzig: Felix Meiner.

Rauhut, F. 1953. "Die Herkunft der Worte und Begriffe 'Kultur,' 'Civilisation' und 'Bildung,'" *Germanisch-Romanische Monatsschrift*, N.F., 3: 81–91.

Raven, D., and W. Krohn. 2000. "Edgar Zilsel: His Life and Work (1891–1944)." In *Edgar Zilsel: The Social Origins of Modern Science*, ed. D. Raven, W. Krohn, and R. S. Cohen. Boston Studies in the Philosophy of Science, vol. 200. Dordrecht, Holland: Kluwer, pp. xxi–lxii.

Reddick, J. 1990. "The Shattered Whole: Georg Büchner and *Naturphilosophie*." In *Romanticism and the Sciences*, ed. A. Cunningham and N. Jardine. Cambridge: Cambridge University Press, pp. 322–40.

———. 1994. *Georg Büchner: The Shattered Whole*. Oxford: Clarendon Press.

Reichenbach, H. 1930. *Atom und Kosmos: Das physikalische Weltbild der Gegenwart*. Berlin: Deutsche Buch-Gemeinschaft.

———. 1933. *Atom and Cosmos: The World of Modern Physics*. New York: Macmillan.

———. 1951. *The Rise of Scientific Philosophy*. Berkeley: University of California Press.

Reichinstein, D. 1935. *Albert Einstein: Sein Lebensbild und seine Weltanschauung*. Prague: Reichinstein.

Reinharz, J., and W. Schatzberg, eds. 1985. *The Jewish Response to German Culture: From the Enlightenment to the Second World War*. Hanover, NH: University Press of New England.

Reiser, A. 1930. *Albert Einstein: A Biographical Portrait*. New York: Albert & Charles Boni.

Renn, J., and R. Schulmann, eds. 1992a. *Albert Einstein—Mileva Marić: The Love Letters*. Princeton, NJ: Princeton University Press.

———. 1992b. Introduction to *Albert Einstein—Mileva Marić: The Love Letters*, ed. Renn and Schulmann. Princeton, NJ: Princeton University Press, pp. xi–xxviii.

Rey, A. 1907. *La théorie de la physique chez les physiciens contemporains*. Paris: F. Alcan.

Ringer, F. K. 1969. *The Decline of the German Mandarins: The German Academic Community, 1890–1933*. Cambridge, MA: Harvard University Press.

Robert-tornow, W. 1883. *Goethe in Heines Werken*. Berlin: Haude- & Spener'sche Buchhandlung.

Roe, A. 1952. *The Making of a Scientist*. New York: Dodd, Mead.

Rohrbach, H. 1967. *Naturwissenschaft, Weltbild, Glaube*. Wuppertal: R. Brockhaus.

Roscher, W. H., ed. 1890–94. *Ausführliches Lexikon der griechischen und römischen Mythologie*. Leipzig: B. G. Teubner.

Rosenthal-Schneider, I. 1980. *Reality and Scientific Truth: Discussions with Einstein, von Laue, and Planck*. Detroit: Wayne State University Press.

Rotermund, H.-M. 1957. "Untersuchungen zu Rembrandts Faustradierung." *Oud-Holland* 72: 151–68.

Rothenberg, A., and C. R. Hausman, eds. 1976. *The Creativity Question*. Durham, NC: Duke University Press.

Runco, M. A. 1997. *The Creativity Research Handbook*. Cresskill, NJ: Hampton Press.

Runes, D. D., ed. 1951. *Spinoza Dictionary*. New York: Philosophical Library.

———., ed. 1983. *Dictionary of Philosophy*. New York: Philosophical Library.

Rürup, R. 1975. *Emanzipation und Antisemitismus: Studien zur "Judenfrage" der bürgerlichen Gesellschaft*. Göttingen: Vandenhoeck and Ruprecht.

Rust, B. 1940. *Reichsuniversität und Wissenschaft*. Berlin: Deutsche Forschungsgemeinschaft.

Sandkühler, H. J., ed. 1990. *Europäische Enzyklopädie zu Philosophie und Wissenschaften*. Hamburg: Felix Meiner.

Sarton, G. 1952a. *A History of Science: Ancient Science through the Golden Age of Greece*. Cambridge, MA: Harvard University Press.

———. 1952b. *Horus: A Guide to the History of Science, a First Guide for the Study of the History of Science*. New York: Ronald Press.

———. 1959. *A History of Science: Hellenistic Science and Culture in the Last Three Centuries B.C.* Cambridge, MA: Harvard University Press.

Saul, N., ed. 1991. *Die deutsche literarische Romantik und die Wissenschaften*. Munich: Iudicium Verlag.

Sausgruber, K. 1962. *Einflüsse der Naturwissenschaften auf unser Weltbild*. Schriftenreihe der Vereinigung Vorarlberger Akademiker, Heft 4. Bregenz: Eugen-Ruß-Verlag.

Sayen, J. 1985. *Einstein in America: The Scientist's Conscience in the Age of Hitler and Hiroshima*. New York: Crown Publishers.

Saz, N. 1990. "Zu Gast im Landhaus." In *Ein Haus für Albert Einstein: Erinnerungen—Briefe—Dokumente*, ed. M. Grüning. Berlin: Verlag der Nation, pp. 494–507.

Schaer, K. F. 1941. *Charakter, Blutgruppe und Konstitution: Grundriß einer Gruppentypologie auf psychologisch-anthropologischer Grundlage.* Zürich: Rascher.

Schäffle, A. 1896. *Bau und Leben des socialen Körpers.* 2 vols. Tübingen: H. Laupp'sche Buchhandlung. (Orig. pub. 1875–78.)

Schaller, H. 1934. *Die Weltanschauung des Mittelalters.* Munich: R. Oldenbourg.

Scheidt, W. 1930. *Rassenbiologie und Kulturpolitik.* 3 parts. Leipzig: Philipp Reclam.

———. 1934. *Die Träger der Kultur.* Berlin: Alfred Metzner.

Scheler, M. 1923a. *Moralia.* Leipzig: Der Neue Geist-Verlag.

———. 1923b. *Nation und Weltanschauung.* Leipzig: Der Neue Geist-Verlag.

———. 1929. *Philosophische Weltanschauung.* Bonn: Friedrich Cohen.

Schelling, F. W. J. 1856–61. *Sämmtliche Werke.* Stuttgart and Augsburg: J. G. Cotta'scher Verlag.

———. 1867. "Introduction to the Outlines of a System of Natural Philosophy; Or, on the Idea of Speculative Physics and the Internal Organization of a System of This Science," trans. T. Davidson. *Journal of Speculative Philosophy* 1: 193–220.

Schiebinger, L. L. 1989. *The Mind Has No Sex? Women in the Origins of Modern Science.* Cambridge, MA: Harvard University Press.

Schiller, F. 1968. *Sämtliche Werke*, vol. 3. Munich: Winkler-Verlag.

Schipperges, H. 1976. *Weltbild und Wissenschaft: Eröffnungsreden zu den Naturforscherversammlungen 1822 bis 1972.* Hildesheim: H. A. Gerstenberg.

Schischkoff, G., ed. 1978. *Philosophisches Wörterbuch*, 21st ed. Stuttgart: Alfred Kröner.

Schlegel, A. W. 1846. *Sämmliche Werke*, ed. E. Böcking, vol. 1. Leipzig: Weidmann'sche Buchhandlung.

Schlegel, F. 1800. "Ideen." *Athenaeum* 3: 4–34.

———. 1806. "An Ida Brun." *Isis: Eine Monatschrift von deutschen und schweizerischen Gelehrten* 2: 431–44.

Schleiermacher, F. D. E. 1839. *Sämmtliche Werke*, ed. L. Jonas, Abt. III, vol. 4, pt. 2. Berlin: G. Reimer.

———. 1911. *Werke.* 4 vols. Leipzig: Felix Meiner.

———. 1912. *Über die Religion: Reden an die Gebildeten unter ihren Verächtern.* Berlin: Deutsche Bibliothek.

Schlick, M. 1925. *Allgemeine Erkenntnislehre.* Berlin: Julius Springer.

———. 1949. *Philosophy of Nature.* New York: Philosophical Library.

Schlunk, M. 1921. *Die Weltanschauung von den Griechen bis zu Hegel: Eine Einführung für Suchende.* Hamburg: Agentur des Rauhen Hauses.

Schmidt, J. 1985. *Die Geschichte des Genie-Gedankens in der deutschen Literatur, Philosophie und Politik, 1750–1945,* vol. 2. Darmstadt: Wissenschaftliche Buchgesellschaft.

Schmitz, O. A. H. 1914. *Die Weltanschauung der Halbgebildeten.* Munich: Georg Müller.

Schneider, G. 1898. *Die Weltanschauung Platos, dargestellt in Anschluss an den Dialog Phädon.* Berlin: Weidmann.

Schneider, I. 1921. *Das Raum-Zeit-Problem bei Kant und Einstein.* Berlin: Julius Springer.

Schoeps, J. H., ed. 1989. *Juden als Träger bürgerlicher Kultur in Deutschland.* Stuttgart: Burg Verlag.

Schöne, A. 1987. *Goethes Farbentheologie.* Munich: C. H. Beck.

Schopenhauer, A. 1916. "Erstlingsmanuskripte." In *Arthur Schopenhauers sämtliche Werke,* ed. P. Deussen, vol. 11. Munich: R. Piper. (Orig. pub. 1814.)

Schöttle, J. E. 1977. *Geschichte von Stadt und Stift Buchau.* Bad Buchau: Federsee-Verlag. (Orig. pub. 1884.)

Schrödinger, E. 1951. *Science and Humanism: Physics in Our Time.* Cambridge: Cambridge University Press.

———. 1984a. "Die Besonderheit des Weltbilds der Naturwissenschaft." In *Gesammelte Abhandlungen,* vol. 4. Vienna: Österreichische Akademie der Wissenschaften, pp. 409–53.

———. 1984b. "Die Wandlung des physikalischen Weltbegriffs." In *Gesammelte Abhandlungen,* vol. 4. Vienna: Österreichische Akademie der Wissenschaften, pp. 600–608.

———. 1994. "Ist die Naturwissenschaft milieubedingt?" In *Quantenmechanik und Weimarer Republik,* ed. K. v. Meyenn. Braunschweig: Vieweg, pp. 295–332. (Orig. pub. 1932.)

Seeberg, R. 1930. "Hochschule und Weltanschauung." In *Das akademische Deutschland,* ed. M. Doeberl, O. Scheel, W. Schlink, H. Sperl, E. Spranger, H. Bitter, and P. Frank, vol. 3. Berlin: C. A. Weller, pp. 163–78.

Seelig, C. 1954. *Albert Einstein: Eine dokumentarische Biographie.* Zürich: Europa Verlag.

———, ed. 1956. *Helle Zeit—Dunkle Zeit: In Memoriam Albert Einstein.* Zürich: Europa Verlag.

Segner, J. A. 1746. *Einleitung in die Naturlehre.* Göttingen: Heumann.

Sepper, D. L. 1990. "Goethe, Colour and the Science of Seeing." In *Roman-*

ticism and the Sciences, ed. A. Cunningham and N. Jardine. Cambridge: Cambridge University Press, pp. 189–98.

Shankland, R. S. 1963. "Conversations with Albert Einstein." *American Journal of Physics* 31: 47–57.

Shea, W. R. 1975. "Trends in the Interpretation of Seventeenth-Century Science." Introduction to *Reason, Experiment, and Mysticism in the Scientific Revolution*, ed. M. L. Righini Bonelli and W. R. Shea. New York: Science History Publications, pp. 1–17.

Shepherd, L. J. 1993. *Lifting the Veil: The Feminine Face of Science.* Boston: Shambhala.

Simmel, G. 1905. *Die Probleme der Geschichtsphilosophie: Eine erkenntnistheoretische Studie*, 2nd ed. Leipzig: Duncker and Humblot.

———. 1964. *The Sociology of Georg Simmel*, trans., ed., and intro. K. H. Wolff. New York: Free Press.

Simonton, D. K. 1988. *Scientific Genius: A Psychology of Science.* Cambridge: Cambridge University Press.

Simpson, J. A., and E. S. C. Weiner. 1989. *Oxford English Dictionary*, 2nd ed. Oxford: Clarendon Press.

Small, A. W. 1905. *General Sociology: An Exposition of the Main Development in Sociological Theory from Spencer to Ratzenhofer.* Chicago: University of Chicago Press.

Smith, N. K. 1962. *A Commentary to Kant's "Critique of Pure Reason."* New York: Humanities Press.

Smocovitis, V. B. 1996. *Unifying Biology: The Evolutionary Synthesis and Evolutionary Biology.* Princeton, NJ: Princeton University Press.

Snelders, H. A. M. 1970. "Romanticism and Naturphilosophie and the Inorganic Natural Sciences, 1798–1840: An Introductory Survey." *Studies in Romanticism* 9: 193–215.

———. 1990. "Oersted's Discovery of Electromagnetism." In *Romanticism and the Sciences*, ed. A. Cunningham and N. Jardine. Cambridge: Cambridge University Press, pp. 228–40.

Snow, C. P. 1979. "On Einstein the Man." In *Albert Einstein: Four Commemorative Lectures*, ed. L. S. Swenson, Snow, H. Stein, and I. Prigogine. Austin, TX: Humanities Research Center.

———. 1980. "Einstein." In *Einstein: The First Hundred Years*, ed. M. Goldsmith, A. Mackay, and J. Woudhuysen. Oxford: Pergamon, pp. 3–18.

Snyder, C. 1903. *New Conceptions in Science.* New York: Harper & Brothers.

———. 1905. *Das Weltbild der modernen Naturwissenschaft nach den Ergebnissen der neuesten Forschungen.* Leipzig: Barth.

Solmsen, F. 1979. *Isis among the Greeks and Romans*. Cambridge, MA: Harvard University Press.

Sombart, N. 1989. "Der Beitrag der Juden zur deutschen Kultur." In *Juden als Träger bürgerlicher Kultur in Deutschland*, ed. J. H. Schoeps. Stuttgart: Burg Verlag, pp. 17–40.

Sombart, W. 1911. *Die Juden und das Wirtschaftsleben*. Leipzig: Duncker & Humblot.

Sommerfeld, A. 1948. *Electrodynamics*. Lectures on Theoretical Physics, vol. 3, trans. E. G. Ramberg. New York: Academic Press. (Orig. pub. 1948.)

Sonnert, G., with G. Holton. 1995. *Gender Differences in Science Careers: The Project Access Study*. ASA Rose Book Series. New Brunswick, NJ: Rutgers University Press.

———. 2002. *Ivory Bridges: Connecting Science and Society*. Cambridge, MA: MIT Press.

Sonnert, G., and H. Brooks. 2001. "The Basic-Applied Dichotomy in Science Policy: Lessons from the Past." In *Science for Society—Cutting-Edge Basic Research in the Service of Public Objectives: A Blueprint for an Intellectually Bold and Socially Beneficial Science Policy*, ed. L. Branscomb, G. Holton, and G. Sonnert. Report on the November 2000 Conference on Basic Research in the Service of Public Objectives. Cambridge, MA: Belfer Center For Science and International Affairs, pp. 75–116.

Sorkin, D. 1983. "Wilhelm von Humboldt: The Theory and Practice of Self-Formation (*Bildung*), 1791–1810." *Journal of the History of Ideas* 44: 55–73.

Spengler, O. 1922–23. *Der Untergang des Abendlandes: Umrisse einer Morphologie der Weltgeschichte*. 2 vols. Munich: Beck. (Orig. pub. 1918–22.)

Spenser, E. 1978. *The Faerie Queene*, ed. T. P. Roche Jr. with C. P. O'Donnell Jr. Harmondsworth, UK: Penguin Books. (Orig. pub. 1590–96.)

Spicker, G. 1883. *Lessing's Weltanschauung*. Leipzig: G. Wigand.

Spranger, E. 1922. *Lebensformen: Geisteswissenschaftliche Psychologie und Ethik der Persönlichkeit*. Halle (Saale): Max Niemeyer. (Orig. pub. 1914.)

———. 1949. *Goethes Weltanschauung: Reden und Aufsätze*. Wiesbaden: Insel.

Stadler, F., ed. 1987. *Vertriebene Vernunft I: Emigration und Exil österreichischer Wissenschaft 1930–1940*. Vienna: Jugend und Volk.

———. 1988. *Vertriebene Vernunft II: Emigration und Exil österreichischer Wissenschaft 1930–1940*. Vienna: Jugend und Volk.

Stahl, F. J. 1845. *Rechts- und Staatslehre auf der Grundlage christlicher Weltanschauung*. 1. Abt. Heidelberg: J. C. B. Mohr.

Steiner, R. 1894. *Die Philosophie der Freiheit: Grundzüge einer modernen Weltanschauung*. Berlin: Emil Felber.

——. 1948. *Goethes Weltanschauung*. Freiburg: Novalis-Verlag. (Orig. pub. 1897.)

——. 1955. *Geisteswissenschaftliche Erläuterungen zu Goethes Faust*, vol. 1. Freiburg: Novalis-Verlag.

——. 1981. *Die Rätsel in Goethes "Faust": exoterisch und esoterisch*. Dornach, Switzerland: Rudolf Steiner Verlag.

Stern, C., ed. 1971. *Lexikon zur Geschichte und Politik im 20. Jahrhundert*. Cologne: Kiepenheuer & Witsch.

Stern, F. 1974. *The Politics of Despair: A Study in the Rise of the Germanic Ideology*. Berkeley: University of California Press.

——. 1980. "Einstein's Germany." In *Einstein and Humanism: A Selection of Six Papers from the Jerusalem Einstein Centennial Symposium*. New York: Aspen Institute for Humanistic Studies, pp. 25–60.

——. 1999. *Einstein's German World*. Princeton, NJ: Princeton University Press.

Stimson, D. 1948. *Scientists and Amateurs: A History of the Royal Society*. New York: Henry Schuman.

Stodola, A. 1931. *Gedanken zu einer Weltanschauung vom Standpunkte des Ingenieurs*. Berlin: Julius Springer.

Stonequist, E. V. 1937. *The Marginal Man: A Study in Personality and Culture Conflict*. New York: Charles Scribner's Sons.

Strack, F., ed. 1994. *Evolution des Geistes: Jena um 1800: Natur und Kunst, Philosophie und Wissenschaft im Spannungsfeld der Geschichte*. Stuttgart: Klett-Cotta.

Strauss, H. A., K. Fischer, C. Hoffmann, and A. Söllner. 1991. *Die Emigration der Wissenschaften nach 1933: Disziplingeschichtliche Studien*. Munich: K. G. Saur.

Studienzentrum Weikersheim, ed. 1984. *Die Stellung der Wissenschaft in der modernen Kultur*. Mainz: Von Hase und Koehler.

Sulloway, F. J. 1996. *Born to Rebel: Birth Order, Family Dynamics, and Creative Lives*. New York: Pantheon Books.

Swoboda, W. W. 1978. "Ernst Brücke als Naturwissenschaftler." In *Ernst Wilhelm von Brücke: Briefe an Emil du Bois-Reymond*, ed. H. Brücke, W. Hilger, W. Höflechner, and Swoboda. Graz: Akademische Druck- und Verlagsanstalt, part 1, pp. xxix–xlii.

Syndram, D. 1990. *Ägypten-Faszinationen: Untersuchungen zum Ägyptenbild im europäischen Klassizismus bis 1800*. Frankfurt: Peter Lang.

Szende, P. 1921. "Soziologische Gedanken zur Relativitätstheorie." *Neue Rundschau* 32: 1086–95.

Tal, U. 1975. *Christians and Jews in Germany: Religion, Politics, and Ideology in the Second Reich. 1870–1914.* Ithaca, NY: Cornell University Press.

Talmey, M. 1932. *The Relativity Theory Simplified, and the Formative Period of Its Inventor.* New York: Falcon Press.

Tänzer, A. 1927. *Die Geschichte der Juden in Jebenhausen und Göppingen.* Berlin: W. Kohlhammer.

———. 1931. "Der Stammbaum Prof. Albert Einsteins." *Jüdische Familien-Forschung: Mitteilungen der Gesellschaft für jüdische Familienforschung* 7: 419–21.

Taylor, I. 1938. *Kultur, Aufklärung, Bildung, Humanität und verwandte Begriffe bei Herder.* Gießen: von Münchowsche Universitäts-Druckerei Otto Kindt.

Terrasson, J. 1732. *The Life of Sethos. Taken from Private Memoirs of the Ancient Egyptians,* trans. Lediard. 2 vols. London: J. Walthoe.

Timm, H. 1990. "Bildungsreligion im deutschsprachigen Protestantismus—eine grundbegriffliche Perspektivierung." In *Bildungsbürgertum im 19. Jahrhundert, Part II: Bildungsgüter und Bildungswissen,* ed. R. Koselleck. Stuttgart: Klett-Cotta, pp. 57–79.

Tocqueville, A. de. 1862. *Democracy in America.* 2 vols. Cambridge, MA: Sever and Francis. (Orig. pub. 1835–40.)

Toennies, F. 1970. *Gemeinschaft und Gesellschaft: Grundbegriffe der reinen Soziologie.* Darmstadt: Wissenschaftliche Buchgesellschaft. (Orig. pub. 1887.)

Toland, J. 1696. *Christianity Not Mysterious: Or, a Treatise Shewing That There Is Nothing in the Gospel Contrary to Reason, Nor above it: and That No Christian Doctrine Can Be Properly Call'd a Mystery,* 2nd ed. London: Sam Buckley.

Toulmin, S. 1970. Introduction to *Physical Reality: Philosophical Essays on Twentieth-Century Physics,* ed. Toulmin. New York: Harper Torchbooks, pp. ix–xx.

Toury, J. 1977. *Soziale und politische Geschichte der Juden in Deutschland. 1847–1871: Zwischen Revolution, Reaktion und Emanzipation.* Düsseldorf: Droste Verlag.

Trahndorff, K. F. E. 1827. *Aesthetik, oder Lehre von der Weltanschauung und Kunst.* Berlin: Maurersche Buchhandlung.

Trapp, D. 1956. "Augustinian Theology of the Fourteenth Century: Notes on Editions Marginalia, Opinions, and Book-Lore." *Augustiniana* 6: 146–274.

Traweek, S. 1988. *Beamtimes and Lifetimes: The World of High Energy Physicists.* Cambridge, MA: Harvard University Press.

Trbuhović-Gjurić, D. 1983. *Im Schatten Albert Einsteins: Das tragische Leben der Mileva Einstein-Marić.* Bern: Paul Haupt.

Treitschke, H. v. 1879. *Deutsche Geschichte in neunzehnten Jahrhundert,* vol. 1. Leipzig: S. Hirzel.

Troels-Lund, T. F. 1900. *Himmelsbild und Weltanschauung im Wandel der Zeiten.* Leipzig: B. G. Teubner.

Trunz, E., ed. 1949. *Goethes Faust.* Hamburg: Christian Wegner Verlag.

———., ed. 1986. *Goethe—Faust.* Munich: C. H. Beck.

Turner, R. S. 1971. "The Growth of Professorial Research in Prussia, 1818 to 1848—Causes and Context." *Historical Studies in the Physical Sciences* 3: 137–82.

———. 1977. "Hermann von Helmholtz and the Empiricist Vision." *Journal of the History of the Behavioral Sciences* 13: 48–58.

Tylor, E. B. 1873. *Primitive Culture: Researches into the Development of Mythology, Philosophy, Religion, Language, Art, and Custom.* 2 vols. Repr., London: John Murray. (Orig. pub. 1871.)

Vallentin, A. 1954. *The Drama of Albert Einstein.* Garden City, NY: Doubleday.

Veblen, T. 1934. *Essays in Our Changing Order,* ed. L. Ardzrooni. New York: Viking Press.

Verein, Ernst Mach, ed. 1929. *Wissenschaftliche Weltauffassung: Der Wiener Kreis.* Vienna: Artur Wolf Verlag.

Vickers, B., ed. 1984a. *Occult and Scientific Mentalities in the Renaissance.* Cambridge: Cambridge University Press.

———. 1984b. Introduction to *Occult and Scientific Mentalities in the Renaissance,* ed. B. Vickers. Cambridge: Cambridge University Press, pp. 1–55.

Vierhaus, R. 1972. "Bildung." In *Geschichtliche Grundbegriffe: Historisches Lexikon zur politisch-sozialen Sprache in Deutschland,* ed. O. Brunner, W. Conze, and R. Koselleck, vol. 1. Stuttgart: Klett, pp. 508–51.

Virchow, R. L. K. 1877. *Beiträge zur physischen Anthropologie der Deutschen, mit besonderer Berücksichtigung der Friesen.* Berlin: Ferdinand Dümmlers Verlagsbuchhandlung.

Visher, S. S. 1947. *Scientists Starred 1903–1943 in "American Men of Science."* Baltimore: Johns Hopkins Press.

Vogel, P. 1915. *Das Bildungsideal der deutschen Frühromantik.* Berlin: Weidmannsche Buchhandlung.

Voigts, M. 1998. "Thesen zum Verhältnis von Aufklärung und Geheimnis." In *Schleier und Schwelle: Geheimnis und Offenbarung,* ed. A. Assmann and J. Assmann. Munich: Wilhelm Fink Verlag, pp. 65–80.

Waal, H. v. d. 1964. "Rembrandt's Faust Etching, a Socinian Document, and the Iconography of the Inspired Scholar." *Oud-Holland* 79: 7–48.

Wach, J. 1930. "Hochschule und Weltanschauung." In *Das akademische Deutschland*, ed. M. Doeberl, O. Scheel, W. Schlink, H. Sperl, E. Spranger, H. Bitter, and P. Frank, vol. 3. Berlin: C. A. Weller, pp. 198–204.

Wachsmann, K. 1990. "Beobachtungen in der Haberlandstraße." In *Ein Haus für Albert Einstein: Erinnerungen—Briefe—Dokumente*, ed. M. Grüning. Berlin: Verlag der Nation, pp. 136–60.

Wagner, J. 1970. *Was Einstein wirklich sagte*. Vienna: Fritz Molden.

Walker, M. 1995. *Nazi Science: Myth, Truth, and the German Atomic Bomb*. New York: Plenum Press.

Ward, L. F. 1903. *Pure Sociology: A Treatise on the Origin and Spontaneous Development of Society*. New York: Macmillan.

Warner, M. 1985. *Monuments and Maidens: The Allegory of the Female Form*. New York: Atheneum.

Waschkies, H.-J. 1990. "Alexander von Humboldts aufklärerisches Weltbild." In *Alexander von Humboldt: Weltbild und Wirkung auf die Wissenschaften*, ed. U. Lindgren. Cologne: Böhlau Verlag, pp. 169–86.

Weber, A. 1927. *Ideen zur Staats- und Kultursoziologie*. Karlsruhe: G. Braun.

Weber, M. 1967. *Wissenschaft als Beruf*. Berlin: Duncker & Humblot. (Speech given 1917.)

Weil, H. 1930. *Die Entstehung des deutschen Bildungsprinzips*. Bonn: Friedrich Cohen.

Weingart, P. 2001. *Die Stunde der Wahrheit? Zum Verhältnis der Wissenschaft zu Politik, Wirtschaft und Medien in der Wissensgesellschaft*. Weilerswist: Velbrück Wissenschaft.

Weingart, P., J. Kroll, and K. Bayertz. 1988. *Rasse, Blut und Gene: Geschichte der Eugenik und Rassenhygiene in Deutschland*. Frankfurt: Suhrkamp.

Weisberg, R. W. 1993. *Creativity: Beyond the Myth of Genius*. New York: W. H. Freeman.

Weisgerber, L. 1953–54. *Vom Weltbild der deutschen Sprache*. 2 vols. Düsseldorf: Pädagogischer Verlag Schwann.

———. 1929. *Muttersprache und Geistesbildung*. Göttingen: Vandenhoeck & Ruprecht.

Weizsäcker, C. F. v. 1945. *Zum Weltbild der Physik*. Leipzig: S. Hirzel.

———. 1979. "Einstein's Importance to Physics, Philosophy, and Politics." In *Albert Einstein: His Influence on Physics, Philosophy and Politics*, ed. P. C. Aichelburg and R. U. Sexl. Braunschweig: Friedrich Vieweg, pp. 159–68.

Welling, G. v. 1784. *Opus Mago-Cabbalisticum et Theosophicum*, 3rd ed. Frankfurt: Fleischerische Buchhandlung (Orig. pub. 1719.)

Wenzl, A. 1936. *Wissenschaft und Weltanschauung: Natur und Geist als Probleme der Metaphysik.* Leipzig: Felix Meiner.

Wertheimer, M. 1959. *Productive Thinking.* New York: Harper & Row. (Orig. pub. 1945.)

Wetzels, W. D. 1971. "Aspects of Natural Science in German Romanticism." *Studies in Romanticism* 10: 44–59.

———. 1990. "Johann Wilhelm Ritter: Romantic Physics in Germany." In *Romanticism and the Sciences*, ed. A. Cunningham and N. Jardine. Cambridge: Cambridge University Press, pp. 199–212.

Whitehead, A. N. 1925. *Science and the Modern World.* New York: Macmillan.

———. 1929. *Process and Reality: An Essay in Cosmology.* New York: Macmillan.

Wickert, J. 1972. *Albert Einstein in Selbstzeugnissen und Bilddokumenten.* Reinbek: Rowohlt.

Wien, W. 1901. "Über die Möglichkeit einer elektromagnetischen Begründung der Mechanik." *Annalen der Physik* 5: 501–13.

Wiener, P. P., ed. 1968–74. *Dictionary of the History of Ideas.* New York: Scribner.

Wigner, E. P. 1950. "The Limits of Science." *Proceedings of the American Philosophical Society* 94: 422–27.

Willems, M. 1995. *Das Problem der Individualität als Herausforderung an die Semantik: Studien zu Goethes "Brief des Pastors zu *** an den neuen Pastor zu ***, "Götz von Berlichingen" und "Clavigo."* Tübingen: Max Niemeyer.

Williams, L. P. 1973. "Kant, *Naturphilosophie* and Scientific Method." In *Foundations of Scientific Method: The Nineteenth Century*, ed. R. N. Giere and R. S. Westfall. Bloomington: Indiana University Press, pp. 3–22.

Wilson, E. O. 1980. *Sociobiology: The Abridged Version.* Cambridge, MA: Belknap Press.

Wilson, W. J. 1952. *Exhibit in Honor of George Sarton.* Displayed at the Armed Forces Medical Library, Washington, DC.

Winckler, H. 1905. "Die Weltanschauung des alten Orients." In *Ex Oriente Lux*, ed. H. Winckler, vol. 1. Leipzig: Eduard Pfeiffer, pp. 1–50.

Windelband, W. 1904. *Geschichte und Naturwissenschaft.* Straßburg: Heitz. (Orig. pub. 1894.)

Winteler-Einstein, M. 1987. "Albert Einstein—Beitrag für sein Lebensbild." In *Collected Papers of Albert Einstein*, vol. 1. Princeton, NJ: Princeton University Press, pp. xlviii–lxvi.

Wittgenstein, L. 1974. *Tractatus logico-philosophicus*. Atlantic Highlands, NJ: Humanities Press. (Orig. pub. 1921.)

Wittich, D., ed. 1971. *Vogt, Moleschott, Büchner: Schriften zum kleinbürgerlichen Materialismus in Deutschland*. Berlin: Akademie-Verlag.

Wolff, H. M. 1963. *Die Weltanschauung der deutschen Aufklärung in geschichtlicher Entwicklung*, 2nd ed. Bern: Francke.

Wundt, M. 1910. *Griechische Weltanschauung*. Leipzig: B. G. Teubner.

———. 1926. *Deutsche Weltanschauung: Grundzüge völkischen Denkens*. Munich: J. F. Lehmann.

Wundt, W. M. 1920. *Die Weltkatastophe und die deutsche Philosophie*. Erfurt: Keysersche Buchhandlung.

Wyneken, G. 1940. *Weltanschauung*. Munich: Ernst Reinhardt.

Yates, F. A. 2002. *Giordano Bruno and the Hermetic Tradition*. Reprint, London: Routledge. (Orig. pub. 1964.)

Yengoyan, A. A. 1986. "Theory in Anthropology: On the Demise of the Concept of Culture." *Comparative Studies in Society and History* 28: 368–74.

Zahar, E. 1989. *Einstein's Revolution: A Study in Heuristic*. La Salle, IL: Open Court.

Zevenhuizen, E. 1937. *Politische und weltanschauliche Strömungen auf den Versammlungen deutscher Naturforscher und Ärzte von 1848 bis 1871*. Berlin: Dr. Emil Ebering.

Zimmermann, R. 1996. "Welche Bildung brauchen Studenten für die Zukunft? Überlegungen eines Juristen." In *Studienstiftung des deutschen Volkes—Jahresbericht 1995*. Bonn: Studienstiftung des deutschen Volkes, pp. 1–19.

Zimmermann, R. C. 1969–79. *Das Weltbild des jungen Goethe: Studien zur hermetischen Tradition des deutschen 18. Jahrhunderts*. 2 vols. Munich: Fink.

Zloczower, A. 1981. *Career Opportunities and the Growth of Scientific Discovery in Nineteenth Century Germany*. New York: Arno Press.

Zuckerman, H. 1989. "Accumulation of Advantage and Disadvantage: The Theory and its Intellectual Biography." In *L'opera di R. K. Merton e la sociologia contemporeana*, ed. C. Mongardini and S. Tabboni. Genoa: Edizioni Culturali Internazionali Genova, pp. 153–76.

INDEX

Moore, Sally Falk, 65–66
Moore, Walter, 255n80
Moos, Hermann, 268
More, Thomas, 50
Morley, Edward, 286
morphology, body. *See* Konstitu-
   tionstypen
Morrell, Jack, 166
Moses's Mission (Schiller), 161
Mosse, George, 259, 260
Moszkowski, A., 308, 311
"Motiv des Forschens" (Einstein),
   223, 280, 292n2
Mozart, Wolfgang Amadeus, 39,
   45n34, 161
Müller, Georg Elias, 240
Müller, Johannes, 193, 247n41
Müller-Freienfels, Richard, 98, 107
Munich Luitpold Gymnasium, 27,
   28, 42–43n12
Münz, Ludwig, 296
music and Einstein, 39, 45n34
mysticism, 319, 320

Napoleon Bonaparte, 57
Nathan, Otto, 348, 348n4
Nathan der Weise (Lessing), 259
nationalism, 99, 135–36, 174n17,
   321
   and Kultur, 57–58, 60, 128
   mystic-racist, 234
   and science, 141–42, 215,
      336–37n2
National Socialism, 115, 169,
   172–73n3, 173n5, 337n5,
   342–43, 344, 345
   anti-Newtonian sentiments, 245n8
   causing scientists to flee, 238,
      329, 330, 337–39n8

and race, 63, 124n52, 131
National Socialist Weltanschauung,
   97, 108–109, 137
   and Welteislehre, 133–34
national styles of science, 129–34,
   165–72, 172n2
naturalism, 104
Naturalismus, 104
naturalistische Weltanschauung,
   198, 245n21
natural theology, 166
Naturanschauung, 245n21
"nature vs. nurture" controversy, 20
natürliche Weltanschauung, 200
Naturphilosophie, 156,
   177–78n42, 186–88, 189,
   191–94, 293n10
   condemnations of, 185, 190,
      195–97, 208, 210, 229, 244n7
"Naturphilosophie des Unorganischen" (Hartmann), 250–51n57
Naturwissenschaft, 102, 122n36,
   135, 235
   See also science
Naturwissenschaft, Weltbild, Glaube
   (Rohrbach), 114
Naturwissenschaftliche Gesellschaft
   Isis, 149
Naturwissenschaftliche Volksbücher
   (Bernstein), 28, 231, 292n3
naturwissenschaftliche Weltbild,
   205, 246n29
Neobarbarei, 236
neohumanism, 24, 233–35
Neo-Kantian philosophy, 84, 303,
   304
Neoplatonism, 176–77n31
Neoromanticism. *See* Romantic
   philosophy